# The Face on Mars
## Second Edition
### Filotto

I0462351

First Edition
Copyright © Giuseppe Filotto, 1995

Published by EXACT PRINT
PO Box 751494
Gardenview
2047
South Africa


Second Edition
Copyright © Filotto, 2014

# The Face on Mars

ISBN: 978-1500725808

Dear P.
Latest reports indicate that my love for you has just expanded past
Radio Galaxy 3C 184 at the furthest reaches of the frontier, and is now
rapidly heading into uncharted space.

It would also seem that Einstein was wrong about the speed of light.

Forever,
G.

# Addendum to the 2014 Second Edition

*To all my family far and near,*
*past, present and future,*
*be you a blood relative or not,*
*you all know who you are.*
*I love you all with all my heart.*

# Thanks

Most of this book was written in solitary darkness. Literally so, since my writing was done mainly between 11pm and 5am.

Similarly, for the main part I conducted my research alone, so in that sense, the number of people that helped me write this book are relatively few. Nevertheless, there is one person to whom my thanks are long overdue and without whom it is truly doubtful that I would have managed to complete the volume you are now holding.

This person is Daniela Sisto; my mother.

My mom is not a person of a scientific bent, nor would she deny this (if she reads this book it will be out of love and not conscious choice). So it is with even more awe that I look back at how she supported me in a very practical sense and without hesitation, in what I wanted to do; regardless of how it may have looked to her from time to time.

Another person to whom I owe a great deal, is my lover and friend Preshilla, whose belief in me never faltered.

If honour is the greatest of virtues, then surely, loyalty must be the greatest of its parts.

Which brings me to: "My correspondent in London".

A man I have been friends with for over ten years, and one of those rare few that knows honour. Paul Bentley, who tirelessly dug up and researched obscure information for me when I could not, regardless of what personal efforts he had to make.

I would also like to thank all the librarians I came into contact with, and particularly Mrs. Ethleen Lastovica of the South African Astronomical Observatory, whom was always extremely helpful.

Dr. Dave Laney, also of the SAAO, educated me a little and worked out the distances to the stars in chapter 7 for me.

Rahman Cassiem, using his architectural skills produced some of the diagrams in chapter 3, as well as the sketch map of Mars on plate 11. Considering his schedule, getting these to me on time was at least a minor miracle, and I (not being partial to deadlines myself) remain in awe.

My thanks too to my friends Robert Myles for finding some useful things for me on the Internet, and Russell Petersen for similarly aiding me in my net-searches extensively and more importantly for performing all the computer enhancements on the Face and City as reproduced in this book, as well as the cropping of this information from the original Viking data supplied by the National Space Science Data Center on CD-ROM disks.

Russell was also responsible for the creation of the cover of this book and for suggesting some excellent ideas about computer imaging techniques which could be performed on the Face in order to further verify it's three dimensional qualities. The contour-from- shading map of the Face found on plate 5 was his idea and one which I believe has never before been undertaken by any other group which is currently investigating this phenomena.

PCE Electronics supplied the video-capture equipment that I required for the images on plate 12, along with technical support by Mark Nowitz, for both of which I am very grateful.

Thanks too to my brother Aldo who found out the relevant information he needed to have, then purchased and mailed me the NASA CD-ROM disks without which most of the images presented in this book would not have been possible. And of course, thanks to NASA too for making such data available to the public.

And lastly to all those whom listened to my ideas as I developed them. Even if perhaps they felt differently, none of them laughed.

Thank you all.

*Giuseppe Filotto*
*Cape Town ,    20. 11. 1995*

# Addendum to the 2014 Second Edition

Once again, my thanks to my mother are due. She stored a ton of books and my old computer for over a decade, which allowed me to retrieve some images of the original book, making this version possible.

The photographer Gerald Cubitt was also kind enough to re-issue me with his photographs for this new edition. He continues to travel and take amazing pictures for a living. The images are viewable at my web-site (with his permission): **www.gfilotto.com/image-gallery**

Gerald remains a gentleman and a man who undoubtedly understands the concept of personal integrity. My thanks to him and his family are long overdue.

Finally, I have to thank my own inimitable Redhead Girl. Her gentle ways have power over me, and she had more than a little to do with this new edition. While my Martian ways are more in keeping with the traditional view of a Martian male (shoot first, then ask yourself why you didn't shoot faster later) her Martian ways are more in keeping with Ray Bradbury's version of a Martian female. A subtle, kind, loving and hopeful creature. As all good Martians we are also both telepathic and happy, meaning that when we eventually decide to reproduce, it will be for the express purpose of annexing your planet quickly and fully.

*Addendum to the addendum:* I wrote the above in very early 2011. We married and had a beautiful little Martian Warrior Princess, but unfortunately everything collapsed. My little girl was taken to Brazil by her mother in early 2013, and despite my best efforts it looks as if I will be spending the next years in courts, estranged from my daughter even by the language she will speak. I don't know how long it will be, but one day I will be near her again. And in all that time, her Martian father will not forget her for a single minute. I just hope her telepathy, which was self-evident from birth, manages to somehow keep a link to me for as long as it takes. Whatever I do from here on, it's all for you Anya. And hang on. Dad's coming.

*London,   11.05.2014*

# Contents

**Tables**

**Figures**

**Plates – Not included in this book version.**
**See the images at: www.gfilotto.com/image-gallery**

# Introduction to the Second Edition

**ooo**

The original *Face on Mars* was first published in December of 1995 in South Africa. At the time, this book was the first one to coherently make sense of all the disparate topics you will find in it.

As a result of the publication of it, several strange things occurred.

Firstly I was asked to speak at conference in London, two years in a row, hosted by the then owner of a new magazine called Amateur Astronomy and Earth Sciences (AA&ES). At these conferences, I met esteemed people I had only written about in this book, such as Mark Carlotto and Professor Stanley McDaniel. I offered them both copies of my book, which they graciously took. At the second conference, Carlotto had obviously read my book because he was enthusiastic about it and over dinner with him, myself and Professor McDaniel, he stated that the scientific community should probably take a page from my own book and be a bit more daring with its conclusions. Carlotto and McDaniel and a few others that certainly did not fit into the "lunatic fringe" of the UFO crowd, had been painstakingly trying to receive official recognition for their work — which from a scientific point of view was impeccable— and were failing. The reason for this was obvious to me, and I think would be obvious to anyone logical who looked at the evidence I present. Carlotto and McDaniel and other good scientists, however, are often used to years going by between discoveries, so it may be that in their view, this was just due to business as usual.

But their work not receiving wider exposure in official channels such as NASA is far from business as usual in the healthy way they might have thought about it. There are very real, very present, and very concrete reasons for how their work has been addressed by the "official" bodies such as NASA, JPL and so on. One *can* say this is due to *business as usual* though, and one would be right. It seems then, we really need to define what we mean by *business as usual*.

What I mean by it is simple. The way you think the world works, is quite removed from how the world *actually* works.

In some ways, the intervening 19 years between the first edition of this book and this second edition, has paradoxically both helped as well as hindered the ability to see just how the world really works.

The spread of the Internet in all its forms has facilitated the spread of information, making it harder to hide secret agencies and organisations. In a way. And yet... it's also made the whole world stage just so much more homogeneous. We all drink the same soft-drinks, we all think the same thoughts about the same TV shows. We all, pretty much, watch the same news channels and soap operas and, for the most part, both types of brain-rot induced by them is interchangeable.

The lone genius that figures out a new scientific discovery is much more easily assimilated, ridiculed or silenced, as well as also much more easily touted as champion, put in charge of new discoveries for large institutions and so on. In other words, the *illusion* of personal freedom is grander than ever before, while *actual* personal freedom, may not be as grand as it was in say the 1800s.

Most people of course don't care. Most people are conditioned not to care. You have great choice of course. Your ever increasing upward mobility (if you can read this book, you are amongst the privileged after all, and are undoubtedly part of the upwardly mobile masses) allows you to choose which house you may live in. Which car to drive. Which fashion to wear. And while you are busy choosing (in complete freedom) what colour of new sports car you want, *because you can afford it, you smart, hard-working, honest, and why not, let's say it: good-looking, guy, you,* it totally escapes you that there is actually no need for us to continue to be a petroleum based economy on this planet. Which means your car could function very differently, and in fact, if only you could afford to spend some time thinking this stuff through, and checking up on it, instead of being free to choose whether you will take out Jill from accounting or Jane from dispatch, after you pick up a new suit to go with the new sports car, you may conclude (correctly) that the whole use of cars is in fact completely obsolete. As is the use of aeroplanes, and anything else that burns petroleum based products.

But surely that's not right? Surely not. I mean... trillions of dollar industries like the vehicle industry, the road building industry, and the war industry, can't be wrong? Surely we need to keep using petroleum based stuff? And... and what does this have to do with weird objects on Mars? Oh surely... *Surely*... this is all just so much nonsense. Conspiracy theorist nutcases and all that. And it gives us a headache. Better to just check the news for the latest stock exchange fluctuations and pick out the red sports car. Red looks best after all. It seems faster doesn't it.

After the *Face on Mars* was published at the end of 1995, my life took a slightly surreal turn. Or I should say *more* surreal turn. I've always had a pretty surreal existence; despite my fervently trying to stick to basic engineering principles of empirical reality testing.

My phone lines were tapped. At first I thought this idea of my phones being tapped was just due to my natural paranoia, but later I had opportunity to verify it. I was working in close protection services at the time and through a contact I had then, I was able to unofficially borrow specialised equipment used by the national security agency of South Africa (which was in more turmoil then than it usually is) to test my phone lines. I was indeed being tapped.

My home was also broken into and footage I had of UFOs I describe in chapter 9 was taken. Cash money that was next to the bed was left alone, as was the stills camera, after it had been checked for what film it had inside it. The breaking and entering was done professionally, as the entry was done through the front door and with either a duplicate set of keys, or some advanced lock-picking. Nothing else was taken except the evidence of the UFO footage. My computer hard drive was also crashed while I was in an argument with a "debunker" in an IRC chat room online. I had friends in the IT business then, as I managed a computer hardware firm I owned with my business partner, and I took my hard drive to a specialist. He returned it to me 2 weeks later and took some time with me, asking me exactly what had happened. He explained to me that the sectors on the drive that had been damaged were destroyed in a way he had never seen before. Effectively, the drive head had been rammed into specific sectors of the drive, damaging both the head as well as the drive sectors. He had managed to retrieve about 80% of the information on it but the sectors in which the drive head had been rammed repeatedly were unsalvageable. He was shocked because that kind of thing is supposed to be impossible. The way he described it to me at the time, one would have to be able to re-write some ROM-based interface code to force the drive head to behave in such a destructive manner. When I got home and linked the drive back to a PC I kept off the net, it was even more interesting.

I could retrieve all my data, except, anything that had the words "alien", "UFO", "Extraterrestrial", "Mars" or "Research" on it. Directories with those words in their names were wiped out in their entirety.

I also met some very strange people as a result of the book being published. Some of these were to my mind undoubtedly mentally unstable people, but a few had rather more interesting stories and situations. Some were professionals in various fields who asked to speak with me privately and then told me their stories. These were men and women not only of obvious good education and fully functioning minds, but they were in some cases also from prestigious pedigrees and came from families that had much influence. In some cases, world-wide influence and in some cases, of a military nature. Their stories had a common thread between them. Some professional astronomers confirmed my own ideas privately but were very specific that they could not do so publicly without severe repercussions to their careers and professions.

In one other instance I met with someone (a professional person that was very well-known nationally in South Africa at the time) who recounted to me the story of a "friend" who had been sworn to secrecy in one of the most effective (if brutal) ways possible, and who was now close to death and had no living relatives left. He described how this friend of his had personally worked on large anti-gravity craft in an underground military base in Southern Africa. The story was fantastic and one could easily dismiss it as too fantastic to be true, except for a few details that the person telling me the story had no way of knowing I had already corroborated years before by random coincidence. I was also certain that the events could not have been "designed" to happen to me just for this purpose because thanks to my eclectic father, my formative years are practically unavailable to any number of Men In Black. The location of where this work on anti-gravity craft supposedly took place happened to be an area I was a little familiar with, and there was no way the person telling me the story (or indeed anyone else that I could think of) could know this. Yet his description was exact. In another instance I met the daughter of a very highly placed officer in the British military (so highly placed in fact I will not even name the rank or the branch of military involved as it would be possible to identify him later if I did so) who described to me in some detail her own story. She had gone from being a straight A student with an already prescribed path into a military career, facilitated by her father's connections, and only a few months of her degree to go, to, in her own words:

"Waking up in a drug-addled state in a field, next to some drug-user I obviously had had sex with, and no idea of what day it was or what had happened to me."

Six months had passed from her last available clear memory. She had inexplicably dropped out of school and she had no clear memory of the last six months, only snippets that made no sense to her and were totally out of character in terms of people she frequented. It would have been easy to dismiss her story as the raving fantasies of some drug-addled young woman looking for an excuse to explain away her failings. Personally however, as I listened to her story and the details she gave me, I remained convinced then, as I still am now, that there is a lot more to it than that.


These are just a few of the things that happened, there are more and I will not discuss some of them, but suffice it to say that personally, I have no doubt that my view of how the world really works, is a lot closer to reality that the view Joe-normal has of how the world really works.

This realisation invariably leaves one with a certain sense of frustration and paranoia. Neither of which emotions is useful for you, so let me explain something else about large cycles.

As individuals, it is only seldom that any of us can really affect the course of human history, and those of us who do, generally do not have a great time of it. They tend to get nailed to trees, shot in public, persecuted for the duration of their lives before being blown up, and so on. Even then, for every one of them nailed to a tree that changed human history, untold thousands, if not millions, get nailed to trees without any success or fanfare.

Slowly however, over the decades, centuries and millennia, humanity evolves. The right of kings is no longer. And whilst the Banker, may now determine your economic future to a large degree, they are not themselves immune from retribution, often from the same mechanisms they subject the rest of us to.

Furthermore, with a little care, and conscious attention, it is not impossible for an intelligent person to live a relatively trouble-free life whilst going about their business, so long of course, as they are not particularly interested in changing human affairs on a large scale, or at least, in a way that affects "the powers that be" directly.

Humanity does evolve, and over the next hundred years or so, it has to do so quite drastically. When I was born we had about 4 billion people on this planet. Some 40 years later, we are rapidly approaching double that figure. In another 40 years, unless humanity learns certain basic fundamental rules of ethics, we are in for some serious situations. And in the next 100 years, drastic ones.

Nor, are we talking about your using eco-friendly light-bulbs and ozone friendly socks. By all means go ahead and do all those things, but if you think the human race will be saved by your boiling only two cups worth of water when you make tea instead of a whole kettle-full, you really don't understand the problem in its proper scale or context.

The only way the human race is really going to "make it" and make it in a meaningful way, is by our eventual acceptance and use of anti-gravity technology. Doing so without understanding our already ancient history regarding this technology will only ensure the human race in this solar system goes out with a bang instead of a whimper. Admittedly, if we have to disappear as a species I think it's preferable we do so in one fell swoop of dramatic Earth-shattering proportions than by the slow and inevitable degeneration into rat-like beings trying to exist on a depleted planet, but ideally, I'd like us to make it.

We have already gone through the Earth-shattering scenario at least once (and probably twice) in this solar system. And I don't want my descendants to live on in some semblance of life on a ruined planet.

Our destiny is the stars, and that is where we need to go.

This book, in its first edition, was the first one to coherently map out our ancient past using factual evidence to discover it. As a reference work it remains largely unsurpassed and still quite unique. This work was in fact (badly, if quite directly) "plagiarised" by at least one famous author, whom I confronted after the fact in person. I was disturbed by the fact that this individual, a journalist by trade originally, had asked for a copy of my book, which I had sent him free of charge, and then went on to use many of its central ideas as

though they were his own, without giving credit. Although any kind of suit in such circumstances is pointless, when seeing him in 1997 in Cape Town at a talk he gave, I said to him:
"Graham, it's customary to give credit when you use other people's ideas" and I at least had the pleasure of seeing Graham Hancock apologise in a rather pathetic way as he explained he had a ghost writer, and he had not been responsible himself for that section of his Mars book, as he'd been oh so busy, and didn't have a chance to go through it all, and how he was sorry, and how my work was "excellent" and I had produced a "brilliant" scientific discovery and so on.

In the intervening 17 years or so since, there have been some discoveries made and some information has come to light that was not available before. Surprising me more than anyone, the facts presented here in this book seemed only to be more solidly verified by practically all new information on the various topics.
        Then I saw a trend. A trend that could be described as: "a surfeit of false or incomplete, or low-quality information".
The few serious (and truly scientific) researchers, faced with such a veritable flood of babbling from mainstream outlets, became drowned out. The "information" being used by these mainstream outlets was in no case new. It was a mere regurgitation of already old, already described, already seen information, but it was touted as great new discoveries. Mars had large bodies of water on it? We have known this for at least 50 years. We have known it without any shadow of a doubt for at least 40 years, since the 1970s. And since a full topographic map of Mars was built, anyone with even just a passing knowledge of geology who bothers to take a look at Mars in any kind of detail would be able to arrive at this conclusion in minutes if not seconds.
        Yet this kind of "useless knowledge" was now swamping the internet and even more-so the more traditional media channels. You could no longer search for Mars information without having to wade through billions of bits of such useless data.
Then something else happened. They started to parse pictures of both the Moon and Mars through automatic "filters" of some kind. The information about this was scant and came from a very few whistle-blowers within the established space agencies. Again, given the current climate of "information", even reading these very words I am

writing now, makes it sound like I am the worst kind of conspiracy theorist. But, you must know, it was not always this way. A few whistle-blowers, by the way, is how some of the most factual information has ever been found. In all fields. Ed Snowden comes to mind, and no one is calling him a conspiracy theorist now, are they? In fact, if you consider that only a small fraction of the documents that Ed has appropriated have actually been published, it takes no real imagination at all to see that the very concepts I am discussing here are not just likely, but in fact must be more than a little plausible, as for possible, well, that is not in any kind of question anymore now that the average man has a better understanding of how pervasively the information age has been hijacked as a whole.

The original Face on Mars images were discovered by people who worked with or at NASA. When they released the information they were quickly sidelined. This is not new. Respected and respectable members of the scientific community who are honest, and who speak out, invariably have drastic changes in careers with very few exceptions. More likely is the route Carl Sagan took. Once a vociferous proponent of the extraterrestrial ancient astronaut theory, Carl Sagan did a pretty sudden about turn and then began to toe the party line. Going so far as to actively lie, in order to try and dissuade the general public from thinking about alien artefacts on neighbouring planets. Lies that have since been factually demonstrated to be lies.[1] In this context, his book (and later film) *Contact*, takes on a rather new meaning, and gives a quite clear message about what Carl really did and why this was the case.

Carl is dead and can't defend himself, so I don't want to malign him; maybe he thought he could do more from the "inside". Maybe he did do "more" too. Who knows. Personally I am not so sure of that, but that's just my (considered) opinion.

Today, the bits of good information are sometimes more difficult to find than 19 years ago, because the amount of "useless noise" is even louder, and invariably, sorting the truth from the fake, the falsified and the intentionally misleading, is a time-consuming process. One that our ever-shortening attention spans (thank TV, e-mail, sms messages, facebook, twitter and "smart" phones for that) cannot indulge in so easily.

---

[1] See The McDaniel Report pages 147-156 (see bibliography).

Nevertheless, some excellent information is still (and probably always will be) available. It is human nature. The good guys may be few and far between, and some get killed, some get turned, and some give up, but they exist. And as long as a few of them exist, some information, somewhere, will always leak out one way or another.

In this second edition of the book, I have decided to keep the first edition's information in its original form. There were a number of reasons for doing this:

1) The way the book was originally written, many of the references are rather convoluted and linked. Meaning a complete re-write would be painful and difficult to do, as each reference or comment in the body of the book may have three or four later or earlier statements also referring to it.

2) In some cases new discoveries would require an addendum or re-write of a specific section, apart from the problem already mentioned in 1) above, this would also require very time-consuming and painstaking research to firstly understand how relevant the new discovery really was (often the "new" discovery in fact added nothing, as it was not new at all, and just as often, the "new" discovery did not relate at all to the subject it purported to relate to; a kind of "mislabelling" of the topic if you like, a clear intention to further confuse and frustrate in other words) and secondly to incorporate that new discovery (in the small cases where it was relevant) into the existing text would at times require an overhaul of a whole section. While at other times this could easily be done with a footnote, doing so would re-label all the existing footnotes and thus make a mess of the cross-referencing inherent in the original work.

3) In some cases, the new information was not actually contradictory of the original information I had in the book, but in fact just added a different dimension to the overall topic. Trying to "slice" the new facet into the existing work would be laboriously difficult and probably unlikely to produce a clear result.

In the end I have chosen to keep the new information separate and present the original work as it was back in 1995 except for the minor corrections of basic errata which I had in any case distributed with the book from 1996 in the form of a single sheet that was added to each book by the publishers. In essence then, the book as you find it after this introduction and up to the section labelled BOOK THREE, is identical to the original except for the fixing of a few relatively minor errors of punctuation or grammar (I hesitate to say the grammar is now "fixed" because in many cases throughout the book I specifically broke the rules of grammar in order to force the thinking process of the reader down certain pathways.) Undoubtedly, some errors probably still remain. Any book of this size and complexity invariably has some, I only hope they are not as many as they used to be, and in any case am satisfied that they do not materially alter the various theses of the book.

The new information, collated in the new section labelled BOOK THREE, taken together with this introduction, more easily explains any new concepts that have surfaced since 1995 than trying to mesh them into the original work. Especially since, as already mentioned, the original work is still relevant in any case and the references found in it for the various topics it discusses remain for the most part still the most relevant ones available even after 19 years.

Creating a second edition took much longer than I would have liked for a number of reasons; these bizarre by mere nature instead of nefarious design of mysterious forces.
After publication and various interviews, at the age of 27 I was suddenly being recognised in the street, approached at bookshops I used to hang out at that now carried my book, and approached by any number of pretty women at book signings. After I appeared on TV in the USA and the UK I was even recognised by someone whilst sitting in the audience of a conference in London.

It was not the being recognised I minded *per se*, but up until that time, I had led a relatively misanthropic life. I grew up pretty wild in Africa and my sense of ethics as well as my sense of people in general remained simple and somewhat primitive (and for the most part it still remains so). What disturbed me was the fact that all of these people viewed me as different or special simply because I had written a book. It didn't just extend to strangers, but to people I had known for years. I sometimes felt like slapping their face and shaking them by their necks and saying:

"You moron! It's me! The same guy you disagreed with daily for the last ten years. Wake up! I'm still that guy."

I literally had to turn women down at book signings, women who it was clear to me, would have been quite happy to go off to the nearest hotel with me if I so chose. It was also equally clear to me that they had not the slightest idea of what my book was about. I had to explain to a few close friends why I invariably turned such offers down, aside from the fact I was in a relationship at the time, it was just against my nature. Even if I had been single, or in an open relationship, I would not have taken those offers. There was something artificial and unnatural about it. A chance encounter fuelled by animal magnetism and lust? No problem. But getting laid as a result of "fame"? It just felt wrong.

Being "famous" was not to my liking.

I threw myself into my karate training and my work in close protection and tried to forget about the book and writing for a while, as other, (surreal) things suddenly occupied my time. A few years passed and then when I thought of updating the book life circumstances had changed, I had broken up with my long-term girlfriend (not as a result of any book signings if you must know!) and moved city and then gotten married (disastrously).

By the time I had settled a bit, the files which held the images of the book were in a computer in storage and it was not known if they were even retrievable. The plates of the book had been stolen by one of the employees of the printers as could be sold for scrap metal.

I pretty much gave up.

Then I moved continent, divorced and had to just get on with the job of surviving. Eventually, I retrieved some of the images from that old hard drive, just before the format for it became completely obsolete, but I was still missing some of the original images and there was no way to recreate a specific one. The one of the UFO I had filmed. The tapes had been stolen and I had no copy of the image on the hard drive as it had been destroyed by the "hard-drive read-head crunching attack" it had suffered. And for that image, as it had taken a lot of computing power to transfer from the old camera tape, I had not been able to store it separately on a stiffy disk as it was too large (some of you reading this will be giggling and wondering what a stiffy disk is. Google it you young padawan, you!).

The cover of the book too had been in a format now difficult to parse. Then there was the problem of all the new information and the idea of spending a couple of years or more to re-write the whole book just seemed exhausting as well as practically impossible. I had written the original over a period of some 3 years of research during which I did nothing else. I had survived without working full time or doing anything other than writing my book and going to karate classes for over a year. I couldn't do that now, more than a decade later. Nevertheless, over time, the relevant information filtered through and I realised that it was not necessary to spend as much time as I had done originally to verify it and sort it. Back then I had needed to go to libraries and meet people in person to get information, now much of it could be verified online. But more importantly, the truly *relevant* information, it soon became obvious, was not that great in volume.

Two events finally decided me to update this book. One was what remains the single largest correction in this work. This largely came about as a result of my reading the works of Joseph Farrell, as you will discover in book three. I do not necessarily ascribe to all his theories (and it is clear that in many cases he merely presents potential options, and he does not necessarily believe such options himself anyway), but concerning the origin of Terrestrial UFO craft, his work was instrumental in revising my overall theory in its specifics. Interestingly, the work of Farrell in no way invalidated my views, in fact, it only served to bring them more sharply into focus and confirm them in more exacting detail.

This pattern, of the original position I took being merely refined by new evidence as opposed to refuted, is one that has yet to be broken. And not for lack of trying on my part or that of others. All of which suggests to me the main theory is a solid one.

The second event was a more practical one, and that is, the ease with which one can now produce a book through print-on-demand publishers, at little personal financial risk and without the painful and random process required of the traditional publishing of books.

A last final point was the work of a man who has been painstakingly looking at the "new" images of Mars online. Michael J Craig (See www.secretmars.com) I reference his work in book three and I add that he is far more well-placed than I am regarding these images to know their true nature. I have not taken the time to investigate them in as much detail as I did the original pictures of the face. To do so would again require more time than I possess at the moment. Nevertheless, I believe his work certainly merits scrutiny.

Finally, I would like to return to the point concerning paranoia and frustration, and come to it in a slightly indirect way, by returning to my experiences regarding the use of this work by Graham Hancock.

It is true that Hancock certainly made a lot more money than I ever did from his (and my) writing. Ask me if this bothers me in any way though, and if you understand why my answer is a very calm, very clear, "No" then you can begin to understand why frustration and paranoia is not the answer once you recognise how this planet actually works in the human context.

There is no amount of money on this planet that I would trade for my own sense of integrity. Of course, perspicacious readers will be aware that if you do not have a sense of personal integrity, or rather, if your sense of it is about as evolved as one of the lower animals, say a small reptile or bacteria, or some "humans" who may at times also be journalists, then you would not even be aware there is a choice. And this of course is perfectly true. And the real issue of this book is that we all *need* to become aware of it.

There are many pitfalls on such an evolutionary road. Not least amongst them is the mistaking of prejudices for ethics or integrity.

Your worst kind of "journalist" does not really have prejudices, other than the imperative of survival. They will write "for" or "against" articles on the same topic as long as they get paid.

A more "evolved" journalist will write a prejudiced article on a topic. No sign of objectivity or factual information is likely to dissuade him though. His mind is already made up and he will steadfastly hold on to his bigoted opinion in the face of any mere facts that oppose it. It is the same mindset of the brainwashed religious fanatic. In essence it is a mindset that is based on fear. Fear of the unknown. Fear of the unthought of possibilities. In short: Fear.

And in the long run, that mindset does not win. The theories of the Graham Hancocks of this world, in the end, are exposed as the products of the erroneous thinking that they stem from.

Hancock's theories on the ancient astronauts have been pretty resolutely shown to be wrong in their main detail, which was that ancient constructs on Earth supposedly align with various stars in the night-sky. While there may be some evidence for this, Hancock went way beyond this and made claims that were easily refuted or even "duplicated" with any modern building you care to use, given how loose his fundamental parameters were.

In fact, if you read Hancock's "Martian theories", despite having obviously read this book in its original form, he still got it wrong. Frankly, I don't know what I was more irritated by, the fact he didn't give credit, or the fact his reading skills and mental processes obviously didn't even allow him to present the information in such a way as to at least get the theory right.

When The film *Mission to Mars*, with Gary Sinise, came out, my brother spent about a month urging me to "...sue them!", by "them" meaning the producers of the film.

I had no such intention or feeling. Firstly because I was sure that in this case my book probably had not been used as a source for inspiration without credit being given. My sense of it being that this was merely a case of Morphic Resonance. Something you can read about in chapter 7. And secondly, even if my book had been the source of inspiration, this was a Hollywood version that obviously had a lot of input and work from many people. I was just happy they were kind of "spreading the word". In some mild form, making the whole thing more "possible" in the general subconscious of the cinema-going population.

The point here is that paranoia is good in small doses. As the saying goes: Just because you are paranoid does not in fact mean they are *not* out to get you! This is a good motto in general to keep at the back of your mind. In jest of course. And remember that the Russians say a joke is only a joke for 25% of the statement.

But! And it's a big but; true paranoia is not good for anyone. Better to go out a little naïve and fighting than to live in constant fear and "survive" in some semblance of life that is crippled by what effectively amounts to a mental illness; regardless of whether it has foundations in reality or not. In short, if they really are out to get you, first of all *they will* get you, and secondly, no point in worrying too much about it.

As for frustration at the fact that you and I both cannot hop into our own little hyperspace capable starship and check out the planets with the lascivious sexual laws and super-hot alien females, as well as go hunting for Tyrannosaurs Rexes on some primitive planet, where they are causing problems for the local farmers, well... that's a tough one. But as far as I can tell, even the guys with the starships are not having all that much fun with them. We are still a long way away from anything like a Star Trekkish kind of Universe. We live in more of a Star Wars kind of Universe, and one in which anyone with any kind of starship can basically "do an Alderaan".

So it may be some time before you get your own little starship without a lot of brainwashing, safety features and self-exploding charges embedded in both starship and pilot, and set to go off when certain manoeuvres are attempted.
In other words, it's a false sense of frustration. The human monkey still has a lot of biological and mental evolution to do before it can be trusted with proper starships. Until then, we will have to satisfy ourselves with good quality graphic novels, science fiction novels and films for a while yet.

# Conventions Used

### Distance, Volume and Weight Conversions

1 Primitive inch = 2.543 Centimetres = 1.00106 Imperial Inches
1 Imperial Inch = 2.54 Centimetres
1 Mile = 1.609 Kilometres
1 Gallon = 4.546 litres
1 Pound = 0.454 Kilograms
1 Ton = 1.016 Tonnes

### Temperature conversions

$C = 5/9 \ (F-32)$
$F = 9/5 \ C + 32$
$K = C + 273.16$

### Metric Prefixes

Kilo    or k  =  $10^3$    = Factor of  1 000
Mega  or M  =  $10^6$    = Factor of  1 000 000
Giga   or G  =  $10^9$    = Factor of  1 000 000 000
Tera   or T  =  $10^{12}$   = Factor of  1 000 000 000 000

So a Kilometre or Km = 1000 meters
and a Megametre or Mm = 1 000 000 metres

### Distances

1 Astronomical Unit (AU) = Mean distance of Earth from Sun
= 149 597 870 Km = 1 AU

1 Light second = 300 000 Km
1 Lightyear = 300 000 x 60 x 60 x 24 x 365.25 = $9.4673 \times 10^{12}$ Km

1 Parsec or pc        =  3.26  Light years
1 Kiloparsec or Kpc  =  1 000  Parsecs = 3 260 Light years
1 Megaparsec or Mpc  =  1 000 000 pc  = 3 260 000 Light years

**Einstein's Time Dilation equation:**

$$t' = \frac{t - vx/c^2}{\sqrt{1 - (v/c)^2}}$$

where:

$t'$ = time taken for the trip as measured by the ship's crew (in seconds)

$t$ = time taken for the trip as measured by people on Earth if travelling at c (in seconds)

$v$ = velocity of ship (use 90% or more of c) in Km/s

$x$ = distance being travelled by the ship (in Km) as measured by the people on Earth

$c$ = speed of light (300 000 Km/s)

If the variables are used as described above, you can use this formula for working out the amount of time (as measured by the crew) a ship travelling at close to the speed of light would take for the completing of an interstellar journey.

**Titius-Bode 'law':**

$$D = 0.4 + (0.3 \times 2^n)$$

where:

$D$ = the distance from the primary star in AU and n varies depending on which planet we are trying to find the distance of.
For planet number one (the closest to the parent star) n is equal to negative infinity, for planet two $n = 0$, for planet three, $n = 1$, for planet four $n = 2$ and so on (for subsequent planets n increases by 1 each time).

**Einstein's Special Theory of Relativity Equation:**

$E=Mc^2$

Perhaps the most famous formula in the world, where:

| | | |
|---|---|---|
| E | = | Energy measured in ergs |
| M | = | Mass in grams |
| $c^2$ | = | The speed of light squared (in cm/s) (30,000,000,000 x 30,000,000,000 = $9 \times 10^{20}$ ) |

This equation of course (along with the atom bomb) is the proof that matter and energy are one and the same thing.

An erg is a very small unit of energy; a kilocalorie is almost equal to 42,000,000,000 ergs, and a person can survive comfortably on about 2500 kilocalories a day. Of course, the energy provided to us by the food we eat, is nowhere near the energy present in the matter the food is composed of. If we could convert matter to energy at 100% efficiency (and without doing it explosively) the amount of energy present in just one gram of matter would be enough to keep one person alive for about 23,500 years.

# 1     **Basic Concepts**

The previous pages entitled Conventions Used are to be of general reference to the person not used to discussing distance in light years or parsecs and unfamiliar with scientific notation.

The purpose of this chapter instead, is twofold. Firstly it introduces some basic concepts by making certain comparisons which our minds can grasp more easily than a sentence such as: "The nearest star lies approximately 4 light years away".

This in turn will serve to highlight just how strange and awesome a Universe we *do* live in. Secondly, I present factual information that shows how (theoretically at least) it is possible for intelligent creatures to travel between the stars despite the enormous distances between them.

By doing this, I hope to create an atmosphere of receptivity in the reader's mind. This is necessary, because unfortunately, most of us, caught up in the everyday running of things have a rather stifled imagination. And since through the course of this book I will be making claims which are, for some people, bound to be classed as completely outrageous, it would help if I first made some attempt to describe some features of our environment (which anyone can look up on their own with minimal trouble if they so wish) which prove that in fact we already live in amazing circumstances as it is, and that what may prevent some of my claims from being accepted as fact, is not the Universe around us, but rather our own subjective view of it.

Wherever possible, I have persevered to try and make my theories follow such a trend throughout the book.

Ideally, a person interested in doing so, should be able to independently check on my statements, and find them to be true and correct, which in turn would ensure that the subject matter at hand is then taken with the seriousness it deserves. I have also endeavoured not only to explain to the reader how I reached my conclusions, by way of giving factual information, but also to present the various parts of this theory in the order which I received them myself.

Whenever creating a theory, it is necessary to base it on some very solid foundations, for this reason, the more accepted and widely known aspects of physics and other sciences is a good place to start. A theory after all is nothing more than the best possible educated guess one can make with all the information at hand and if the foundations are shaky to begin with, it's all the more likely to place the whole structure in jeopardy.

In addition to the basic principles of science though, there is available to us, a very useful tool for discerning the possible validity of a new theory. So useful a tool in fact, that without it, our very sound and most brilliant scientific knowledge would not only be meaningless; it would not exist in the first place. We all possess this quality in varying quantities. It is sometimes referred to as the analytical mind.

A logical analysis of a problem, by its very nature, is a mathematical process. Since maths is not affected by our opinions or beliefs, it could be said (and has been) that maths is the only truly universal language. It could also be said that maths is a facet of ultimate truth and therefore, the better our analytical equipment, the closer our minds can come to the truth of a situation.

Ultimately, in arriving at a decision of whether a statement is true or false, we play a game of probability.[1]

If we can arrive at the same conclusion in a number of different ways then chances are, that it is the correct one.

Although the process of logical analysis sounds simple, in effect it is not, because unlike mathematical operations, our minds tend to be prejudiced in a number of ways. A simple example might illustrate the problem.

It is usually easier for an atheist to hypothetically consider the existence of a God for the purpose of an argument than it might be for a devout Christian to consider (also hypothetically) that there is no God. Why is this?

Generally because an atheist has usually considered the problem a little more at length.

Most of us are born into some type of religious upbringing, so for a person to say he or she is an atheist would suggest they have given the matter at least some marginal thought and arrived at a *personal* conclusion which differs from the one accepted by the majority.

The devout Christian on the other hand, has probably been brought up to believe that doubting the existence of God is blasphemous, sinful and productive of punishment in the after-life. It can be harder for this

---

[1] Incidentally, It has been noted by casino owners worldwide, that if all other factors are equal, the best gamblers are good mathematicians. Edward Thorpe perhaps being the best example of this.

person, and sometimes very much so, to actually consider the non-existence of God for a few hypothetical moments.

And yet, it becomes impossible to argue about the existence or non-existence of God in a *logical* manner if one is incapable, at least for the purposes of the argument, to start from either premise. And this would be necessary of course if we are arguing *logically*, since neither starting point should have an advantage over the other and therefore both possibilities need to be examined.

The above is only one of many examples that come to mind of instances where we allow our emotions or beliefs to prevent us from logical thought.

I would like to ask of the reader then, that in the process of reading this book, he or she try as much as possible to avoid making a judgement on impulse. I invite you to consider the propositions I put forward and critically analyse the logical steps taken to arrive at the various conclusions. Where you have doubts about the veracity of my comments, I would urge you to check up on them for yourself. The local library has been of invaluable help to me in my research and I would suspect it would be just as useful to a person trying to disprove my theories.

Lastly, the subject of this book is so vast and far-reaching that I do not expect to be correct about every detail of my assumptions. Generally, I have tried to steer clear from detailed particulars for this very reason and have painted my canvas with a rather broad brush. This was not because I hoped to escape criticism, but because one man cannot be expected to tackle the myriad details of a theory which, if correct, has the potential to change the way every human being on this planet thinks about him or herself. Neither does it help to confuse an already vast issue with countless suppositions, guesses and in the main irrelevant titbits of information which do not really alter the broader outline one way or the other.

It is left for others (may they be plentiful in number) to explore the details of the situation since it is not an exaggeration to say that a number of lifetimes can be spent analysing any one of the many facets of this topic.

# Distances in Space

To most of us a light year is a distance so vast it is meaningless in giving us an idea, a feel, of just what such a dimension means.

By building some conceptual models, however, it becomes easier to visualise distances in terms we are more familiar with.

Imagine that we shrink the Earth until it is a mere 3 millimetres across (just under $^1/_8$ of an inch).

The whole of planet Earth is now only a tiny dot about the size of the head of some coloured pins one might use on a map. Everything you hold dear lives there, on that little ball. Our oceans cover most of the tiny sphere so it looks mainly blue. Details of the continents are too small to see. Countries like the UK or Italy, despite distinctive shapes would be difficult to spot and would be represented by no more than a dot.

Now place this little ball somewhere in space near you. Imagine it floating in the air a few centimetres from you perhaps. Using this as our scale, we find that the Sun is just over a foot in diameter (327 mm) at a distance of 35 metres.

If you take a beach ball and place it roughly 35 largish steps from a pin with a coloured head, you have a pretty good idea of how far we are from Sol, our primary star. It helps to do this in your mind. Add a fireball the size of a beach ball to our imaginary Solar System. Place it at a distance that you estimate to be 35 metres (114 feet) or so from where you are reading this book now; you might need to look out of a window to find a suitable structure near which you can visualise the Sun. Continue doing this, adding each new element to the model in your mind by using points of reference you are familiar with.

Mars would be just over half the size of Earth at about 1.6 mm and it would be placed 54 metres away from the beach ball, being only 19 meters away from Earth at perigee (the point at which its orbit is closest to Earth).

Jupiter and Saturn would be about the size of ping-pong balls and they would be placed 183 and 336 meters away from the Sun respectively.

Uranus and Neptune would be a grape at about 675 metres and another grape at 1 Km from the beach ball, while Pluto, the furthest member of the Solar System would lie at an average of 1.4 Km of distance and would be less than 1 mm in diameter, perhaps the size of the tiny ball one finds in the tip of certain fine-line ball-point pens.

The Moon, which is the only body in the solar system we have made a manned landing on to date, lies 9 centimetres (roughly 3.5 inches) away from the Earth and has a diameter just a little bigger than Pluto.

How far do you think the nearest star to our solar system lies on this scale? To test yourself on how well you judge interstellar distances (distances between stars) make a best guess based on what you've just read without resorting to a calculator, then check your answer at the bottom of this page[2].

By now it is obvious that even if we shrank the entire planet earth to 3 mm it would not be possible for us to talk about even a small part of our Galaxy. The scale would once again become composed of numbers too huge for us to comprehend.

Now imagine what kind of technology one needs if we are to attempt to travel by means of some type of spaceship to these nearby stars. At first the task before a prospective interstellar[3] traveller seems impossible.

Yet, in theory at least, this is not at all so. When we consider interstellar travel, we have a number of alternatives, before discussing these however, we should take into consideration a man that has single-handedly advanced the most widely accepted (and arguably the most significant) theory of physics to date.

Albert Einstein.

We don't need to do all the maths. In any event, the number of people even today who can honestly say they fully comprehend the theory of relativity are very few indeed. For our purposes, it's enough to know that according to this theory, it is not possible to travel faster than the speed of light. We can take this as pretty good advice considering that most modern physicists have yet to find a better model for the universe than the one proposed by Einstein. In fact, advances in science up to now have seemed to vindicate Einstein to the nth degree.

---

[2] The nearest Star to our Sun is Proxima Centauri (also known as Alpha Centauri C) and its distance is estimated as 4.3 light years. Translating this distance into the scale being used for our model of the solar system, Alpha Centauri C would be another beach ball 9500 Kilometres from the Sun. If we placed the first beach ball, representing Sol in the centre of London, then Alpha Centauri C would be a beach ball in Cape Town at the Southern tip of Africa, and remember that there is nothing between these two fiery balls but dark, empty space.

[3] Interstellar travel means travelling from star to star. Interplanetary travel means travelling from planet to planet of the same solar system. It is thought that most stars have some form of a solar system, with a number of planets and gas giants orbiting them. An example of a gas giant is Jupiter in our own solar system.

If we take this fact into account, it becomes obvious that even travelling at the speed of light, it would still take us 4.3 years just to reach the nearest star. If we wanted to make it a return trip we would have to spend another 4.3 years in a ship to do so. Assuming we don't stop for longer than a few days at arrival, (unlikely when it took us so long to get there!) a round trip would take 8.6 years, and all of that would be spent inside a ship!

This however is not telling the whole story. The theory of relativity also proved that time and space are not two separate entities but rather different aspects of the same dimension. After Einstein, it has become customary for physicists to refer to the *space-time continuum* rather than just time or space.

All that space-time continuum really means is that space and time are inextricably linked, but neither is immutable. Translated into English, this means that time does not pass at the same rate for everyone, but its passage is relative to your movement through space.

One analogy that works quite well, is to consider time as being ripples in a pond. These ripples emanate from every object in the universe. Now if that object stays still, then the ripples move further and further out and leave the object further and further behind. You can think of this as time passing and the object ageing.

If the object moves in a straight line[4] at the speed of light or close to it though, the ripples are still moving away from the object, but not as quickly as before. It's a little like running after a car. You're going to be left behind, but you're closer to it than you would have been if you'd stood still. The same principle applies to time, and the closer to the speed of light you travel, the less you age compared to someone who stands still.

For the people back on Earth however, time passes at the normal rate we are all used to. How does this affect Interstellar travel?

It makes it a little easier.

---

[4] For the purposes of relativity it does not matter what direction you move in as far as I know, but in this analogy, travelling in a straight line helps.

# Interstellar Travel

Without taking relativity into account, if we wanted to reach a nearby star, we could only do so in what some people call generation ships. These ships would be closed, self-sufficient environments and would act as tiny reservoirs of life. Most of the people on board these ships would live their entire lives inside them, being born in them, having children in them and ultimately dying in them. Their children's children's children might be the ones to reach the target star. Such ships would have to be enormous and able to survive for many years in space. Depending on the distance to the star, they may have to survive up to hundreds of years in space.

For this very reason, such ships would have to virtually be tiny planets. Since they are unlikely to come across raw materials in the vacuum of space, they must take all they need for the entire duration of the journey. It goes without saying that there is no room for waste on such ships; everything would have to be recycled throughout the life of the ship.

There are obviously a number of serious problems for such a theoretical ship. For a start outgassing[5] would eventually deteriorate the ship's hull unless we found a method of preventing it.

Assuming some form of technology to prevent this was discovered, we are still left with some major dilemmas. On a trip of this length it would be wise to have some type of gravity, since calcium deterioration is the eventual outcome of staying in low gravities for long periods of time. This in itself might not pose such a big problem, because a ship of this size could be given a spin and the inhabitants would then experience the centrifugal force generated as a form of gravity. Effectively, the people aboard would walk around on the inside of this type of ship.[6] All well and good, but the gravity would

---

[5] Outgassing is a process where the vacuum of space slowly sucks out individual molecules and atoms from the hull of a spaceship or satellite. While modern satellites are less prone to malfunction even after theoretical long periods of time, this type of loss is bound to cause problems for a ship that is meant to be in space for a few centuries.

[6] Imagine a small cylinder rotating about its axis in a zero gravity environment. Ants inside the cylinder would experience the centrifugal force generated as gravity, and would be able to walk around on the inside of the cylinder, the floor under them becoming 'down' for them regardless of their orientation.

only be constant while you are at the same level in the ship. The closer to the centre you get, the less the effect of gravity. What this means is that there would still be a lot of ship volume with low gravities and its axis would of course still be at zero G for practical purposes. Effectively this would result in the following generations of pioneers born on the ship to become progressively weaker, and the average crewman would be weaker than the average Earth bound person. By the time the ship actually arrives at its destination the colonists might find they're constitutionally severely disadvantaged, especially if they were to try living on a planet with a similar gravity to Earth's.

I suppose though, it's not impossible to envision different spin cycles being run regularly, so that over certain periods of time, the ship gravity is perhaps a little more than 1 G in order to compensate for the crew's inevitable wanderings into the upper parts of the ship.

These problems and many others of a similar nature however would only be the tip of the iceberg, because the real problems would come from cosmic radiation and social disorders.

Space is alive with a multitude of deadly rays. If unprotected, human beings travelling in ships for long periods of time would eventually fry. Cancerous growths, leukaemia etc. would be run of the mill in the best of cases, most would die from direct radiation sickness.

The way to prevent this is by shielding. The first type of shielding would simply be as thick a mass of matter as one can place between himself and the rays. But the walls of a ship would be pathetically thin even if constructed of solid lead a meter or so thick. Another alternative would be to create some type of constant magnetic field in order to protect the ship occupants.

Such a field would have to be many times stronger than Earth's natural one in order to protect the ship since Earth's is much thicker. Providing the energy for such a field (which has to be on all the time) would certainly pose a large problem.

It would mean that several fusion generators would have to be in operation on the ship. One for the propulsion system and a second to run the magnetic field. It would also probably be necessary to carry at least one more for backup purposes. A minimum of three would be the least one could envision on such a ship.

It is the sheer dimensions of such a craft however which ultimately are the most damning characteristic. A project of this type would require the combined efforts of an incredibly vast number of people, untold trillions of dollars and years of complete co-operation in order to build it.

Lastly, who would actually set off in the thing? It would be like being sentenced to life imprisonment with a number of other people. Psychological problems would be bound to arise; don't forget we have wars here on Earth, where we have a lot more space than would be available on such a ship. Even given the fact that the men and women selected for such a trip are bound to be extraordinary individuals, one cannot guarantee that over a long period of time they will maintain their discipline, nor can one guarantee that their offspring will be as disciplined. In fact, knowing human nature, it would be wiser to bet that there *will* be a war on board sooner or later.

Thanks to relativity however, all of these problems would be greatly diminished by the time dilation effect which the crew would experience if we could travel at a large percentage of the speed of light.

A person inside a ship that travels at let's say 95% the speed of light towards Alpha Centauri would reach it in 4.53 years as time is measured from Earth ($4.3 \times {}^1/_{0.95} = 4.53$), *but for the person in the spaceship less than 1 year would have passed.* For that person, the trip would have taken only about eight months.

       If this person where to explore the Alpha Centauri system for one year and then returned to Earth, for the people on Earth a total of 10.06 years will have passed ($4.53 + 1 + 4.53 = 10.06$) but for the space traveller, only two years and four months will have passed (one year and four months of total travel time and one year of exploration).

If the astronaut sets off from Earth at age thirty and had left a wife of a similar age and his one year old child behind on Earth, on his return his wife would have aged by 10.06 years, making her 40 and his child would now be 11 years old, while the astronaut would only be 32. Strictly speaking, the figures are not completely accurate because there are many factors to take into account, such as the fact that such a ship would have to accelerate gradually otherwise the occupants would be crushed by the G forces. Similarly, at the other end of the trip, it would have to turn around and decelerate.

Suffice it to say however, that Carl Sagan, a respected figure in most aspects of astronomy, is on record as having stated that even the centre of the Galaxy, which is 50,000 light years away, can be reached during a human lifespan thanks to the time dilation effect experienced at velocities approaching the speed of light.

If we assume a ship capable of travelling at 99.9999 % the speed of light, then a trip to the centre of the galaxy would take 36 years (as time is measured by the crew members). Earth however would have undergone 50,000 years of evolution by the time the ship gets to its destination and would be totally unrecognisable to the elderly crewmen if they returned to their home planet another 50,000 years later (another 36 years for them. Assuming they took no stops along the way or on arrival and they set off at the age of 20, they would be 92 years old on their return).

The human race might have blown itself up in the 100,000 year interval, or the entire solar system might be colonised with technology reminiscent of magic to the ship travellers, the Earth itself could even have been utterly destroyed by being slammed into by some wild asteroid.

Because a ship travelling at such speeds would reach its target in less time (as measured for the ship) it would permit the building of smaller ships since the crew only has to live in them for a few years at most. While the biggest problem of such a design would be the controlling of the incredibly high temperatures generated by its propulsion system, the other problems would be lessened by several magnitudes. Artificial gravity could still be induced in the same manner, and calcium degeneration would occur to some degree, but would be acceptable. Astronauts have already lived for several months in a zero G environment and recovered fully once back on Earth. Supplying them with a minimal amount of gravity and regular exercise should make a few years in space a possibility. As for the radiation problem, a ship that had a nuclear engine system capable of propelling it at a theoretical 95% the speed of light for a period of a couple of years would not be short of energy for powering its magnetic coils. The key to such ships would lie in the development of fusion and anti-matter rocket engines.

Even so, we still have the problem of actually building a ship which can travel at such speeds right? This would certainly be a challenge, to put it mildly, but actually, travelling at an appreciable percentage of the speed of light might not be as difficult as one might suspect.

## N.E.R.V.A.

Nuclear Engine for Rocket Vehicle Application.
It is a little known fact, like so many things to do with space exploration, that as long ago as 1959, tests were being done by NASA in order to determine the feasibility of using nuclear power for better propulsion of spaceships. The first nuclear test engine was fired on July 1, 1959.

By the spring of 1969 several reactors had been tested, and preparations were being made to try the first test flight of a rocket ship powered by a nuclear engine. There are a few worthwhile results of these tests which should receive a little more exposure.

Consider the technology of the sixties. Think of the kind of home appliance you had then compared to the ones you have now. The writing of the very text you are holding would have been done on a typing machine, resulting in endless corrections and changes, all done on paper, unlike what has actually occurred, which is the writing of it on a word-processing program. Corrections are easily handled by simply shifting words about, adding, deleting, and then storing the information on a magnetic floppy disk. The only paper used is for the printing of a few copies of the manuscript and ultimately for the printing of the book you are holding. If you were even around in the sixties, try to recall in some detail what it meant then to take a plane to a foreign country. It was still a relatively major undertaking. Today there are people that commute to work every week by aeroplane. Some even commute two or three times a week. Personal computers did not exist. The information network to which millions of users are linked, and in which any user can dump a question and is assured of several replies from all over the world within a few hours, did not exist.

Making a long distance phone call to, say, Nigeria from Europe was an ordeal if it succeeded at all; today, a person in Europe sending a fax from his personal computer to Hong Kong is very likely to succeed on the first try.

Try and get a little appreciation of the technology back then. Then think of the Space Shuttle.

That's right, that amazing piece of engineering.

It takes off like a rocket and comes back to Earth like a plane. The computers on board are staggering. How much more advanced such a craft is than *anything* that could have existed in the sixties.

Well, not quite.

The last NERVA engine to be tested, early in 1969, burned 1 pound of fuel in one second for 850 pounds of thrust. The Space Shuttle Main Engine (SSME) which the Shuttle uses today, burns 1 pound of fuel in one second for 455 pounds of thrust. The SSME engine, unlike the NERVA engines, is being used and can therefore be said to have reached its peak, especially since it is considered that chemically powered engines have a limit of about 500 pounds of thrust for 1 pound of fuel burning in one second.

The NERVA-XE engine which produced 850 pounds of thrust was a fission reactor and an experimental prototype. Nuclear engines were still in their infancy back then, and yet already they had proven superior to even present day chemical rockets.

In fact, this particular prototype had already been recognised even then as no great achiever. The NERVA-XE was a solid core reactor and this had already been recognised as having limitations. It was thought that it would probably not produce more than 1000 pounds of thrust per pound of fuel per second.

Gaseous core nuclear engines were to be the future, where theoretically, up to 20,000 pounds of thrust per pound of fuel per second might be reached.

Gaseous core reactors however were more difficult to construct due to the great heat they generated. Studies therefore were limited to experiments and analysis of these in order to get a better idea of the workings of such reactors. This process began at around the same time that the NERVA project started. By 1967 research on gaseous core reactor engines had determined that they would function well if the high temperature problems could be overcome. Testing of materials and design ideas at reactor temperatures was the next step taken. This too was a success. The biggest hurdles had been overcome. Gaseous core reactor engines for spaceships could be built, now it was just a matter of building them and testing them. The next step would in fact have been the building of miniature reactor engines. After testing these tiny reactors, the first full blown gaseous core reactor would have been built.

As it turned out, the funds NASA had for further research with respect to this project were drastically cut. Over the first ten years of research for NERVA engines, approximately 1.1 billion dollars were spent.

Now, suddenly, this amount had been reduced to 38 million dollars.

One of the two most promising designs of the first gaseous core reactor engine, was studied by United Aircraft Research Laboratories. The actual creation, despite its feasibility, was cut short by the lack of funding of NASA's budget, but the project director of United Aircraft at the time, estimated that the first of these reactors would produce a thrust of 1800 pounds for 1 pound of fuel used in 1 second. This value is nearly four times the SSME value of 455 pounds. NASA engineers, who worked on a different, if similar design, estimated a first generation gaseous core engine with a thrust of 2000 pounds per second for each pound of fuel used, well over four times as powerful as the SSME.

No doubt, with today's technology such values could be improved upon.

While a NERVA engine might not make trips to the nearby stars an everyday occurrence, it would permit us to send unmanned probes to explore these stars and their possible planets and receive results during the lifetimes of the scientists who saw them off in the first place. It would also place the entire Solar System within our reach for colonisation purposes, by reducing travel times between planets from periods that would span months or years when using chemical rockets, to weeks or even days. And eventually, with the undoubted advances in technology which would arise, future versions of the engine would permit man to reach the nearest stars.

Keep in mind too, that while the first NERVA engines would use nuclear fission, it is only a matter of time before man masters nuclear fusion, which is far more efficient, and after that, anti-matter propulsion is the next logical step. These last two technologies are still in our future, but it has been proved quite conclusively, at least on paper, that nuclear fusion and anti-matter propulsion would make the stars truly a reachable goal for mankind. These are facts which anyone so inclined can check on.

◆ ◆ ◆ ◆

It's only logical that if mankind can one day reach this sort of technology, another intelligent race, perhaps much older than our own, existing on some distant planet, might have gone a little further and have done so before us.

The chances that we come in contact with such a race might seem remote given the vastness of space. In fact so remote that we may be forgiven for saying that it would be almost impossible.

If we assume some race at some point in time did develop such technology, given the complexities involved, the more likely outcome, is not necessarily that they come into contact with alien races, but rather, that in the spreading of their own race between the stars, occasionally, some of their number get cut off or lost from their home planet.

And this may be somewhat closer to what I propose is the case for our own human origins in the course of this book. In this case then, we need not have come into contact with an alien race that built the structures on Mars,[7] but rather, we are the descendants of those aliens, which of course would only mean we have lost contact with our origins. Considering the many false ideas the human race has had and continues to have about itself in relation to its Universe, such a proposition stands up to much better scrutiny, and, as you will discover in this book, it is also a theory for which we have undeniable proof. Written in stone one might say. Several tons of it.

## Other Theories

While Einstein placed the speed of light as an upper speed limit, in mathematical terms, there is a tiny loophole in the theory of relativity. Mathematically (and only mathematically so far) it can be proved that there is some possibility for the existence of what G. Feinberg (their "discoverer") called Tachyons.[8]

Tachyons are supposed to be particles which only travel faster than the speed of light. Such hypothetical particles have never been detected in reality and it's arguable as to whether they exist at all, but if they do, they may provide some means for us to travel faster than light.

A final alternative for interstellar travel would be the concept of hyperspace. A ship with a hyperspace drive would be able to somehow "flip" into an alternate dimension or universe. In this theoretical universe one could then move a small distance and then "flip" back to our own universe where lo and behold, you have moved billions of kilometres. The theory is that the alternate universe could be a shrunken mirror image of our own dimension, and moving one inch there might be equivalent to moving one light year here.

---

[7] See the next chapter.

[8] Feinberg popularised the notion of tachyons. The credit for actually postulating them in the first place belongs (as far as I can tell) to O. M. Bilaniuk and E. C. G. Sudarshan.

Tachyons and hyperspace drives however are usually the realm of science fiction and not seriously considered in most cases.

Hyperspace theories aside, the point has nevertheless been made that despite the enormous distances involved, interstellar travel is possible by more conventional means. In fact, if we were truly inclined to do so, we currently already have the technology required to send at least the first probes to the nearest stars. In fact we had the beginnings of this technology as far back as 1969. A sobering point to keep in mind when we become inclined to think it impossible that some alien race may have visited this Solar System long ago.

# 2      The Face

In 1976, the now famous Viking probes arrived in Mars orbit.
Although earlier probes by both the Russians and Americans had been
sent to the red planet before, and taken many pictures already,[1] the
Viking probes performed what remains to date the most extensive
survey of Mars,[2] returning over 52 000 photographs.

Both Viking I and Viking II comprised an orbiter and a lander. On
arrival, the orbiters would go into Mars orbit and return pictures of
the surface, while the landers would provide more detailed
information of the planet's soil composition.

The Viking I orbiter took thousands of photographs of Mars
and these were relayed back to Earth by electromagnetic waves. Due
to the sheer number of pictures, it was some time before all the
photographs had been thoroughly analysed by the Jet Propulsion
Laboratory staff of Pasadena California.

Amongst the pictures received, were a few which should ensure that
Viking I is remembered as the most important probe in space
exploration for a very long time.

These photographs[3] showed an enormous structure, about 1.5
kilometres in length which the Viking imaging team came to call "The
Face".

Despite the fact that the resolution of the pictures was not of
the best quality, this structure looked uncannily like a human face.
Clearly visible, even in the original picture, is one eye, the hint of a
nose, and half of the mouth. The other half of the structure though is
hidden in shadow. On enlarging the picture a bit, one can also make
out that the Face is wearing some type of helmet, or perhaps, the
feature is supposed to represent the hair, and the right nostril also
becomes more visible. A larger portion of the mouth is also clearer.

Although immediately noticed by the imaging team,[4] the Face was
almost as quickly dismissed as a natural rock formation. The concept

---

[1] The US probe Mariner 9 for example, returned over 7300 photographs between
November 14, 1971 and October 27, 1972.

[2] The Mars observer probe, launched on 25 September 1992, was supposed to perform
a complete topographical survey of the planet, but according to NASA, the craft was
lost during its last manoeuvring stages on August 21, 1993.

[3] Received on July 26, 1976.

of an artificial structure being present on the surface of Mars was just too incredible to accept.

Not all were so hasty though. A group was formed by technicians and staff that did not believe so readily that the Face was a natural rock formation. They called themselves the Independent Mars Investigation Team. It's probably no accident that the initials of the name spell the acronym IMIT. I'm it.

The Independent Mars Investigation Team made some additional shocking discoveries.

About 15 kilometres or so to the south and west of the Face is a group of pyramids. The shadows formed by the light of the sun on these structures is identical to the ones formed on aerial photographs of the pyramids found in Egypt. Another (very large) pyramid structure is found further south and west of the Face; this has four regular sides descending from the top, but one of the sides seems damaged, resulting in what looks overall as a five-sided pyramid.

It was postulated that this pyramid had suffered a partial collapse, probably as a result of meteoroid impact, since several craters are clearly visible in the region.

Some smaller and less well defined structures are also present near the group of pyramids, and although these are not really considered as evidential of anything, their shapes although incomplete in most cases, have a distinctly regular look about them, suggesting further artificial constructs were probably in the area.


Being scientists after all and thus of a cautious nature, the IMIT group did not release this information right away, they wanted to be sure that they were not wrong.

When they did go public in the early 1980's, their discovery was met with ridicule and hostility from most of the scientific community. The idea that extraterrestrial life could be found so close to home was unthinkable for most of the so called "experts". In addition, the architecture of the buildings was so obviously related to Earth that most people could not accept that the IMIT group was being serious or scientific. To the obvious dismay of the group, the news of a Face and pyramids on Mars was printed not on the front page of every newspaper around the globe, but rather, mainly in the middle pages of tabloid magazines which are well-known for stories about ghosts,

---

[4] The man responsible for first discovering the Face was Thomas Owen.

alien abductions and whether Elvis was spotted in Times Square recently or not.

The reasons for this are many and we shall cover some of the most important ones in a later chapter, but for the moment we can limit ourselves to say that two main factors prevented this event from becoming the most public and important of discoveries.

Firstly, the atmosphere of ridicule generated by the media placed the discovery quite firmly in the field of 'Unbelievable-claims-made-by-people-trying-to-sell-news'.

The fact that in this case the news were real could not be appreciated. In large part this is because since the 1940's or 50's the world at large has been very carefully taught to meet stories about UFO's, alien abductions and encounters of the third kind with a lot of scepticism if not out and out disbelief.

Most people however don't readily recognise that their perceptions about this subject are not really their own. The media, subtly influenced and encouraged by the various governments of their respective countries, has made it almost impossible for a respected scientist to come out and say: "Listen here, the UFO's really are out there." Even before anyone listened to the evidence (or lack of it) for such a claim, the scientist in question would become the laughing stock of the scientific community, with the result that no-one will really take him seriously no matter how brilliant the proof.

Secondly, the average person in the street didn't (and to a large extent still doesn't) really care about UFO's, extraterrestrials and all that "stuff". The general reaction of people, if they were to find out that aliens existed on Mars at some point in time, would be to say something like "Gee, that's really amazing isn't it? Would you pass the salt please?". To be fair though, once again this is at least in part the result of years of subtle brainwashing. How many classes did you take at school with a name like Aliens 101? How often are you allowed to mention the possibility of alien life in a serious manner at an educational institution?

Personally, I distinctly recall a primary school teacher asking the class during a geography lesson if we knew how planets and stars were created. When the child she'd decided to call on began to express the concepts of swirling gasses contracting into denser materials, he was cut off short by the teacher.

"No. Where did you read that?" she asked him.

The boy became a little intimidated and began to reply he'd read it in a science book he'd seen in the local library.

"Well, that's a lot of nonsense." Said the elderly woman "God created[5] all the planets and stars of course. We all know that." This was the first and last time any aspect of astronomy was mentioned in that class for the entire year I was in it.

While most people's experiences in relation to astronomy and things of an extraterrestrial nature may not have been as radical as that, most of us never really had a chance to learn much about the topic or even discuss it generally. Talking about UFO's and such things is considered a waste of time and impractical by well over 95% of the planet's "professional" population, or so at least goes the perception.[6] This however is not because such things are unimportant or irrelevant to our lifestyles, but because we have been *taught* that they are.

On the contrary, the existence of artificial constructs on Mars has implications which are so vast and far-reaching in their effect that they will not leave any sphere of human endeavour untouched. I am not merely referring to philosophical musings or the expansion of mind which would naturally occur from such knowledge. A man with barely enough to eat for himself and his family is bound to be as disinterested in his true origins as he's going to be of discussing Picasso's painting technique. It's irrelevant to his life.

But the Face and pyramids (as I hope to demonstrate in a clear and logical manner) will have an intensely practical and very real effect to our lifestyles. These changes however, cannot come about until a sufficient number of people are made aware of the possibilities which lie before us.

---

[5] It's not the concept of God which is damaging of course. I can readily accept that some people believe in a God and I have no objection to their saying God created all the planets and stars, regardless of whether I choose to believe it myself or not. What I do object to though, is that the way in which planets and stars are formed is completely ignored, in fact the very subject is completely denied. In short, why is it inconceivable for people such as this teacher to think that while God created the planets and stars, he/she/it did it by making masses of gas swirl and condense due to gravitational attraction?

[6] We shall see later that this is in fact far from the truth.

For these two main reasons then, ridicule and a sense of impotency when faced with the disinterest of the 'professionals', the Face and nearby pyramids were largely ignored.

As the years passed however, imaging techniques and computers improved, and soon, by using a new method for interpreting the original data, various investigators were able to achieve an unprecedented resolution without altering the original information.

Perhaps the most efficient of these processes is referred to as CSI (Cubic Spline Interpolation) and it is a computer process whereby the original blocky image is re-interpreted to a finer degree.

The pictures which resulted from this computer technique left absolutely no doubt as to the nature of the Face or its nearby 'City' of pyramids.

I have included in this book, the original NASA pictures of the Face as received by Viking I (plates 1 and 2). When I first saw these along with the nearby City I compared the Mars pyramids to aerial photographs of the Giza pyramids found in Egypt. On doing this I became almost certain that the pyramids on Mars were very real and definitely artificial in origin.

Then one day, while I was busy buying a vacuum cleaner in an electrical appliances shop I saw on the many television screens that were displayed there, a programme showing images of the Martian Face at unprecedented resolutions. The Face had a clearly defined mouth, nose complete with nostrils and eye sockets. The pyramids were unquestionably pyramids and not angular mounds of soil or rock. Some looked to be in better condition than the ones we have in Egypt. On seeing these images any shadowy doubt I might have still harboured vanished instantly.

Unfortunately, as I was in an electrical appliances shop I was unable to videotape the programme, nor have I been able to get a copy of these super-enhanced pictures from the local broadcasting company or through the Internet.

My interest however was definitely piqued, and I resolved to find out all I could about this structure and its neighbouring anomalies on the Martian surface.

# A Summary of the Face Evidence

The dimensions of the structures on Mars can only be estimated rather roughly, but still adequately enough to give us a reasonable idea of their size. On doing this, we find that the Face is estimated to be 1.5 kilometres (almost a mile) in length and some 400 metres (about 1300 feet) high in elevation. The largest pyramid of the group, shown in Plate 7 and known as the D&M pyramid, is estimated by Vincent DiPietro and Gregory Molennar (the two persons credited with first discovery of this pyramid) to be about 1.6 Km (1 mile) in length by 2.57 Km along the diagonal (1.6 miles).

About 15 kilometres to the south-west of the Face (9.3 miles) we find a collection of pyramidal structures which in addition to having artificial properties of structure individually, also seem to have a coherent layout as a group.

As has already been stated, the Face was discovered on July 26, 1976. Since then, perhaps the most influential exponent of the view that the Face is of artificial origin has been Richard Hoagland, who for the past ten years or so has led the independent investigation of this feature. Hoagland has had exposure to the space community for a long time, having been under contract as a consultant to NASA's Goddard Space Flight Centre and generally being involved in a consulting capacity with several space ventures. It would seem though, that Hoagland has mainly been involved from a media point of view, in fact he was a member of the JPL press corps when he first saw the blurry image of the Face. Perhaps for this reason, Hoagland has largely been ignored, despite being quite vociferous on the subject, having published the book *The Monuments of Mars* and produced at least four videos based on the theme.

Hoagland has done his own research concerning these structures, but was originally influenced by the work done by (and he is quick to give credit to them for it) Vincent DiPietro and Gregory Molennar,[7] who

---

[7] DiPietro and Molenaar are both engineers working at Goddarrd's Space Flight Center, who conducted their work in a responsible, technical, and honestly scientific manner, but as a result of their work being of a 'controversial' nature they have been forced to publish their discoveries independently of NASA and the orthodox 'experts', and without endorsement from the organization that employed them for their efforts.

searched the entire Viking data file and found a second picture which revealed more of the part of the Face which had been in shadow.[8]

Using this new image they then enhanced the original data. Sceptics may correctly argue that a computer can indeed be used to change a photograph to make what was not there in the first place, seem a reality[9].

The difference here lies between the terms *computer generated* and *computer enhanced*. Admittedly, the distinction is sometimes a fine one, and in certain cases so blurred as to be difficult to find, yet, as always, we must let logic be the final judge. There are several computer imaging techniques which would be useful in reinterpreting the data of the Viking camera. These would be, in order of importance:

1) Magnification.
2) Filtering out of background noise by averaging out the difference between adjacent pixels[10].

---

[8] Interestingly enough, this second frame was originally misfiled in the NSSDC (National Space Science Data Center) archives.

[9] As an aside, this book's cover, which was largely created by my friend and business partner Russell Petersen, is an example of what computer enhancement can result in. Starting with the original Cydonia region image, we proceeded to enhance its quality by several means. Firstly it was 'cleaned' of the numerous computer errors (single points of bright white) by the same process of pixel interpolation used by Carlotto, secondly it was magnified slightly and then contrast enhanced in order to give all the local features more definition. This in turn brought to the fore a few of the remaining pixel errors by making them once again clearly visible, requiring a second pass of pixel interpolation. Lastly false colour was added by the computer, which reinterpreted the grayscale image using appropriate colour shades chosen by us, to approximate the orange to rust colour of the Martian surface. While extensive computer enhancement took place on the cover then, apart from the false colour, no data which was not there has been added, in fact, if anything, through pixel interpolation, some data was averaged out, resulting in a net loss. This is clearly evidenced by the fact that several passes of pixel interpolation result in a gradually progressive loss of features. The remaining black dots are camera registration marks which form a grid-like pattern throughout the Viking images.

[10] When an image is magnified enough, eventually, each grain or pixel becomes large enough to be individually visible. If we now compare one pixel with the ones immediately adjacent to it, we will usually discover a minimal difference in shade. Where there is a computer error however (as a result of 'noise' in the transmission) the difference in shade of the erroneous pixel is dramatic from that of its immediate neighbours. If we average the difference between two pixels by interlacing them, we have now created in effect a third shade which combines elements of both pixels and effectively gives a smoother result. The interleaving technique which gives the best results is one that is known as Cubic Spline Interpolation (CSI) and uses pixels from

3) Contrast enhancement of the photograph, which in raising the play between light and shadow can raise the level of detail shown by an image.

4) The addition of false colour for a more realistic effect.

By using just such techniques, DiPietro and Molennar have produced pictures which show amazing detail when compared to the original. If we assume that the process was undertaken in an unbiased manner (a fair assumption since the results produce a consistent image)[11] we would then come to the conclusion that the new pictures must have indeed improved on the original, and the story told by them is a better one.

DiPietro and Molennar are not the only ones to have researched the Face and nearby structures. Mark Carlotto, began examining the Viking data in 1985 after reading about Hoagland's theories.

Since then, Carlotto published a book in 1991 entitled *The Martian Enigmas*.[12] As I have not had access to DiPietro and Molenaar's work, I cannot compare it directly with Carlotto's but without a doubt, *The Martian Enigmas* must remain an outstanding work of true scientific investigation and most definitely, until better pictures of the Face become available, remains an indispensable text for the serious researcher.

Dr. Mark Carlotto is probably best known for the fact that he analysed the Face by using a method that re-creates a three-dimensional shape

---

all around the erroneous one to produce a new shade for it. This is also the technique mainly used by Carlotto.

[11] While to tell a computer to interlace each pixel in a starburst (circular) pattern may not be easy, to tell it to do so in such a way that the final image produced resembles a human face, from a supposedly natural rock formation, is a *lot* harder (and in any event this is *not* what DiPietro and Molennar did). The reason why the CSI image processing is so relevant, is because it is performed by a computer in an unbiased manner and is applied to the whole picture. The computer does not care in other words, whether the final result is relevant to a human observer or not, it's not "trying" to make the picture shown any particular feature, meaningful or otherwise. For this reason, if the feature being investigated does end up showing meaningful (and already suspected) attributes, it's more likely that they are indeed *real* attributes. In any event, the very same image processing techniques are used by NASA on other, more natural features of Mars for the very reason that they give a better representation of such geographical features.

[12] See bibliography for details.

from a two-dimensional shaded picture. In other words, the process tries to interpret a shaded image as a three-dimensional object. His findings indicate that the impression of a human face remains consistent even when the structure is viewed from several different angles.[13] Considering that we now have two sources of computer analysis independent of one another which both indicate that the Face indeed does look extraordinarily like a human visage, we should begin at least to seriously contemplate this possibility.

Especially since the pyramids were also examined by Carlotto and once again found to be consistent with the interpretation of DiPietro and Molennar. Furthermore, Carlotto investigated in more detail a structure which Hoagland named the Fort in 1983. This, according to Carlotto is a polyhedral object with very straight sides and regularly shaped markings or indentations. According to his 3-D shape from shading technique, the object appears to have been some type of building which now has a damaged upper portion.

Carlotto is a Division Staff Analyst at the Analytic Sciences Corporation (TASC) an analytic service corporation that performs satellite-based image processing, and has been involved in computer image processing for almost fifteen years now, so he should know what he's talking about. In addition, using techniques developed at TASC to detect *man-made structures in satellite images*, by fractal analysis, Carlotto and some of his colleagues concluded that the Face does not share characteristics of its surrounding terrain, which in itself is at least partly indicative of artificial origin. When combined to the fact that the Martian pyramids seem to be aligned in distinct geometric patterns, it becomes difficult to disregard the subject as being a natural rock formation or trick of the light as originally suggested by NASA.

This last analysis by Carlotto of the Martian structures has been somewhat underplayed, but in my opinion it is in fact very relevant, because the Martian features appear to have a non-fractal origin according to the computer interpretation. This may seem unimportant, especially if we are a little fuzzy as to the meaning of the word *fractal*, but in fact, it's far from it.

---

[13] It is particularly interesting to note that natural rock formations on Earth which look like a face or other meaningful shape when viewed from a particular perspective, *do not* retain this consistency of character when viewed from other angles; further indication that the Face on Mars is unlikely to be natural in origin.

Chaos theory,[14] which is at the front edge of presently used mathematical theory, has proved quite conclusively, that nature follows a mathematical pattern in all of its endeavours. A leaf may look like an irregular, random shape, and yet there lies in that leaf a precise mathematical shape. Fractal images are pictures composed by mathematical formula, which create shapes we find in nature, such as leaves or snowflakes. Remember that no two snowflakes are alike! Yet there seems to be an undeniable and extremely complex order underlying the apparent chaos of nature. It is this very concept, along with quantum mechanics that is forcing more and more previously agnostic or even atheist physicists and mathematicians to re-evaluate their position on the existence of God.

If the Martian features in question are non-fractal in origin it can only mean one of two things, either they are artificial or our computer made a mistake. While Chaos theory, by its very nature, is an immensely complex subject, and therefore not absolute in its predictions, the theory that nature uses fractal shapes is pretty much ascertained by most mathematicians. Once again then, this seems to indicate that the Face is indeed an artificial construct.


Additionally, according to work done by Hoagland, the cartographer Erol Torun and if to a lesser extent, by Carlotto too, it would also seem that the City has definite geometric consistencies which are very unlikely to happen naturally.

It is interesting to note, that Erol Torun began his investigations concerning the claims made by Hoagland of geometrical consistency in order to challenge him on this point. On completion of his detailed study however, Torun changed his mind and in fact began to agree with Hoagland.

This tendency to at first view the Face and nearby City in a very sceptical light is one that is actually almost universal amongst those who are now exponents of the idea that it is artificial in origin.

Carlotto on reading his first report of the Face thought it was some sort of joke, while Hoagland was initially sceptical until he became acquainted with DiPietro and Molenaar's work. As for myself, I initially considered the Face to be in the same sort of class as alien

---

[14] Chaos theory is examined in more detail in chapter 5.

abductions and sightings of Hitler clones living in Argentina: good material for a science fiction novel.

In the picture section, using computer techniques similar to those used by Carlotto, I also present enhanced images of the Face and City, but must stress that his work is in no way superseded by mine in this regard, in view not only of his more extensive expertise which are related in particular to computer imaging, but also due to his more extensive imaging processes with respect to the three dimensional shape of the Face. Having said that, the contour map of the Face shown on plate 5 is a technique that although similar (if somewhat inferior) to Carlotto's shape from shading technique has to my knowledge never been done before[15].

Similarly, the false colour enhanced images are to my knowledge the first of their kind made widely available to the public, although the work done by some members of the Mars Mission (the Hoagland led organisation) may have included false colour photographs of which I am not aware. The merit of false colour is of course more of an aesthetic nature than anything else, but I feel colour gives the whole image a more realistic perspective, and thus its inclusion should be presented at least on occasion.

Those that remain sceptical as to the origin of the photographs are invited to check up on them for themselves.[16]

There is one more way in which the existence and artificial origin of the Face can be somewhat demonstrated. If in analysing our own

---

[15] This process basically relies on the computer to make contour maps of the features shown in a photograph by using the shadows as a form of representation of heights at different levels. Plate 5 incorporates contours taken from three different levels of 'altitude'.

[16] The original Viking Orbiter images of Mars are available on CD-ROMs from the National Space Science Data Center. Contact the Request Co-ordination Office at:

NSSDC, Code 633
NASA Goddard Space Flight Center
Greenbelt, MD 20771              Tel: (301) 286-6695 FAX: (301) 286-1635

Internet: request@nssdca.gsfc.nasa.gov

The Face images are found on CD-ROM volumes 10 (frame 35A72) and 11 (frame 70A13) of the Viking I Orbiter Images of Mars, which span at least 28 CD-ROMs and have the National Space Science Data Center ID of 75-075A-01c/d.

terrestrial pyramids we begin to uncover evidence that they have extraterrestrial origins, this would once again suggest that it is not only very probable, but possibly inescapably obvious, that the Face and Martian pyramids are indeed artificial.

From this point on though, my arguments throughout the remainder of this book are based on the assumption that the Face on Mars along with the nearby pyramids is real and artificial in origin.

If you do not yet accept this basic premise, you might yet change your mind when in the next chapter we investigate the Great Pyramid of Giza found in Egypt.

◆ ◆ ◆ ◆ ◆

If we accept that the Face is artificial in origin, then the next logical steps seem to be:

1) The looking to our own pyramids for some clue as to their true nature and origin, while at the same time keeping an outlook towards our history which is free of preconceived ideas. This naturally would lead to....

2) ....a reconstruction of history based on this new approach.

Given the size and complexity of such a task, we are left with only two approaches: we could analyse in detail any evidence which tends to support the idea of our extraterrestrial origins, or we could try and get a more broad view of the subject, looking for large if somewhat blurry pieces of the puzzle, instead of detailed but tiny ones.

The first method seems to me to be narrow minded and I fear I would be accused of only showing those details which fit with my theory. The second approach of course has the hidden trap of allowing critics to say that the arguments are so general and lacking detail as to be inconclusive of anything. For these reasons then I shall adopt the practice of looking at the problem mainly from a general point of view, but taking some time to outline key details which should validate the 'joins' I make between my sometimes large puzzle pieces. It seems to me also a good idea to introduce at the end of each chapter, a summary of the salient points just discussed.

Before we embark on the above-mentioned points though, it would be wise to try and make at least a preliminary deduction of just how far the relationship between the Martians and Earthlings extended.

Although somewhat speculative, this exercise serves two purposes. Firstly it may shed additional light on the origin or intent of the Martians/Aliens, but perhaps more importantly, if our conclusions in this exercise are later confirmed in a number of ways independent of the ideas outlined here, we then have a check for our theories and this would further indicate that we are most probably on the right track.

# The implications of the Face and City

There are several obvious conclusions (and not a few less obvious ones) we can arrive at in a logical fashion, simply by knowing that the Face and City on Mars exist. The rest of this chapter is dedicated to pointing these out as well as explaining my reasoning, so that the reader can judge for himself whether or not he agrees with the logic.

## A Connection Between Mars and Earth

Let us first of all examine the possibility that the pyramids on Mars are *not* connected in any way to the ones we have on Earth. At first this may seem to go against the generally accepted flow of logic, but by using this approach, what we are in fact really doing, is saying that we assume nothing. We are thus forced to reconstruct events up to their present conclusion, and in doing so we can indeed often find out more of the truth than not if we'd simply accepted the *prima facia* [17] evidence. Our motto should be the one outlined by Sir Arthur Conan Doyle's famous fictional character, Sherlock Holmes: "When you have discarded the impossible, whatever is left, however improbable, must be the truth."

Sherlock's statement however, is only correct of course, if our logic is faultless. It is for this reason that it's better to begin the investigation by not making any *a priori* judgements.

Some people might argue (quite convincingly and possibly correctly so) that humans are a far cry from intelligent life, but for the purposes of this argument, let's assume that by intelligent, we mean a form of life capable of constructing the Face and nearby pyramids.

The first question we might ask then is: What are the chances of *intelligent life* evolving *independently* on two planets of the same solar system?

There is no hard and fast maths rule we can use to answer this question in order to arrive at a precise percentage of probability; nevertheless by approaching the problem in an intelligent manner, we

---

[17] A slight pun on words here, since *Prima Facia* is Latin for "first face". The meaning implied obviously, is that first appearances are not always the correct ones.

can certainly make an educated guess, and this should not be so far from the truth.

All that we know about life indicates that it is an evolutionary process which takes thousands and thousands of years —in the best of cases— to evolve to the point where it's capable of building structures like pyramids. Current scientific theory in fact, would say it took about 2 billion years on Earth, give or take a few million.

This very fact would place the chances of intelligent life occurring *twice* in the same system as remote to say the least.

This 'remote' becomes very, very remote indeed when you analyse our solar system in any detail, because it becomes obvious that only about seven bodies in it (apart from Earth) might, by any stretch of the imagination, have at one time supported intelligent life.

These are: Mars, Venus, Jupiter's satellites Europa, Ganymede and Callisto, the Saturnine moon Titan, and the Uranian satellite Triton.

Any remotely competent astronomer will add that I have been extremely kind in naming all of the above as possible life originators. Venus has a surface pressure of about ninety to a hundred times that of Earth (equivalent to being submerged in about a kilometre of water on our planet) and it's atmosphere is composed mainly of Carbon Dioxide (about 97%) and Nitrogen (about 2%) with Sulphuric acid traces. The Jovian satellites are all inside the gas giant's radiation belt and are far enough from the sun to be rather chilly, while Titan and Triton are certainly of a frigid nature to put it mildly. The lower the number of possible alternatives to Earth as a springboard to life, the less likely it is that we are going to have two types of it evolving and *both* reaching an intelligent status.

It is also true however that the Universe is indeed infinitely large, and therefore, if there is even the remotest chance of something happening, we can pretty much say it's bound to have happened somewhere, at some time in the Universe. Since we are in such an optimistic mood, why not in our very own solar system?

It should be obvious by now that this is not a logical way of explaining the Face on Mars, but let's persevere a little longer.

Since Mars could (and in view of the Face, most certainly has) at one time, have had an atmosphere and running water,[18] which would be prerequisites for the kind of life we are envisioning, let's assume for

---

[18] The ample geographical evidence for this is discussed further in Chapter four.

the moment (against our better judgement or indeed any logical process) that intelligent life evolved separately there and built the pyramids found by the Viking probe. If we were to do this, our theory would instantly become so obviously wrong to be ridiculous, because we would have to somehow explain why creatures which developed completely separately from us, and hence have no reason to look or think even remotely like humans, ended up having a similar sense of architecture.

How could we explain this? Why would Martians want to build something so similar to our own structures? In fact why should they have buildings at all? Their whole social structure could just as easily revolve around the burrowing of vast tunnel complexes in a fashion more akin to termites than humans. The limits of coincidence here would have to be stretched to breaking point, and finally well beyond it when we take into consideration the Face, since this is undeniable proof that the Martians resembled us in appearance, or at the very least had knowledge of our features.

These points then should make it obvious even to the most sceptical of persons, that *if the Face and pyramids on Mars are artificial, there is an obvious connection between those pyramids and the ones we have on Earth!*

I said at the beginning of this section that I shall consider the Face and pyramids to be real and artificial in origin, yet I have chosen to structure the above sentence as I have for the benefit of those that on reading my claims find them too fantastic to be based in reality. While some may be inclined to doubt the reality of the Face, I don't believe though, that one can argue with the logic outlined above, therefore, the only area of doubt which one can have at this point, should be as to the veracity of the Face.

I would beg such sceptics to read on, because it has been my experience while researching and discussing this topic, that while some people tend to initially disbelieve my theories when approached from the Face as a starting point, they are a lot more willing to accept them once they have been made aware of certain facts pertaining to the Great Pyramid at Giza. This is an important point, because the Great Pyramid is not a hazy photograph. Its massive reality resides silently on the outskirts of Cairo, as it has done for thousands of years, and the facts concerning it are not only well documented, but can still be physically checked today.

I chose to approach this subject from the Face on Mars as a starting point for the simple reason that I was led to investigate the pyramids of Giza as a *result* of it, but if some find it easier to be led to an investigation of the Face by first examining the Giza pyramids, then such a route should prove to be just as valid.

Indeed if this approach did not prove valid, we might assume that there is something questionable about my assumption that the Face is real and artificial in origin. Similarly, if looking at the Giza pyramids leads us to look for evidence of a connection with extraterrestrial life, then it would seem to validate the theory that Mars may indeed be harbouring alien artefacts.

The conclusion that there must be a connection between the Mars pyramids and the Earth ones, of course brings us to several realisations. Paramount amongst them is the fact that if there has been extensive contact with extraterrestrial life, then we have some serious gaps in our history as it stands at present. Logic would indicate, that the construction of several pyramids of the size that we have here on Earth and doubly so of the larger ones on Mars, is a rather major undertaking, and would affect a large part of the society present in the location of their construction.

This in turn implies a rather extensive contact with such 'aliens'.

I use the quote marks because obviously these aliens must at least superficially resemble us. At first this may not be completely obvious, but it will gradually become so on a more detailed investigation of this fascinating subject. At any rate we already have some evidence for this. In the first place, it is quite likely that the people who built the Martian artefacts were themselves in possession of human faces, otherwise why built a structure with such features?

Secondly, although we are now aware that our history is at least suspect, we do not really have plentiful legends describing blob-like beings or at least creatures that do not resemble us in some way. We *do* however have an *incredibly vast amount* of legends describing God-like beings of human appearance and possessing fantastic powers, which affected the lives and environment of mere mortals in many and varied ways. In fact there is not a culture on this planet that does not have some form of legend along these lines in their history.

For these reasons then, it's pretty safe to assume that contact between the Martians and Earthlings did exist at some point. We can also assume that the building of the pyramids involved at the very least some instruction by the aliens, and quite probably their direct involvement.[19] Further investigation of history in fact, shows that it's quite likely that we have some genetic relationship with the Martians, as I shall point out later in this chapter, and this of course raises even more interesting conclusions.

At any rate, with the multitude of probes that should soon be headed towards Mars, The Face cannot be kept hidden forever, (I hope anyway) and in the near future it will be announced for what it is by 'respectable'[20] scientists the world over.

The Mars Observer craft could have cleared this whole question up since it was to map the entire Martian surface, and would have made this book largely irrelevant, but this probe was lost for reasons still unclear,[21] just before it arrived in its final parking orbit.

To loose one probe is not unthinkable, of the 24 Mars-bound missions undertaken up to 1980, eleven have been outright failures, either failing to take off, to leave Earth orbit, or contact with them being lost before arrival.

Of the remaining thirteen, five were only partial successes, accomplishing only some of their goals, examples are flyby and lander mission combinations where the lander failed to survive planetfall or make contact with the planet (in one instance at least, the lander *missed* the planet altogether!).

---

[19] Further evidence of this is given in Chapter 3.

[20] I use quote marks, because if even only *half* of what I have said about the Face so far is true, then anyone whom is deserving of being called a scientist, should be calling for a resolution of this matter once and for all. So far, the people that have been asking for clearer pictures of the area by using another probe are not NASA's top brass, but individuals like Hoagland and Carlotto. NASA in fact seems reluctant to even mention the Face at all. This state of affairs leads me to question who the real scientists are. It would seem that most scientists at NASA seem more preoccupied with being 'respectable' than with discovering the truth. Then again, there may be reasons why NASA, among others, might not want the truth about the Face to come out too soon. See part II of this book.

[21] According to NASA, the fault may have been related to some sort of fuel leak. Before writing this book in fact, I had started a science fiction novel where the main theme was that NASA had not actually lost the Mars Observer, but had instead, been forced by its government to keep its findings secret. In the course of writing that novel I undertook research which eventually convinced me that there was a far more important book to write on this subject than a science fiction novel. Some related ideas are examined in Part II of this book.

Of the successful eight missions, one was a communications test, three were flyby missions, two were orbiter missions, and only the last two were orbiter/lander missions. These two orbiter/lander missions were Viking I and Viking II.

With this sort of track record it's perfectly plausible that the Mars Observer *was* lost, along with both of the Russian *Phobos* craft.[22] And yes, it is certainly possible that even the next satellite or two get lost on the way, but with both Russia and NASA having committed themselves to sending more probes to Mars, sooner or later, one *will* arrive intact. Russia's perhaps hastily named Mars '96 mission should, similarly to the Mars Surveyor spacecraft[23] clear up the question of the Face once and for all.

When this happens, any doubt as to the artificial origin of the Face will be dispelled, and following the arguments outlined above, we will also be able to conclude decisively that the pyramids on Mars indeed do have a connection to the ones on Earth.

## The Revision of History

It is not possible of course, in a book of this size to reconstruct the whole of human history in a very detailed fashion, I am forced then, to talk in general terms while pointing out some key examples which highlight the arguments being discussed.

Firstly, it seems appropriate to give a very brief summary of what present historical and archaeological thought tells us about ourselves and our origins. After having done this, it might be useful to try and

---

[22] Incidentally, it's interesting to note, that *Phobos 2*    took one picture three days before failing. The visual image showed a bright, white, potato-like shape with wisps of grey hinting at markings on it, surrounded by darkness. Below this was another shape, similarly bright and with less markings, but its shape was slender and long; a vertical cigar below the potato.

The potato object was Phobos, which orbits Mars at a mere 5980 kilometres of distance. The cigar shape was what the press has alternatively called a UFO, unknown object, or flying saucer.

The image offered no sense of scale, making it impossible to determine whether the cigar shape was relatively small and close to the probe when compared to Phobos, or whether it was near, or even behind, Phobos from the probe's perspective; which would make it very large indeed.

*Phobos 2* was finally lost on March 27 of 1989, while *Phobos 1* had already been lost by this stage, while en route to the Martian satellite.

[23] To be launched supposedly in 1996 and 1998 respectively.

reconstruct these theories without taking much notice of accepted forms of thought, and focusing instead on whatever factual evidence we have. If in doing this we discover gaps or inconsistencies with the presently accepted history, we should then investigate them and see if perchance they might be explainable in terms of a connection with extraterrestrial contact. If this turns out to be the case, it would seem to further advance the theory that our origins might be very different from what we have up to now presumed.

At first the idea may come as a shock or seem impossible, but please recall, that the very same man which could be said to be responsible for our present theory of evolution, Charles Darwin, was at first ridiculed and ostracised by the scientific community of his time for the theories he put forward in his book, *The Origin of Species.*

As it turns out, it may be that while getting the basics right, Darwin only uncovered part of the picture, which is why people with a religionist view of evolution, have been able, off and on, to take the occasional sound shot against Darwin's theories.


The generally accepted idea of our history begins about 3500 million years ago, when it is believed the first form of life first appeared from the primeval soup. For a long while life was pretty boring and involved mainly the slow and gradual evolvement of primitive unicellular organisms into more complex creatures. About 3000 million years later, the first known shelled creatures where around, swimming in the seas. Eventually, life made its first land sorties and over the next few millennia all sorts of creatures had evolved. The famous age of dinosaurs is thought to have taken place between 300 and 70 million years ago, according to the fossil record. The great lizards ruled for well over 200 million years, then something happened that wiped them out almost instantly from a geological point of view. The theory which currently seems most accepted by palaeontologists is that a large meteorite (or perhaps several) hit Earth and the resulting climatic changes were fatal for the unlucky reptiles. This gave mammals, which had developed about 120 million years ago and thus appeared only in the last 50 million years of the dinosaur age, the chance they'd been looking for.

Without the dinosaurs to eat them and generally make life miserable by hogging all the food, the mammals evolved into a multitude of species, several of whom also became extinct. The primates first appeared around 70 million years ago, just around the time that the dinosaurs disappeared,[24] and have basically been evolving ever since. It's interesting to note, since we're primarily concerned with this facet of history, (what with being primates ourselves and all) that not all our distant cousins made it to the present day.

Several species of primate in fact did not survive, amongst them a giant, yeti-like creature known as *Gigantopithecus*. In the book *Other Origins, The search for the giant ape in human pre-history*, by Russell Ciochon, John Olsen and Jamie James,[25] it is estimated that the *Gigantopithecus* reached a weight of up to 1200 pounds (544 Kg.) and a height of over 10 feet (3 metres).

The most well known example of an extinct hominid however is perhaps Neanderthal man, who lived predominantly between 120,000 and 80,000 years BC

Also of note is that from about 14 million years ago, the diversity of primates seems to have been very wide indeed, with many different species scattered all over the globe. Yet prior to this, the fossil record does not seem to show a clear evolutionary trend. Some scientists believe that the various species of primates must have been in evidence long before this, since the differentiation of species is a process which takes a long time, but the fossil record prior to 14 million years ago markedly lacks evidence of the plethora of primate species we find at around the 14 million year mark.[26]

*Homo Sapiens Sapiens* makes his first appearance about 100,000 years BC and is a distinct and quite separate race from Neanderthal man; it is this fact which prompts Biblical scholars to mention that as no clear link between the primates and modern man seems to exist, Darwin's theory of evolution must be in error, hence the 'Missing Link' theory.

With the appearance of early stone age man, begins human history, and if we concentrate on Homo Sapiens Sapiens, that history is only 100,000 years old or so.

The predominant trend amongst historians, anthropologists and archaeologists, is that man evolved from a primitive state to the presently technologically advanced era. While different races and tribes evolved technologically more rapidly than others, each tribe or

---

[24] Significantly (with respect to extra-terrestrial involvement) perhaps?

[25] See bibliography for details.

[26] Like the dinosaur disappearance, this point may have some significance, which however may only become apparent after considering the entire book.

race is said to have improved technologically only gradually and through time. Occasionally, a more primitive tribe coming into contact with a more advanced one would suddenly make use of the new technology, but generally, it is accepted that our present state of affair is technologically superior to anything that our ancestors might have had in their day.

Although ancient Egyptian society, like many cultures of this period, is known to have been very structured and complete with several social castes as well as laws, rules and customs, it is a rare person indeed that would dare suggest that the people in those eras were not backward compared to us.

In fact, while it is generally accepted that once a culture reached its peak, the one that immediately followed it would be somewhat inferior to it at first, the very concept of a gradual rise from the stone axe to the computer guided missile is fundamental to our concept of history.


Fire is thought to have been known to early man from about 350,000 BC and yet, until 30,000 BC archaeologist seem to think man only used stone tools. The first evidence of metalwork does not appear before 9,000 BC or so and then only for decorative purposes; at around the same time farming is thought to have begun.

By 4,000 BC metal tools have become more widespread and written language is a discovery attributed to the Sumerians as recently as 3200 BC, while the Egyptian pyramids are thought to have been built around 2,700 BC

Iron makes its appearance for the making of swords and tools in Greece only from 900 BC or so. Rome was becoming a city by about 600 BC and in 221 BC the Great Wall of China was begun.

From this point on, history is recorded in more detail, and technology generally improved mainly in irrigation techniques and in transportation. Columbus discovers the first islands off the American coast in 1492, and 5 years later Tomas de Torquemada, the infamous leader of the Spanish inquisition dies, although the practice of burning people at the stake does not. In 1642 Galileo Galilei dies still under house arrest for having suggested that the Earth is not the centre of the universe but rather rotates around the sun. Neil Armstrong sets foot on the moon in 1969 and the Vatican finally apologises posthumously to Galileo in 1992.

This brings us to the present and thus to the end of this brief summary.

If plotted on a graph, a small part of our history, as recounted above, might look something like figure 1 below.

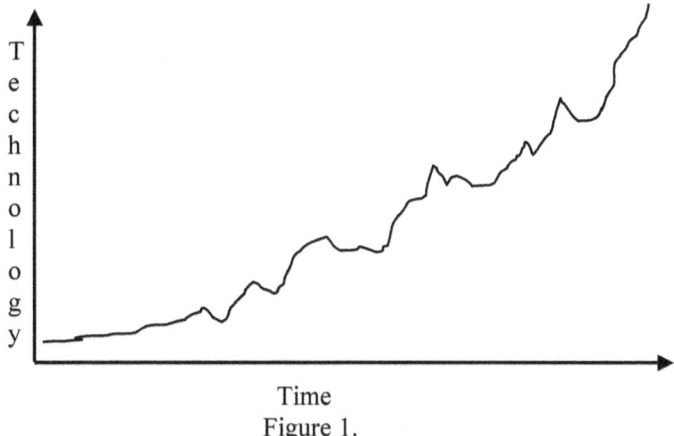

Time
Figure 1.

While there are bumps along the way, which are the result of war, famine, disease or natural disasters which set humanity back for a time, the overall trend is upward. The line above though, does not accurately depict the fact that most of the scientific advances we have ever made have occurred only in the last 100 years of our history, and most of *those* discoveries have only been made in the last 50 years and so on. It is estimated that at present, our level of technology doubles approximately every 5 years.

If we were to plot what is generally accepted as human history, starting from about 100,000 BC the graph would more accurately resemble figure 2 as shown on the next page.

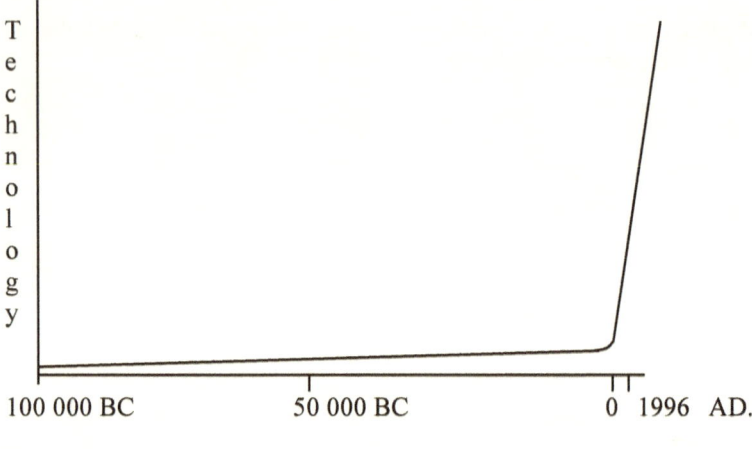

Figure 2.

It is interesting to note that technological progress makes a veritable leap over the last 2000 years or so. In fact, the year zero[27] would be an excellent point to choose if we were inclined to pick a crucial time in our history.

Something indeed must have happened around this time which was of widespread importance, because it is the point from which we date our present history. On investigating as to the reason for our current method of year-numbering, we discover that the event in question seems to be the birth of Jesus of Nazareth.

Whatever one's religious beliefs, it cannot be denied that this man had a pivotal role in history.

Unfortunately, the study of the actual events which took place concerning this historical figure is anything but an easy one. Because the personage of Jesus is so tied up in religious dogma, it is quite difficult, perhaps even impossible, to determine with precision exactly who he was and why his birth should have prompted the world to adopt a new way of recording history.

---

[27] The year zero never actually existed according to our present system of dating. The year which followed 1BC was not the year zero, but the year 1 AD. For this reason, calculations of time across the BC / AD line produce a result which seems a year short.

The only things we can say with some certainty of the period in question are:

1) A person that we refer to as Jesus of Nazareth was indeed born at this time.

2) By all accounts this man seems to have been some type of spiritual leader with views so original that his passing revolutionised human thinking and behaviour. In fact so monumental was the trauma, that even today the echo of it can still be felt in almost every aspect of Western society. Consider that the whole of Christianity and all of its many off-shoots exist solely because of this personage. The impact of Christianity on the world, regardless of our views of it, cannot be denied.

3) It would also appear true to say that Jesus' teachings were fundamentally of a non-violent nature. This is significant, because most of the historical figures we have immortalised have been leaders of great armies and instruments of war. While Jesus' passing certainly resulted in widespread bloodshed, some of it continuing right into the present, it seems from what evidence there is, that Jesus himself, was basically opposed to the use of violence and instead preached the brotherhood of man.

4) The major change in thinking would seem to have been a shift from polytheism to the concept of a single deity or God.

To discuss why the belief of a single deity would give rise to an unprecedented advance in technology would be an exercise which could take several volumes by itself, and at any rate we have no concrete evidence that this factor alone was the key element.

It is fairly safe to assume though, that this point was at least instrumental in developing mankind. Consider that the average ancient Egyptian citizen, regarded his Pharaohs as gods or at least demigods. In addition to this, he also had a number of non-corporeal deities which required worship. The concept of a single God would be very freeing to such a person. All of a sudden his place in the universe has changed. From a poor peasant at the mercy of his many and varied gods, he suddenly becomes aware that there is only one God, and he does not even reside on Earth in human form! Such a person surely would feel that he could now do anything.

Especially since, according to Jesus, this was a loving God and not a vicious or vengeful God as so many of the semi-idolized kings of the time obviously were.

Some chaos and anarchy would no doubt also arise as individuals and groups begin to assert themselves and begin the gradual ascent from a classist or feudal system to a more democratic one. Certainly, this period is indeed fraught with such evidence.

The point which needs highlighting though, with respect to the Jesus incident, is that it is the first of several holes one can shoot in the theory that man evolved gradually and steadily.

As stated a little earlier, fire is thought to have been known to early man from about 350,000 BC and yet, written language appears only in 3200 BC.

If these datings are correct then, it took man well over 300,000 years to evolve from fire to the written word. At this rate it would seem impossible to think that from the written word to having the first manned landing on the moon would take only an additional 5170 years or so. Even if we only consider history from 100,000 BC — which is the alleged time during which Homo Sapiens Sapiens first appeared— from fire to written language would still have taken 96,800 years.

At the very least, then such a rapid rise in technology over the last two thousand years or so, would make the theory of gradual evolution suspect.


More evidence of a fragmented history is written in the very geology of our planet. The legend of the flood, accompanied by 'fire falling from the sky' is one so widespread in human cultures that it is difficult indeed to find a group of people, regardless of geographical location or technological level that does not incorporate it into either their mythology, religion or history. There is in fact geological evidence that in the past ocean levels have varied at times by well over 100 metres, but generally this has been thought of as a gradual change arising from the melting or freezing of the ice caps and glaciers as ice ages came and went.

The reality of the matter however is that Earth has been struck on a number of occasions by asteroids or meteorites, and the resulting changes have been drastic and sudden. The evidence for this, quite apart from any myths or legends, is plentiful. One only need look for it, and a few suggestions as to where one might begin looking for it are presented below.

**A)** The reason that the dinosaurs' disappearance has been attributed to a meteorite impact is because if we look at a cross section of the Earth's crust, we find a persistent but thin layer of iridium at about the time the dinosaurs made their mass exit from the scene. If memory serves me correctly it was Walter Alvarez that made that discovery, and since iridium is a very rare substance on our planet, having a persistent layer of it in the 70 million year old strata is extremely suspicious. It is postulated though that it would be consistent with the type of deposits one could expect if a large meteorite hit the Earth, the resulting explosion would quite likely result in a phenomenon which has been called nuclear winter by prophets of doom.

In the event of a full scale nuclear war, the amount of warheads exploding is so large that the sun would be obscured by the ash thrown up into the atmosphere. The result would be catastrophic to say the least. In a way this has already happened in Kuwait on a smaller scale. Although no nuclear devices exploded in the Gulf war, the burning of hundreds of oil wells darkened the sky to the extent that the sun was at times completely obscured. The loss of life in terms of the environment was enormous, and sure enough, thousands of years from now, the deposits which now lie at the surface will leave a tell-tale trace for future palaeontologists. If they are smart, they might deduce what happened, otherwise, they might put it down to a natural catastrophe.

**B)** No geologist would dispute the fact that Iceland, which today, is a bitterly cold place, as its name implies, once used to be a tropical forest. This may imply that a sudden and violent tilt of the Earth's axis has taken place, had it been gradual, the vegetation would have had time to adapt.

**C)** Although not conclusive, there is some evidence that would seem to suggest that if we were to drain the oceans, what we would find under the Arctic ocean is an enormous crater left by some ancient flying rock which landed rather heavily in the area thousands of years ago.

**D)** Mammoth meat has been served in Russian restaurants. That is because several of the creatures were found frozen in the Northern region of the old USSR.

The interesting thing is that the contents of these creatures' stomachs were not fermented. For an entire mammoth to have frozen and be perfectly preserved, without its flesh rotting for hundreds of years, two things must be true: The temperature must remain very low throughout that period of time, and *the Mammoth must have frozen very quickly*. For the food inside its stomach to still be in an unfermented state, *the entire mammoth would have to be frozen solid to the core in a matter of a few hours at most*.

I can speak with some authority on this point in particular. I come from a family of hunters, and I have done my share of it before changing my ways. I know from personal experience, that even an animal as small as a *Steinbuck* (a small antelope less than a metre tall at the shoulder) will bloat in the stomach within an hour if left under the African sun or even under the shade of trees. Considering that a Mammoth is more akin to an elephant in size, and thus that freezing it to the core would take considerably longer than for a small buck, it becomes obvious that something very unusual must have occurred to freeze that Mammoth in its tracks so to speak. Especially since the Mammoth is a creature known to have adapted to life in cold climates.

One way in which one can explain this is if a meteorite impact caused enough of a disturbance to suddenly flood the area with freezing cold glacier water, or perhaps even tilted the Earth enough to suddenly place the unfortunate mammoths in a freezing cold climate, along with a sudden rise in water level.

How else could one explain the fact that the creatures did not seem to otherwise be harmed or diseased?

E) I am not the first to draw attention to the mammoth phenomena, Immanuel Velikovsky, author of the controversial (and largely erroneous) *Worlds in Collision*, was perhaps the first person to bring it to the attention of the population at large as far back as the 1950's; while Charles Berlitz, in his book *Mysteries From Forgotten Worlds*[28] adds that these mammoths have all been found with flora in their stomachs and mouths which has not grown in Siberia and Alaska (where the creatures have also been found) since it has had an Arctic climate.

---

[28] See bibliography for details.

In one case buttercups were found in one's mouth, clearly proof of a sudden change in climate, since buttercups in Siberia can comfortably be said to be rather difficult to find today!

In addition, one mammoth was found completely frozen, with half-chewed food in its mouth, but its bones had been suddenly broken, apparently immediately before death.

Charles Berlitz, among others, also points out that we have several sites around the world, where the remains of an incredible number of different species of animals are found together. Something which can only happen if a catastrophe of large enough importance forced together animals that would normally never be found side by side. This phenomenon can still be seen today during forest fires, at which times animals will run, hide and sometimes perish together although they may be of species which naturally prey on one another under normal circumstances. Some particularly interesting sites are the asphalt pits of La Brea, near Los Angeles, California, where hundreds of ancient animals were found together, among them, were saber-tooth tigers, horses, camels, mammoths, mastodons, bisons and peacocks. Clearly an unnatural mix of creatures, unless some sudden catastrophe drove them all together towards some point which gave them a few extra minutes of life; perhaps higher ground if the danger was a flood, or a shallow lake in the case of a fire.

Similar sites have been found throughout the world, where lions, deer, bears, rhinoceroses and smaller animals have all been found together. Berlitz mentions a hill near Chalon-sur-Saône in France as another case in point.

In China, many of the old ivory carvings were performed on ancient ivory extracted from "ivory mines" where the material had been preserved by the cold or sudden mudslides that engulfed the animals. This is particularly interesting, because like bone, ivory becomes brittle if exposed to the weather for some time, and it requires to be frozen rather suddenly indeed for it to be usable for carving thousands of years later.

**F)** A meteor of large enough size and ferocity would cause world-wide earthquakes and tidal waves. Entire civilisations could be wiped out in instants, and the few survivors, would no doubt see to it that their offspring were told about these happenings, ensuring they eventually became legend and myth over the millennia. But if the art of writing was known, the events would surely be recorded, making very ancient texts or hieroglyphs a possible source of documentation of large meteoric impacts. Of course, even if writing *was* known, the texts themselves may not have survived in their original format, since the survivors may have had more pressing matters at hand than the saving of their history books.

The evidence that the Earth is periodically struck by celestial bodies large enough to make a difference, is indeed there, like I have already said, one need only look for it. But it would seem that each branch of human knowledge is becoming more and more compartmentalised, and in doing so, the left arm is sometimes little aware of what the right arm is doing. It seems to me, that history, archaeology, palaeontology, geology and astronomy have to be integrated a lot more than they are at present before we can construct a clearer picture of our planet's (and consequently our own) history.

♦ ♦ ♦ ♦ ♦

If technology can rise to a peak in only two thousand years or so, and if the Earth has been subjected to great catastrophes, then it is plausible that humans may have at some time in the past developed technology equal to, or even superior, to our present one.

That technology can rise from primitive to space-faring in a couple of millenniums is most certainly a fact we are all well aware of. Similarly, the evidence showing that Earth has been bombarded by relatively large meteorites, and/or underwent other types of catastrophes which were serious enough to affect human development is also enough to dispel any doubt as to its occurrence.[29] And if the above two statements are true, our next step should then be to look for some evidence of past technology which is superior or at least equal to our present one.

Before we do this, it is wise to first compile a short list of what our expectations in this regard might be.

1) Any civilisation which has reached our present level of technology or higher is bound to have colonised the globe, and therefore we should find evidence throughout the planet.

---

[29] The reader who wishes to pursue this evidence further can refer to Appendix A at the end of part I of this book.

2) If however we do find isolated cases of superior technology we can only conclude one of two things: Either it is a unique incident, perhaps an individual or small group discovered some powerful technology but were soon destroyed and thus had no opportunity to make widespread use of it, or it is not of native origin. A simple analogy might explain this better. If we discovered a planet composed of chocolate, and while scouting its surface we came across a diamond structure, and if further, this diamond structure is unique, we can conclude that either something very strange happened to place that diamond building on this chocolate planet, or that *someone from another planet* placed it there.

3) Considering that we are talking of periods measuring thousands of years, it is unlikely that we would find the remains of any metallic substance. A computer for example, if left alone for a few hundred years, would decompose to a pile of rusty flakes and geometric shapes of silicon. Add a little nature (wind, rain, earthquakes, curious hominids etc.) and pretty soon even those geometric shapes will have been broken and scored so that at best, they perhaps resemble small fragments of some ceremonial pottery. The material least likely to succumb to the ravages of time is not steel, but stone. Stone structures, such as homes and temples however may not necessarily give an indication of technology. Our concrete towers might seem impressive to us, but over the years, the steel reinforcing corrodes. The mighty Empire State buildings of the twentieth century will come crashing down long before the old stone cathedrals if all other factors are equal. If enough time passes, even the rubble of those buildings would degenerate to the point of barely giving a hint that it existed at all. An archaeologist of the future might identify the rubble as being part of large and complex structures, but it is very unlikely that he would expect those structures to have reached half a kilometre into the sky.

Once we begin to look for evidence of a higher technology than previously suspected in our past, we are suddenly faced with some amazing results.

Firstly, there is an incredible amount of what a lawyer might call 'circumstantial evidence'. While it may not be enough to prove anything, some of this is certainly worth mentioning. Our entire history is riddled with descriptions of people and events which seem unexplainable, even dream-like in their complexity and inconsistencies. Broadly speaking, we are faced with a vast number of Gods which seem to go about their business in very human-like ways. The Gods of old went around fighting wars and feuds against each other, marrying, having sexual encounters with just about anything that moved and generally behaving suspiciously more like humans than divine beings. It is for this very reason that we view these stories as being just that: elaborate and fanciful stories told to either amuse or control the people of the time and to explain away a Universe which must have seemed irrational, chaotic and frightening to primitive man. How else can we explain such stories as Zeus, king of the Gods transforming himself into all sorts of animals in order to impregnate some sexy-looking human female?

If, however, we imagine for a moment that our ancestors were not the dumb troglodytes we make them out to be and take some of their legends and stories to be based in fact (and what legend isn't?) we soon discover some really amazing 'coincidences'.

Perhaps not any one or two of these taken alone would be enough to sway us to believe that the 'Gods' were in fact simply more advanced humanoids, but taken as a whole, their number is staggering. There are literally hundreds and hundreds of legends *in every culture* which could be interpreted as being not the fanciful imaginings of a backward people, but the only way in which a 'primitive'[30] person could describe modern concepts such as genetic engineering, atom bombs, rifles, electricity, flying machines and interstellar travel, to mention a few. Appendix B at the end of part I is dedicated to showing some of the more interesting examples of this sort of mythology.

---

[30] Primitive here implies a lower technological level, and not necessarily a stupid or particularly superstitious person, but one who quite simply has no words or knowledge to describe certain phenomena.
If we came into contact with beings that were capable of mentally controlling us and who survived on the red nectar of some fruit native to their planet, we would almost certainly come to call them vampires despite our supposedly superior technology and broader view of the Universe. How much more impressive then, must a race capable of interstellar travel be, to a race that has yet to discover how to smelt iron? Today we have words for things we cannot yet do, such as teleporting, or hyperspace, but what if we encountered creatures capable of doing things which we simply cannot even *imagine?* I for one am not at all sure that we would not refer to them as Gods or angels even today.

As for more concrete evidence, we indeed have some. The problem in this regard lies mainly in deciding exactly what we can consider to be evidence of a superior technology. If for example we had a picture carved in rock of a modern jet plane, would we consider that to be evidential or not? A lot depends on the carving does it not? One person's jet plane carving is another person's ornamental object carving.

Many of the ideas I present here are not in fact new. A number of people before me have indeed come to similar conclusions. Among them, and perhaps most famous for it, is Erich Von Däniken.

In his books, Von Däniken gives a number of examples of objects, carvings, and buildings which he thinks are the result, either directly or indirectly, of extraterrestrial life.

Sceptics have labelled Von Däniken a sensationalist fraud and generally his work is not taken seriously. I have not read all of his books, but in the two I have read, *Chariots of the Gods?* and *The Gold of the Gods*,[31] he makes no claim that he is absolutely correct in all of his theories. While Von Däniken almost certainly has got certain facts wrong, either ignorantly or by design, the basic message is sound in my opinion. Critics have said (and some have gone so far as to write books on the subject) that Von Däniken is a pathological liar, that he's been in jail, that he has embezzled money to finance his trips, that he plagiarised much of his work from other writers and so on. I have not bothered to investigate this beyond the point of being made aware of such claims, and in my opinion they are irrelevant with respect to the question at hand.

In reading not only Von Däniken's but anyone's books, my attitude has never been one of blind acceptance. The reader who simply takes the writer for granted is a fool. One has to critically analyse any book or indeed any information. Whether we realise it or not, we all do this, to a greater or lesser extent. The criticism Von Däniken received, is at least in part due to the fact that some people, whose powers of perception and/or criticism are above average, realised Von Däniken may have been less than candid in his approach to the subject. Despite this though, his basic precept that we have extra-terrestrial origins needs further investigation than it has received up to now.

---

[31] See bibliography for details.

In light of the discovery of the Face, the most obvious objects our attention should focus on would seem to be the pyramid structures. This has been done in two ways in this book. In the next chapter I look in more detail at what could be argued to be the best example of a pyramid structure on Earth. While here, I will briefly and in a general manner point out some easily verifiable facts with respect to cultures which built pyramid structures.

Pyramidal civilisations, were all sun worshippers at a time when most of the rest of the world was composed of fire worshippers. Similarly, both the ancients of South America and of Egypt are known to have had extensive astronomical knowledge. So vast and precise in fact, that the full extent of their understanding of things astral is not fully appreciated today. The ancient Mayans had a calendar which was far more accurate than the one used by the Westerners who eventually destroyed them. The Westerners in fact were ten days or so behind the *real* astronomical date when these two cultures first met.

Hieroglyphs of a complex nature are also evident in both Mayan and Egyptian cultures, and most significantly, especially when thought of in context to sun worship, both cultures worshipped their kings as Gods or descendants of Gods. The Mayan calendar began at year one (3113 BC) when *a white, bearded man, descended from the Sun* along with several others who were said to be scholars, astronomers, architects and priests of great skill.

Both the Mayas and Egyptians built large structures depicting a human face encased in some type of head garment reminiscent of a helmet. The Sphinx is probably the most famous of these for the Egyptians, while the large, helmeted stone heads of the Mayas weighing from 15 to 30 tons and from 1.5 to 3 metres in height (5 to 10 feet) are the South American counterparts. It's interesting to note that the Mayan heads represent headgear which is remarkably similar to a motorbike helmet, complete with chin guard and even what may be a visor that could possibly slide down in the original but which the stone sculptures (obviously) merely depict.

There is also considerable evidence among all pyramidal cultures, of objects, carvings, or drawings which could easily be construed as being representative of flying machines and their respective pilots. Most significantly perhaps, is that the methods by which the pyramids themselves were constructed remains a mystery despite numerous attempts to explain the process by a number of 'experts'.[32] Their erection seems utterly impossible by the technology our historians and archaeologists attribute to these people.

Parallels in culture, legends, and philosophical thought, are so striking as to lead one to the conclusion that it is certainly possible that all these pyramidal cultures originated from a common group of people; whom possessed technology and/or knowledge far in advance of the one exhibited by the races whom eventually became associated with the pyramid structures. And that in time, these 'refugee gods' for want of a better term, were assimilated into the originally more primitive societies. In fact, such propositions have been put forward by a number of people, whom however, rather than pointing to the stars for the origin of the more technologically advanced race, mainly point instead, to the supposedly mythological land known as Atlantis.

It is not possible, in the scope of this one book, along with the other already avant-garde theories, to also justify the possible existence or inexistence of Atlantis in some detail. An ample number of books concerning the subject of Atlantis has been written, and although generally speaking, no definite conclusions regarding the theoretical inhabitants of Atlantis is arrived at, in the main, geologists and geographers do agree that indeed, such a continent or large island could, and most probably did, exist somewhere in the vast expanse of Ocean known as the Atlantic.[33]

Persons who doubt that Atlantis *did* exist are advised to research a number of topics in particular, which may change their mind, but which I unfortunately cannot dedicate more than a few sentences to, in view of the already complex task before me. Especially since, as I mention later, the main objective of this work is one that will become

---

[32] No one can be said to be an expert at ancient pyramid construction, since no person alive today has any experience of it, and while some of the theories proposed by archaeologists in respect of the pyramids are worthy of note, no-one has yet satisfactorily explained how such structures were built in reality. What does become rather obvious on closer investigation, is that even the best orthodox theories put forth as to how a building like the Great Pyramid of Giza was built, fall far short of the mark, as we will see in more detail in chapter three.

[33] While some eminent geologists deny that Atlantis ever existed, the ratio of scientists that believes a large island may have existed somewhere in the Atlantic to that of scientist that believe no such island ever existed, is approximately 2 to 1.

clear only towards the end of the book, and which does not necessarily concern itself with such details.

1) Dr. Manson Valentine, in 1968, discovered and explored what has sometimes been referred to as the "Bimini Road". A sunken wall built of giant stone blocks that lies about 36 feet (12 metres) below sea level, off the North Bimini coast. Some of these blocks are supported by columns and the sheer size of the construction is such to leave one at a loss for words. Further out off the Bimini coast, pilots of commercial and private planes, have spotted a number of features reported as vertical walls, arches, and *pyramids* or pyramid bases. Concentric circles of monolithic stones, forming a sort of underwater Stonehenge are also present. It is interesting to note that fossilised mangrove roots growing over the stone blocks that compose the great wall or 'road' have given carbon datings of 10,000 to 12,000 years of age, which places them not only in the proper context for Plato's dating of the sinking of Atlantis, but also in the correct time-frame, recognised by the scientific community, during which the last ice age glaciers melted, resulting in a rapid rise of the ocean levels world-wide.

2) When the Canary Islands were discovered by Europeans, the inhabitants showed surprise that anyone else in the world had survived the great catastrophe. They thought that all of humanity except for a few survivors that had lived on mountains, like themselves, had perished. Curiously enough, they had customs that fit suspiciously well with Plato's account of Atlantis. They had a ruling body composed of ten kings, worshipped the sun and had an order of priestesses dedicated to it. They mummified their dead, built houses of close fitted stone with walls coloured red, white and black as well as building circular fortifications and having a canal system of irrigation. They also possessed a written, alphabetical language, although this has since been lost. It was also evident, that at one time their technology must have been higher, but that it had somehow degenerated.

What if, for the sake of pure speculation we went so far as to say that the land of origin of such a technologically advanced race was indeed Atlantis?

If we do assume that such Atlantean 'Gods' did indeed exist, and that the survivors of the catastrophe that destroyed Atlantis spread to other continents, we find that this fits rather well with the known facts. The Incas of Peru, the Guanches of the Canary Islands and the Egyptians, all performed a crude form of brain surgery, going even so far as to install metal plates of silver or gold into the skulls of people whose heads had been damaged. A practice which if slightly different is continued even today in the hospitals of the world.

If initially the Atlantean survivors ruled by being worshipped by their subjects as Gods (no doubt due to their extraordinary technology, which must have given them very impressive 'powers' when compared to the average person of the time) it would also make sense that by marrying queens of a more primitive nature, their lineage was eventually watered down further and further and their origin eventually became legend rather than history as humanity underwent the catastrophic changes which very nearly wiped it out and certainly returned it to a primitive state.

The fact that the Mayan civilisation had come into contact previously with white skinned, bearded men, is an indisputable fact.

Especially since it was this very mistaking of the Spaniards for Gods which resulted in the destruction of the Incas and Aztecs.

The word Maya itself means "not many" in its own tongue, and would in fact support the theory that a few survivors of some ancient catastrophe, such as the biblical flood, settled in parts of South America.

Furthermore, while the people of South America pointed to the East when indicating the origin of the land of the Gods in their legends, the Egyptians pointed to the West in theirs.

Regardless of what source one quotes for this supposedly mythological Atlantis, probably all modern researchers would have to start with Plato's description of the fabled island, which could also be said to be the most detailed. But Greek mythology is not alone in representing Atlantis. The Mayas had similar stories if somewhat less well recorded, as do the Hindu religious works of ancient origin.

As its very name implies, Atlantis has always been depicted as having existed somewhere in the Atlantic, which of course would coincide nicely with the Gods originating in the East for the Mayas and in the West for the Egyptians and Europeans. It is also generally accepted that the 'myth' of Atlantis goes something like this:

An ancient civilisation, which had reached technology equivalent, and possibly superior to our own in some ways, was suddenly destroyed by some catastrophic event. The general theory as to the nature of the catastrophe is mixed. About an equal portion of stories attribute the sinking of Atlantis to a natural occurrence as others do to some Atlantean experiment in technology gone wrong. Some legends go so far as to speculate that the Atlanteans may have had a war among themselves or been destroyed as a result of atomic explosions, why or how these might have come about is unknown.

The fact that the Aztec kings were known as Sons of the Sun and are attributed as having had blonde hair and blue eyes as well as a fair or reddish skin coincides interestingly with the fact that the Egyptian male Pharaohs were always depicted as being red in skin colour but never their queens and that the word Phoenician is based on the Greek word for red. While in the bible it says that the *sons of the gods married the daughters of men* in Genesis 6 : 1-3.

Why would there be a reference to *gods*, the plural of the word,[34] when the central theme of the bible, as well as the first commandment, is that there is only *one* God?

Ronald Story, who dedicated at least two books to the debunking of Von Däniken has no clear explanation for this and arrives at simplistic and unreasonable conclusions in a number of other cases. One such example[35] is found where he states that Von Däniken claims the Great pyramid contains about 2 600 000 blocks of stone which weigh 12 tons each. He corrects him by stating (correctly) that the pyramid is estimated to contain about 2 300 000 blocks, and not 2 600 000 and further that the average weight of the blocks is only 2.5 tons and not 12. He neglects to mention however that some blocks are thought to

---

[34] The original Hebrew text, has a reference to the Elohim (Gods) in the plural of the term, although orthodox biblical scholars have tried to deny that the word Elohim refers to a plural term for Gods, and it is sometimes translated as 'Lords and Judges' or 'Angels'. But even so, it is clear that these beings, are of a supernatural and 'more-than-human' nature, otherwise why differentiate between them and the rest of mankind? In addition, their grandchildren, born to them by the union of their sons and the 'daughters of man' are the biblical Nefilim, or Titans. Literally, translated as 'giants' by the orthodox scholars, but originally meaning 'Those Who Were Cast Down'. And although presumably weaker than their fathers (and their grandfathers the 'Gods') the Nefilim were still vastly 'superior' in the hierarchy of things, to the rather primitive 'sons and daughters of man' (normal humans). See also Appendix E.

[35] Pg. 79 of Ronald Story's *Guardians of the Universe?*

weigh up to 17 tons and that certain stones inside the pyramid are much heavier than this. One in the King's chamber complex is estimated to be 72 tons in weight. Nor are these the most severe of Story's oversights. As we shall see in chapter three, the Great pyramid cannot be so easily explained away.

If we summarise briefly such theories as we may construe with regard to the pyramidal civilisations we conclude that:

1) They had technology and knowledge superior to that which is presently attributed to them by most orthodox scholars.
2) Such are the parallels in their legends, religions, buildings, writing and philosophies as to suggest some common origin.
3) Such common origin may have been the lost continent of Atlantis where a supposedly technologically advanced race lived.

Placed in the context of the Face this reads as follows:

The (possibly) Martian visitors had a base on an island or continent called Atlantis. They did not freely mix with the rest of humanity on Earth until some catastrophe (as yet undetermined) forced the few survivors to set up new homes in lands inhabited by less advanced cultures. The ample evidence as well as written description in the Hindu Vedas of flying machines suggests the Atlanteans may well have had the capacity for flight, and in view of the pyramids on Mars this ability would have to be a foregone conclusion anyway. Also, if Atlantis was merely an outpost, it would help explain why the evidence for a superior ancient technology is not so widespread, but rather localised in a few particular sites, while at the same time, evidence of the existence of gods which travelled in *chariots of fire and descended from the sun* is more widespread than their technology.
This also makes sense since they were 'not many' but their technology would give them the ability to visit and thus impress, most of the scattered tribes of a more primitive nature which inhabited all parts of the globe.
The idea of an outpost though would seem to indicate that the Atlanteans, if they originated on Mars, were somewhat different from the humans to be found on Earth. Yet, despite the flying machines,

helmets and space suit type of depictions to be found in the carvings and drawings, these Atlanteans obviously could breathe our air, since they lived among us.

This of course would suggest that if the Atlanteans were also Martians or at least related to the Martians (and considering that they are responsible for the construction of the pyramids, this must indeed be so) then Mars must have had an atmosphere similar to Earth's. This concept would seem to make sense since, as I have previously stated, the construction of pyramids is generally believed to have taken many men and an undetermined number of years, which would make construction of them impossible for space-suited people. This would be true regardless of the technological level of the Martians. It has been estimated for example that it would take six years and about a billion dollars to reproduce the Great pyramid with present technology.[36]

But giving Mars an atmosphere similar to Earth's, which in view of the Face we must do anyway, regardless of any Atlanteans, gives rise to more interesting ideas still.

---

[36] Personally I find this claim an optimistic one when the measurements, precision and alignment of this structure are considered, never mind the fact that the quarries from which the stone was taken lie on the opposite side of the Nile. More significantly though is the fact that we may not even be able to find the correct place for construction, since we would need a solid underground rock base which would limit settling and damage caused by earth movements. With present soil surveying techniques, it would be a major undertaking which would by no means be sure of success. Even today, geological surveying is to some extent a matter of guesswork. The best of geotechnical engineers cannot always eliminate damage to structures due to settlement. While in most cases this is reserved to some cracking, on occasion it results in condemnation of a building, and keep in mind that the Great pyramid is about thirty times heavier than the Empire State building. How the ancients performed their geotechnical surveys remains a mystery, like so much of their technology.

## A Question of Atmosphere

At this point in our generic reconstruction of human history we have found evidence that suggests the pyramids were built under the instruction of a more advanced race. Indeed this reasoning cannot be said to have been too heavily influenced by the discovery of the Face because it is one which a select number of individuals had arrived at before any knowledge of the Martian pyramids was known, and in fact, a number of books on the subject exists.

We have also concluded that if the architects of the pyramids were Martians, while possessing perhaps some slightly eccentric features, such as possibly a fairer skin and lighter colour of eyes and hair[37] they resembled humanoids to the extent that they could mingle with them and pass as reasonably normal humans.

This being the case, the Martians and Earthlings must have been able to breathe the same air and of course this suggests that Mars had a similar atmosphere. We have also tentatively identified the previously mythical Atlantis as a possible outpost or base of operations for the Martians.

In order to substantiate such claims we should next investigate Mars, to find out whether it might indeed have had an Earth-like atmosphere. The planet Mars is discussed in more detail in chapter four, here I will limit myself to say, that there is indeed evidence which suggests that the planet could have supported an atmosphere as well as an ecology similar to the one we have on Earth today. While the amount of vegetation and animals would be a matter of considerable speculation, it is relatively safe to assume that Mars could in the distant past support human life in quantities large enough to enable the construction of the Face and nearby pyramids.

There are canals on Mars which were created long ago by what astronomers think must have been running water, and in this case, a much thicker atmosphere must also have been present. It only follows then, that if Mars once had water and a thicker atmosphere, and it also has pyramids, that that atmosphere was similar enough to Earth's to allow humanoids to go about their business on its surface without having to wear vacuum suits.

---

[37] It is also suggested, if somewhat less so than the other features, that they may have been somewhat taller.

The fact that Mars had an atmosphere similar to Earth's does not surprise us unduly we might say. After all, with a Face and pyramids there, it only follows that the two planets must have had a similar atmosphere at some point in time, right?

Not really. In fact it doesn't follow at all.

The coincidences required for Martian and Earth pyramids to have evolved separately have been considered and found to be too many and of too great a significance to have been coincidences, and for this reason, logic forced us to conclude that there must have been contact between Mars and Earth.

We did not however speculate as to where humanity originated first. Was it on Mars or on Earth?

Given our Atlantis/Outpost theory we would have to say Mars. If we had evolved on Earth first, it means we would at some point in our past have had to develop the technology to travel to Mars in relatively large numbers. There is no evidence at all to substantiate this in our history, but there is instead, evidence to substantiate 'Gods' descending from the sky who then helped us to grow technologically. This however implies two things:

  1) There already were humans on Earth to receive the 'Gods' or Martians.
  2) If we accept that we originated on Mars first, this is inconsistent. It implies either that the more primitive Earth-humans were the ancient ancestors of a first attempt on the part of the Martians at colonisation, or that there were humans here on Earth originally, which we have already discarded since it is a virtual certainty that two types of intelligent life which evolve separately would not resemble each other so closely.

But an earlier attempt at colonisation by the Martians is also inconsistent because we are once again left with some large questions which are very difficult to answer, such as why would the atmosphere be the same? There is no reason why it should be! And besides that, why would the second lot of Martians wait thousands of years until the original colonists descend into barbarism before making contact with them again?

The only alternative which make sense, is that the human race did not originate in this solar system at all.

At this point, the intelligent reader may have a query. If a planet's atmosphere is very, very unlikely to resemble that of a neighbour, how much more unlikely that it resembles that of another planet which circles a completely different star! But the thought that we originate from a different solar system in fact explains the problem of atmosphere rather than complicating it, because a civilisation capable of interstellar travel would also have at its disposal, the technology to terraform a whole planet.

Terraforming, is the process where an entire planet is engineered to have conditions suitable to sustain humans. It is a term used mainly by science fiction writers, but has received serious consideration from respected scientists, and it is a recognised fact that while the process would take up to several thousand years, (depending on technological level) it would certainly be possible if our technology was a little superior to our present level. Studies have been done to approximately estimate how long it would take us to terraform Mars for eventual proper colonisation instead of mere outposts, and a conservative estimate arrived at was 100,000 years.

Because it is so unlikely that Mars and Earth had originally similar atmospheres, we must conclude that either one or the other, or both, were engineered by aliens of humanoid origin who then colonised the planets. It is also very unlikely that people from Mars terraformed Earth, because of the great time this would have taken.

If however you remember the concept of interstellar travel outlined in chapter one, you will recall that travelling to another star would entail only a few years for the crew of the ship, but many more for the people who stay on their respective planets. In this way, probes or scout ships with organic bacteria suitable for the manufacturing of oxygen and perhaps some genetically engineered plants, could be sent off from the home planet to a suitable, nearby system, where they would begin their work and then come back home.

By the time the ships have returned, a great deal of time could have passed for the planets. The ships could now load up with the first batch of explorers who would then travel back to the planet where the terraforming would be well under way or possibly even complete by the time of their departure and even more so by the time they actually arrive there.

We'll see later, that the theory of our origin being outside of this solar system also explains some other concepts (which would otherwise have no answer) in a very neat and appropriate way, further suggesting that this assumption is correct.

I will discuss these ideas later in more detail, for the moment I am merely concerned with giving the reader a foundation with which to work. At this stage I do not expect these theories to be immediately approved of. As the book progresses, however, the reader should begin to discern a pattern and it is in this fitting together of pieces of a vast puzzle that must come a degree of certainty.

As I have stated before, this subject is incredibly vast and covers areas of thought and details of subjects as diverse as biology is from history or palaeontology from astronomy. It is impossible therefore to form an airtight, completely accurate, reconstruction. I am limited to showing patterns of relationship. This is by no means an invalid or inaccurate way of discovering the truth about something. In fact, for the purpose of making us generally aware of our true origins, this method is more than adequately accurate.

Remember that our entire existence is based on patterns of relationship. We recognise a cup as such only because of it relationship to containing liquids. We are able to read, not because of the individual words, but because of the relationships they make with one another, and we indeed have words themselves, only because of the relationships that letters make with one another. And so it is with anything.

The higher we lift our eyes, the larger the relationships involved. If some of the concepts I advance seem strange or unbelievable, it is not because they are not real or based in fact, but more likely because we have been busy with examining the relationships at a much lower and detailed level. If we take the analogy of human history being represented by a book, we have been involved with the piecing together of letters to make words, and we have perhaps composed a few paragraphs and struggle already to relate even those few into a coherent whole. By examining the possibility that we originate from another solar system, we have lifted our eyes from a worm's eye view of events to a bird's eye view, and piecing together large chunks of history becomes easier. From analysing letters or words, we have jumped to analysing chapters. Of course, it is still necessary to investigate each chapter in more detail and this is a task I leave to people more competent than I in the various fields, my aim here, is merely to point the way, and my expertise, if any, could be ascribed to being in the recognising of patterns of relationship rather than in any one branch of the various disciplines.

## Interstellar Origin & Terraforming

If we accept the possibility of a race so advanced that it had interstellar travel facilities and terraforming capacity, we are left with other unanswered questions which at first seem to invalidate the theory of an extra-solar origin for the human race.

A race this advanced surely would not be so easily wiped out by some natural catastrophe? Where, then, are they? Why have we descended in technology and lost these ancient secrets?

Especially since the Ancients had colonised not one, but apparently two planets in this system.

Mars is a dead planet now, and only traces of this once technologically advanced culture exist on our own planet. What could have happened?

After all, with our present day technology, if we had a little advance warning of a meteorite headed on a collision course with Earth, chances are that we could prevent it. It would not be a simple affair, but the technology to send a shuttle or space vehicle carrying a nuclear payload in orbit is certainly with us, and from there, all that would be required is that the missile be guided to the threatening satellite. Depending on the size of the asteroid/s a few missiles could be dispatched. The remaining rubble might still crash on Earth, but having been reduced in size, most of it would burn up in the atmosphere on re-entry.

Any damage to the inhabitants of this planet would have been greatly reduced at the very least, and quite probably completely eliminated.

How would we explain then, the apparent fact that the Atlanteans and/or Martians were unable to prevent a similar catastrophe? An Earthquake does not explain the situation any better, because that would still leave the problem of explaining how Mars came to be the way it is today.

There is one answer which fits rather well though.

What if the catastrophe that destroyed the Ancients was not natural, but just like the pyramids, man-made?

A war of a scale so large that it very nearly destroyed mankind altogether in this system.

As you will find out through the course of this book, this idea is not as fantastic as it sounds. Indeed we already have plenty of that circumstantial evidence I talked about before.

Every culture that was aware of Mars, labelled it a war God.

Ares, Mars, Harmakhis, Nergal. Whatever the name, the purpose was always the same. God of War.

What is more, the ancient Hindu scriptures, the Vedas, describe a war in which weapons of mass destruction are used. Some are described as weapons capable of producing explosions brighter than ten thousand suns.

A war would go a long way to explain why the Ancients are no longer with us today, but so far the 'evidence' is relatively circumstantial.

There is in fact ample concrete evidence for such a war having taken place, but the discussion of this is done in a later chapter as it is somewhat involved and for the moment I want to limit myself to laying before you a general concept of my theory as to our origin. The first step in *presenting* a new theory, is always the general laying down of a broad concept. This is then refined and crystallised by analysing first its key concepts in some detail and later its minute particulars. As I have already stated, the last step is one which I leave to hopefully many and more qualified persons than myself.

It is important for me to point out though, that the process whereby a new theory is *formed* is the reverse of this.

In other words, one detail will lead an investigator to another detail and so on, until eventually a coherent picture emerges from the amalgamation of these details. This is the only honest way in which any new theory can be formulated, because if we start with an overall concept of how things should be, then we are already biased to look for relationships which in truth may be secondary in importance or not even existent. In this respect I feel that without a doubt I have been honest in my arrival at certain conclusions, because at the beginning of my investigations, like most people, I viewed the Face on Mars, Atlantis, UFOs and such concepts with what I thought was a healthy and rational degree of scepticism.

Each step led me to look into a new facet of this vast topic, and it was only about two years after I first began my investigations that I came to see an overall picture. It would be fair to say that the thing which ultimately led me to my conclusions was not any personal preference regarding any 'outcome' I may finally arrive at, but rather an insatiable curiosity, that sometimes caused me to look into places I would not normally have given a second glance to.

# Summary

Many ideas have been presented in this chapter in what is sometimes an admittedly rather vague and undetailed manner.
In addition, although the reader can probably follow a tenuous logic between the various steps, this is certainly still far from being very clear. It might be useful then, to give a summary of the more salient points so that a skeleton for the theory of our true origins might be constructed and from which we can then proceed.

1) The Face and pyramids on Mars, lead us to conclude that some type of humans lived on that planet in a distant past.

2) These humans must be either directly or indirectly responsible for the building of the pyramids we find here on Earth.

3) On examining the civilisations that built pyramids on Earth we find large amounts of 'circumstantial' evidence to support the theory that a more advanced race of humans once roamed the Earth.

4) We also arrive at the conclusion that these more advanced humans could be the 'Gods' referred to in many of our legends.

5) Point four further leads us to conclude that there must have been humans on Earth prior to the arrival of such Gods. This is an important and confusing point. How did the original humans arrive on Earth? Why were the 'Gods' so much more advanced than the Earthlings if they were related to them?

5.a) The evidence points strongly to a separateness between the 'Gods' and the less evolved humanity of Earth, possibly indicating that the 'Gods' may have only had an outpost on Earth.

5.b) This outpost has tentatively been labelled as Atlantis. Investigation into the myth of Atlantis seems to reveal that it could well have been a real continent or island which perished in the Great War (see point 8 below).

5.c)  The 'Gods' are generally referred to as being helpful to mankind on Earth, in general educating our ancestors by teaching them how to write, demonstrating various forms of social structure etc. Despite this there is also evidence that the Earthlings were at times treated in a similar manner to the way we treat animals today.

5.d)  There is no evidence that suggests the 'Gods' came from Mars, so for all we know, the Martians were at a similar stage of development as the Earthlings.

6)  By assuming we had a similar atmosphere on both Earth and Mars we must allow for the possibility that Mars or Earth, or both were terraformed. The technology required for this can only exist if the race in question had an extremely long period of (probably) peaceful existence at a very high technological level. There is no evidence in our history to suggest this and it is very unlikely that the Martians terraformed Earth since Mars is so much smaller than Earth. The ability to terraform would also indicate the race is beginning to expand in space and therefore interstellar travel is at the very least being contemplated and quite possibly already in existence.

7)  Point six indicates that the human race must then have originated on a distant planet of some solar system other than our present one. This implies the technology for interstellar travel.

8)  We cannot explain the disappearance of the 'Gods' and of the Martians unless we postulate that a great war took place in which Mars was made into a dead world and Earth's outpost of the 'Gods' (Atlantis?) destroyed along with most of the native Earth population.[38]

9)  The surviving 'Gods' once again helped mankind to rise from their ashes by educating them and leading them. Over time and as perhaps more catastrophes (of a natural origin this time?) befell mankind on Earth, the 'Gods' were assimilated into the rest of Earth Humanity.

---

[38] It is important for me to once again point out that I did not start with this as a premise, but rather I came to believe it later as a result of certain features described in more detail in Chapter four.

9.a) Evidence of this is also quite good since the first Pharaohs and Mayan kings were worshipped as fully fledged Gods, but later generations are only worshipped as demigods and later generations still are said to be human.

9.b) A first alternative is that after getting humanity back on their feet, most of the survivors left Earth in any ships they might still have. Evidence for this is less than for 9.a. above but still present in many myths and legends.

9.c) A second alternative is that a second Great War followed later, fought between the various 'Gods' on Earth, and in this, once again, humanity was nearly destroyed and thrown back into barbarism, developing writing again only thousands of years later.

9.d) A mixture of all the above (points 9 to 9.c.) resulted in the eventual dispersion/obliteration of the 'Gods'.

10) Points which would substantiate or disprove this general theory can be found mainly in:

10.a) Legends, myths and ancient scripts of such a 'mythological' nature. These of course could be interpreted in many ways and therefore any evidence of this type can only be taken to be circumstantial or corroborative and not conclusive. Nevertheless, enough of this evidence exists that it cannot be ignored, and detailed translations of ancient texts by expert and unbiased linguists could still be extremely useful and perhaps even conclusive to a great degree.[39]

10.b) Drawings, carvings and statuettes or jewellery that represents, or depicts technology of an advanced nature. This type of evidence can also be said to be circumstantial although particular examples may be quite conclusive.

---

[39] The discovery of an ancient Sumerian tablet which describes the workings of a space ship or the origin of the 'Gods' would indeed be very valuable! Incredible as this sounds, we appear to have just such writings. See appendix B at the end of part I of this book.

10.c) Items from the distant past which are technologically advanced. Such items would have to be conclusive, but often they may be so corroded or ambiguous in nature that their true function is unrecognised and alternative explanations may be plausible.[40] While this type of evidence is rare, there are examples of it. See Appendix C.

10.d) The pyramids of Earth. A study of this has been done in more detail in the next chapter.

10.e) Mars. A study of Mars and its pyramids will not be complete until we land a manned craft there and establish a permanent base, but until then, much can be concluded from high resolution photographs of the structures in question. Mars is analysed in more detail in chapter four.

10.f) While points 10.c, 10.d and 10.e seem to me to be the most important, there are also a number of other, less easy to prove, but perhaps not insignificant theories, we should at least mention. One such idea is the concept of race memory, which Rupert Sheldrake in his work concerning 'Morphic resonance' has basically proven to be valid. The extent to which this is relevant though is unclear. Some examples are mentioned in Chapter seven.

Additional points of interest are:

A) There seems to be some evidence (in mythology and legend form only, as far as I know) that would indicate that the 'Gods' were involved in genetic manipulation of some sort or other with respect to the Earthlings. This in itself is not so important since genetic manipulation would have to be a fact if we accept the possibility of terraforming. The interesting part is that it ties in rather well with part two of this book and suggests the possibility

---

[40] If it could be proved for example that Stonehenge was a recharger for alien spaceships or a matter teleporter (this is not to imply the author sees it as such!) and that, such a function was its intended purpose, then we could say this would be conclusive evidence. As a matter of interest, Stonehenge is thought to have been a type of astronomical computer, and the calculations it is capable of being used for, are remarkable indeed, especially since it was built by supposedly ignorant primitives of a druidic sect. It is also quite possible that it may be used to perform many more functions which we may not have yet discovered.

that the 'Gods' may resemble us only rather distantly and have gone to the trouble of genetically engineering generation after generation of Earthlings in order to reach a final product which is intelligent and capable of greater awareness.

Either due to alien chauvinism or perhaps because they may have used genes similar to, or even deriving, from their own, we resemble them; but in fact we could be said to be an artificial race, having been created by the 'Gods' which visited us many times over the millennia in order to steer us down the correct evolutionary path.

B) Why a race of aliens would want to populate a planet with a race of beings it has basically engineered, is open to speculation, but some thoughts on this rather vague subject are given in chapter seven and later in part two of this book.

C) Point A above would raise some additional confusion as to the origin of the Great War. A subject which is of course already dim and confused due to its great distance from the present and lack of exact reference points. Some speculations on this point are undertaken in chapter six.


In the following chapters then, we shall investigate the various points outlined above from several different angles and at the end of this exercise we shall then hopefully have the necessary information required to revise this theory outline as required.

# 3  The Great Pyramid of Giza

## Prelude to the Chapter

The time has come, figuratively speaking, at which I would like to challenge the reader to a simple agreement.

The ancient Romans used to say that to keep silent was equivalent to assent. It is a practice I would like to adopt at this point, since in effect, chapter three could be said to be an alternative starting point for this book.

The challenge then is this: If by the time you are finished reading this chapter you are still unconvinced as to the existence of any extra-terrestrial connection, then I urge you to get yourself either to your library or to a well-stocked book shop and to research this topic more fully, until you are satisfied one way or the other. If however, you choose to do nothing while still being uncertain about what you believe, then I would like the reader to be bold enough to face the consequences of such inaction. In this context, the consequences would amount to your accepting, for the purposes of the remainder of this book at least, that the possibility of extra-terrestrial involvement is now a fact and no longer a mere possibility.

This is important, because while I have endeavoured to show the validity of this point of view throughout the remainder of this book wherever possible, it is NOT my primary concern. This work, is not merely an attempt to rewrite human history in the context of alien origins. My aim ultimately is in fact quite different.

I hope to show that the human race is not only at a point in its development which is critical, but that the reader him or herself can play a major part in not only the well-being of humanity at large but also in their own self-development.

Ultimately, what difference would it really make if we were the result of an alien genetic experiment or the culmination of millennia of random evolution? Monday would still follow Sunday, the phone companies would still want their ridiculous bills paid and life would carry on pretty much the same.

Luckily though (and possibly quite frighteningly so) this is not where it ends. In part one of this book, I try to reconstruct history in a very general way, while at the same time giving concrete evidence of our extra-terrestrial contact, involvement or even origins, but it is only in part two that I show why this knowledge has practical applications of truly paramount importance for the future of the human race.

◆ ◆ ◆ ◆ ◆

The Great Pyramid of Giza is probably the best example of a pyramid ever built which survives to the present day. It also has the added bonus of being one of the most surveyed buildings in the world. In fact it has often been referred to as *the* most surveyed building in the world. This is a great advantage, because it means there is no ambiguity as to the size and shape of the pyramid. Even today, anyone that wishes to check on the measurements found here can do so without having to resort to a trip to Cairo.

Your local library probably has several books on the pyramids of Egypt, and while one might have to refer to textbooks of an occasionally drier nature than some, it is a simple matter to check the veracity of the claims you will find in these pages.

It is important to state, and occasionally reiterate, that I have by no means made a complete study of any one aspect of this broad subject, and the pyramids in general, are no exception to this. There are various pyramidal constructs in South America which are just as interesting and startling as the Great pyramid, and perhaps even more so. I have chosen to focus on the best known pyramid simply out of convenience. Since I am already well off the beaten track, I saw no point in further ostracising myself by choosing to examine an obscure example of pyramid construction.

Other examples of pyramids certainly worthy of close scrutiny might be the Temple of the Sun and the Temple of the Moon, both located in Mexico; the pyramids of Angkor in Cambodia and the pyramids said to lie in the Shensi province in China. Whether this last complex of pyramids is true or fictional has not been investigated by this author, but it is said that the largest pyramid of the group is over one thousand feet (304 metres) in height.

A person whom I feel deserves much praise with regard to pyramids, is Peter Lemesurier, whose book *The Great Pyramid Decoded* was written in 1977, long before I entertained such fanciful ideas as I do now with regards to pyramids.

Lemesurier freely admits that in the writing of his book he made use of the dimensions and measurements provided by Doctor Adam Rutherford. He goes on to say that these measurements almost without exception fall well within the tolerances of the best available surveys. This is not surprising, since Rutherford's own measurements are largely based on such surveys, but I have taken some pains to check on this, and as with all the statements of Lemesurier I have bothered to research, his claim is indeed correct.

In the writing of this chapter, I in turn have made extensive use of the figures and facts provided by Lemesurier with respect to the Great Pyramid. Those readers whom feel they would like a truly complete book as far as the Great Pyramid is concerned are advised to purchase Peter Lemesurier's book,[1] he does a far more thorough job than it is possible for me to do here, and I would add that until new discoveries are made concerning this monument and its surroundings, one could do worse than labelling Lemesurier's text as definitive.

The Great Pyramid of Giza is also referred to as the Great Pyramid of Cheops or Khufu. A less well known fact is that the real pronunciation of the name is buried in the mists of time, for the word Khufu in reality is derived from the letters HWFW. No vowels were provided. Just like YHWH is normally translated as Jehovah.

Lemesurier makes the interesting proposition that the word YHWH may in fact be a Hebrew version of the older and original word symbolised by the letters HWFW.[2]

It is always a good idea, I feel, to give wherever possible, the true origin of a thing. While this may not strictly speaking be possible with regard to the Great Pyramid, letting it be known that even the name of the Pharaoh whom it was supposedly built for is not known in its

---

[1] No, this is not a plug for a book which exclusively supports my ideas. I think anyone that reads *The Great Pyramid Decoded* will agree that Lemesurier seems to be quite detached and unbiased with respect to the outcome of his predictions, nor is it a plug for a friend of mine. I have never met Peter Lemesurier or even spoken to him, although I would certainly look forward to the opportunity of doing so.

[2] Page 251 of *The Great Pyramid Decoded*.

entirety, is a start. Especially since so much today is simply taken on faith by so many. Allow me to demonstrate what I mean.

I shall set out a general idea of what most archaeologists think with regard to the pyramid and its construction. In conjunction with this, I shall also state some facts about the Great Pyramid which should show even the most sceptical of persons that the first view becomes untenable. It will be interesting to note then, that right up to the present day, the orthodox ideas with regard to this building still prevail, despite their shortcomings being so severe that they cannot be based on fact.

When the Great pyramid is talked of at all in scientific lectures, universities and schools, the story goes something like this:

In about 430 BC Herodotus completed his history of the wars between Greece and the barbarians. While undertaking this admirable work, Herodotus travelled widely throughout the ancient world. He was born in present day Turkey but he met Egyptian priests and from them learned what today is taken only in part (and quite arbitrarily so) as fact, concerning the building of the Great Pyramid. Keep in mind that the Great Pyramid was at least as old to Herodotus as he and the people of his time are to us. He certainly did not get the facts directly from the horse's mouth is what is implied here.

At any rate, Herodotus' account of the building of the pyramid is not adhered to in all aspects. The reason for this is unclear, but may be tied to the fact that Herodotus, in the writing of his history, referred to various Gods quite liberally.

Interestingly enough, the further back in time you go, the more overbearing is the presence of these fantastical beings we shall refer to as 'Gods' for now. It is evident to me, that the ancient Greeks believed in the Gods as a fact of life. I would have thought this clear enough if one but reads the surviving works of the day, Homer's possibly being the most famous case in point.[3]

---

[3] Homer's Iliad and Odyssey were always considered to be merely the writing of legends and myths, with no historical value (other than that implicit in the plays themselves) until the supposedly mythical city of Troy was unearthed by Heinrich Schliemann, a German amateur archaeologist that decided to take Homer's stories a little more seriously. Even so, the practice of looking at myths as being devoid of factual historical content prevails in the vast majority of cases, right up to the present day. Incidentally, it is very interesting to note that Homer's plays have plentiful references to numerous Gods.

And when you consider that Herodotus' purpose in undertaking the project in the first place was not in fact to record the time of his day for us, but to record his own distant past for the people of his own time, it becomes obvious that he would have been unlikely to knowingly insert legends or myths as part of his history.

In other words, the references to 'Gods' in the time of Herodotus were to be taken literally, perhaps in a similar fashion to the one we have today of referring to the dinosaurs, whom although extinct, were very real and are taken as a historical fact by us today.

Why we continue to assume that the ancients spent their time wandering from place to place, no doubt at great personal cost and even greater risk, in order to write fanciful stories of imagined beings, is quite beyond me.

In any case, Herodotus writes that he was told by the Egyptian priests of the time that the Great Pyramid was built in twenty years by 100,000 men. Several other factors are mentioned too.

King Khufu for example was not buried in the King's chamber according to Herodotus' account, but rather under the pyramid in a vast cavern where an artificial underground lake had been created. The water supposedly came from the nearby Nile and the King's final resting place was on the cavern's central island.[4]

Ignoring this, and the fact that when the King's chamber in the Great Pyramid was entered the sarcophagus was not only empty, but even missing a lid, many archaeologists up to the present day insist that the 'giant tomb'[5] must have been robbed, and no allowance for a subterranean cave as a final resting place is made.

This despite the fact that tombs which *have* been robbed of course showed some concrete evidence of such a disturbance, and normally the mummy and sarcophagus lid were nearby, even if severely damaged.[6]

---

[4] Any faithful translation of Herodotus' work in book II will attest to this, as for example on pg. 179-180 of: *Herodotus, The Histories*. Or pg. 75, par. 124 & pg. 76, par. 127 of *The History of Herodotus*. See Bibliography for details on these books.

[5] The pyramids in Egypt, even today, are usually regarded as tombs in their primary function. Only recently, is the Archaeological community beginning to accept that the main purpose of the pyramids may have been something other than an elaborate gravestone.

[6] Since the mummies often had jewellery hidden in the wrappings, grave robbers did tear or remove these, but the actual mummy was normally left behind.

The historians of our day then proceed to state that the Great Pyramid was built by using the ramp, which Herodotus says, along with the underground complex of channels and passages, took ten years to build.

This causeway is 915 metres in length, 18 metres wide and was 15 metres at its highest point. It also incorporates numerous intricate carvings.

Although Herodotus says wooden machines were used to place the blocks of stone, it is taught to school children (if mentioned at all) that the great blocks of stone were dragged up this ramp by slaves.

The passage of such stones was made easier either by the use of logs that acted as temporary wheels or by some sort of wooden sledge under which the Egyptians poured oil or some other fluid to alleviate friction. The period of 20 years is taken as being correct generally speaking, as is the number of men used in the project.

I have come across certain books however, that claim the Great pyramid was built in as short a period of time as five years by only 80,000 men[7].

But let us now consider certain facts with respect to the Great Pyramid; so that we can assess these claims based on a mathematical or logical basis, rather than on that so vastly abused and misused factor: 'expert opinion'. Which of course, sometimes successfully masquerades as scientific fact.


First of all, the Great Pyramid has been built on an underlying strata of rock. This point, which is often overlooked even by more objective persons than the 'experts', is of paramount importance.

The Great Pyramid is thought to weigh nearly 6 million tons,[8] and any structure with this sort of weight is bound to undergo some rather large settlement unless it's built on a rock strata.

Even then, *rock classified as soft or very soft by geotechnical engineers might crack under the weight.*

To be on the safe side, the rock needs to be at least of *medium/hard* quality. This type of rock has a minimum compressive strength of

---

[7] In the interest of fairness though, I shall refer to this last proposal only for the purposes of comic relief.

[8] Lemesurier gives an approximate value of 5,955,000 tons (Pg. 8 of *The Great Pyramid Decoded*).

10 MPa (Mega Pascals) which is equivalent to a pressure of 1000 tonnes per square metre.

On average, the Great Pyramid exerts a pressure on the underlying strata, of only a little more than 1 MPa. But at its centre the pressure on the underlying rock comes to over 3 MPa, which is above the minimum compressive strength of rock known as *soft rock* to soil engineers.

Furthermore, the rock layer has to be thick enough to support the pyramid, and, more importantly, extend for a considerable distance throughout the area in order to be considered stable.

In order to find a suitable site, present day engineers would have to dig or drill quite a number of bore holes in the general area.

In order to ensure that the underlying rock is a consistent strata, a large number of these bore holes would have to be drilled, extending over an area several times larger than that eventually taken up by the finished building.

This procedure is not as simple as it sounds, because the depth of the rock stratum also has to be consistently thick, and there is no sure way to determine this without the use of equipment which, apart from being rarely used due to its prohibitive cost, is considered sophisticated even by our own standards.

The determination of consistency is important. Particularly when the incredible precision of the finished pyramid has to be taken into consideration.[9]

We'll look at this precision in some detail later. For the moment let's just take a look at the basic structure of the Great Pyramid.

The Great Pyramid today stands at a total height of about 137 metres and has a square base with sides which measure some 232 metres. These are not the *precise* measurements, (we will get to these later) but for now they'll suffice. It is composed of about 2,300,000 limestone blocks,[10] which on average weigh some 2.5 tons. Some of

---

[9] Incidentally, if the original resting place of the King is as detailed by Herodotus, considering that presumably quite a large amount of rock would have to be excavated at a point which lies under a thick layer of the rock stratum and ultimately under the pyramid itself, this would in part go some way towards explaining how the builders could be sure of the thickness of the rock strata.

[10] Some sources quote this figure as 2,500,000 but I have stuck to the more conservative measurements wherever possible, in order to give the 'benefit of the doubt' so to speak, to the more sceptical reader.

these blocks though, are much larger, weighing up to an estimated 16 tons or so.

Internally, the pyramid has some passages, and various rooms or chambers, and here we find blocks of granite as well as limestone. Interestingly, some of these slabs of rock weigh a lot more than 16 tons, at least one in the King's chamber complex is estimated to weigh some seventy-two tons or so, and several others weigh only a few tons less.

Let's just pause here and deal with this information for the moment. Assuming that Herodotus had been given the correct information by the Egyptian priests, let's say that the Great Pyramid really did take 20 years and 100,000 men to build.

Now let's further assume that the ancient Egyptians were far more diligent at their work than *any* present day building crew. We'll assume that in that whole period of twenty years, not a single man was sick, died, or slacked off during the building of the pyramid. Given their diligence, obviously, strikes are also completely out of the question.

We'll also assume that the weather was always perfect during every single day of construction and in no way adversely affected building time. I think given these circumstances no one would be so bold as to say we've not given these theoretical builders every possible advantage.

But let's really stretch ourselves here. Let's also say that they worked not a five day week like us, but a seven day week. They take no holidays or breaks whatsoever for the entire period.

Additionally, just to make sure the sceptics are truly satisfied, we will endow these (by now superhuman) workers with the ability to work not eight hours a day (pulling huge rocks up ramps mind you) but a full twelve hours each day.

We assume of course that the heavens are going along with us on this and provide at least twelve hours of daylight during which to work every single day for twenty years.

Right then, let's begin. Take a calculator and type in 2,300,000 (number of limestone blocks) and divide it by 20. This will leave a total of 115,000 blocks to be placed every year. Now divide this number by 365.26[11] and you are left with a result of 314.84 blocks which have to be placed every working day. Since the working day is composed of a

---

[11] The length in days of the Anomalistic Year. A value slightly longer than the conventional 365.242 days for a Mean Solar Tropical Year, but hey, we don't want to be accused of cheating.

back-breaking twelve hours in our case, divide this number by twelve. This will leave you a value of 26.24 blocks to be placed every hour which of course is very close to one block every two minutes.

It doesn't matter whether that block weighs 2.5 or 16 or 70 tons mind you, a block is a block. We are not to squabble over such details you understand.

I wonder if at this point there are still people out there that think this is feasible by any stretch of whatever fevered imagination.

Especially when we add a few other points which we have conveniently neglected up to now.

According to the archaeologists, experts and historians, the people who built the Great Pyramid did not even have knowledge of the rudimentary pulley. It is also difficult to see how a ramp which is fifteen metres high at its highest point can help in building a structure that is estimated to have been about 147 metres in height when completed.[12]

Even if we choose to ignore such 'details', there is still the matter of explaining how the blocks were aligned with such precision.

Even today, after thousands and thousands of years, at least two major earthquakes for which we have records, and quite possibly a few more for which we don't, earth tremors and severe human scarring, the casing stones of the Great pyramid are so perfectly aligned that the joints are hairline cracks, and that other than the places where obvious destruction has taken place (either man-made or due to natural causes) it is not possible to slide a thin knife blade between the blocks.[13]

These casing blocks, which placed over the stepped sides of the pyramid are what gave the finished structure its smooth finish, were cut and polished to something akin to optical precision standards.

Additionally, the artificially levelled ground the Great Pyramid is built on, is still perfectly level, the error of shift in the horizontal plane of the

---

[12] Additionally, it has been mathematically calculated, that the building of a ramp of the required size, would entail even more work than the construction of the Great Pyramid itself in terms of material used.

[13] Lemesurier attributes the average join width as less than a fiftieth of an inch (0.51 mm) (The Great Pyramid *Your Personal Guide* Pg.13) This of course is approximately the size of the lead in a mechanical pencil. It is worth pointing out, that in order for this sort of precision to be possible, each block must have been perfectly cut with tolerances for error of about half this distance (0.25 mm). It is difficult to see how this type of precision could be achieved given the theoretical technology of the time, and if each block had to be 'sanded' down to size after having been carved out, this would add to the building time by several degrees of magnitude.

structure being less than a tenth of an inch (2.54 mm), with a total settling shift over a period of thousands of years, that amounts to about a half inch; and this for what remains the heaviest building in the world! A feat that would be considered more than outstanding even today, since despite the use of modern equipment, it is one we cannot hope to achieve, and certainly one that would not last for thousands of years.

By now, it should be clear that the theory of the blocks being dragged by slaves on sledges or over wooden rollers[14] does not have much basis in fact, or that if it does, then the Great Pyramid took a lot longer to build than 20 years, and probably a lot more than 100,000 men.[15]

Incidentally, the longer the building period, the less likely that the pyramid would have been completed at all, since the chances of war, famine, revolt at such a megalomaniac scheme and similar problems would all be much more likely to occur. And if we are to assume that King Khufu really was responsible for the building of the Pyramid, we cannot really take the entire period for construction to be much more than 100 years, even if we assume that he 'stole' the unfinished building of the Pharaoh which preceded him.

If you thought however to explain away the pyramid by allowing the wooden machines described by Herodotus, then it's time to look at a few other factors which should promptly dispel you of the idea that the Great Pyramid was built by a superstitious bunch of ignorant primitives, and at any rate, by allowing the wooden machines, you have already taken the first step in leaving behind the orthodox view of its construction.

As we progress though, you'll find that even the explanation of some type of primitive wooden machine will become quite inadequate.

---

[14] Incidentally, wooden rollers would be crushed by the weight of some of the building blocks and in any event Egypt has never possessed the kind of trees that would be in any way useful for this kind of work, meaning that if trees were to be used they would have to have been imported, with all the additional trouble which that entails.

[15] It is interesting to note that if the Great Pyramid was to have been built in five years by 80,000 men, then the number of blocks which would have to be laid down each minute is 1.74, or just under 105 blocks per hour. Conversely, if we assume a building period of 100 years, then just over 5 blocks per hour or 1 block every twelve minutes or so would have to be placed. Other factors would still have to be as described for a twenty year building period (twelve hour working days, no breaks, etc.).

The total area of the four faces of the Great Pyramid adds up to about 85,000 square meters. An area that is equal to a 'road' 85 metres wide and one kilometre in length (or if you prefer, 57.76 yards wide by one mile in length). This whole area, was dressed and finished in smooth limestone. One can see why the Egyptians used to refer to the Great Pyramid as *The Light*. It must have been visible as a huge shining mountain for miles and miles.

Present day stonemasons in England feel that completing a fifth of a square metre of smoothed limestone a day is a good day's work.

At this rate of work, one thousand men would have had to work every day for 425 days, just over a year, to complete merely the smoothing of the outer facing of the casing stones of the Great Pyramid, to say nothing of the smoothing of their other sides, which in view of their precision of placing, would have certainly been required and which would of course have further added to the construction time.

And of course, even the value of 425 days assumes that the ancient Egyptians had access to present day methods for stone working.

But perhaps they could make up for what they lacked in technology by using the same type of tireless masons that we used for the block laying. You know, the ones who work twelve hours each day, seven days a week, for the whole period and never get sick, strike, complain, waste time, take a break for lunch or have the bad luck of having nasty weather before the end of the project.

Oh yes, did I mention where most of the limestone used in the Great Pyramid's construction was mined from? It was done at a place called Tura (hence the name Tura limestone) which lies only fifteen miles or so from the location of the Great Pyramid.

It *is* on the other side of the Nile of course, but you can't expect *every* little thing go your way now, can you.

Besides, the black granite used in the King's Chamber, which is estimated to weigh a total of about 1500 tons, some of it composed of blocks weighing between 60 and 72 tons, was imported from Aswan in Upper Egypt, which lies some 1000 kilometres away in a southerly direction.[16]

Egyptologists tell us that the Nile was used to transport these stone blocks from their quarries to the location of the Great Pyramid.

---

[16] See *The Orion Mystery*, pg. 46 and footnote 53, on pg. 310. This distance assumes travel by the Nile; in a straight line, Aswan lies about 640 Km from the Giza Pyramids.

Just what kind of boats, and how many of them were supposed to have been used for the moving of these stone blocks, the experts do not say, but then, this is a trait we're starting to become familiar with by now.

And yet, despite all this rather damning evidence to the contrary, I would still be ready to accept that the Great Pyramid *could* have been built by regular humans with old technology.

Not by pulling on giant blocks of stone and using rollers and ramps of course. I would have to *at the very least* allow for some ingenious contraption to facilitate the placing of the blocks as well as their mining, and their precisely cut shapes would suggest to me some additional technology in this field too.[17]

Given that much as an absolute minimum, and stretching my rather elastic imagination considerably, I would be ready to accept that the Great Pyramid has nothing to do with any Martians.

But then, even if we just put aside for the moment the fact that we would still have the messy prospect of explaining away the Martian pyramids,[18] there is still an uncomfortable problem before us:

Up to now, I have not mentioned any of the *truly* disturbing facts in relation to the Great Pyramid.

The really interesting stuff is still to come.

The Great Pyramid was built to be correctly aligned with the true points of the compass. Its slight deviation from true North today, is not the result of poor workmanship, but rather due to the gradual shift which the Earth's axis undergoes over long periods of time. This shift is so minute, that even after thousands of years, the error amounts to just a little less than a twelfth of a degree. Pretty amazing accuracy then, considering the ancient Egyptians supposedly had no compasses at the time (experts tell us so).

---

[17] It is normally accepted that the blocks were cut out of the quarries by the use of saws and the placing of wooden wedges in the rock face which would expand on wetting and/or hammering. Personally I find the idea of cutting out well over two million blocks in this fashion ridiculous. See also footnote 13 earlier in this chapter.

[18] "Never mind them" I can almost hear the 'experts' say in unison, "it's a trick of the light, or of the computer imaging process, or of some other factor...."
Sadly reminiscent of a Catholic priest from the middle ages trying desperately to hold onto the principle that the Earth is flat when faced with satellite photographs to the contrary, we would still have to ask why NASA seems biased in its treatment of this topic. See Part two of this book for evidence of this as well as possible reasons for this type of treatment.

In order to appreciate some of the finer points of the Great Pyramid though, it is necessary for us to understand a little about the units of measure used by its builders. Once we do this, proper investigation of the various properties of this monument can take truly a lifetime of study. I am limited therefore, to simply show *some* of the more easily expressed points of interest, but I once again urge the interested or still sceptical reader to get a hold of Peter Lemesurier's book, *The Great Pyramid Decoded*[19] for a more complete treatment of this subject. I cannot stress enough the inadequacy of this chapter when compared to Lemesurier's more complete effort.

The metric system of measurement is based on the unit known as metre. This unit was taken to represent one ten-millionth of the distance from one of the Earth's poles to its equator, as figure 3 below shows.

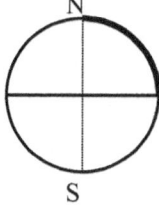

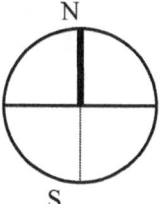

Metric distance                                    Collet distance
(cannot be precisely determined)      (can be precisely determined)

Figure 3.                                                  Figure 4.

The problem with this type of measurement, is that it cannot be accurate, since the Earth is not a perfect sphere. For this reason, the metre could be said to be a quasi-arbitrary unit of measure and thus imperfect.

The mathematician Collet, as far back as 1795, stated that a better unit of measure would have been the ten-millionth part of the distance from the pole to the centre of the Earth (See figure 4 above). This

---

[19] See bibliography for details.

distance has been established quite accurately by the use of satellites and it comes to be 3949.89 miles, which also equals 250,265,030.4 British, or Imperial inches.

One ten-millionth of this length would then equal 25.0265 Imperial inches, and this length could be said to be the perfect unit of measurement, since unlike the metre, it is not arbitrary.

The unit of measure known as the Sacred Cubit, used by the Egyptians in the building of religious monuments as well as in the construction of the Great Pyramid, is *precisely* equivalent to 25.0265 Imperial inches (correct to four decimal places).

If we divide the Sacred Cubit into twenty-five equal parts, we are left with a distance that is equivalent to 1.00106 Imperial inches. This distance is once again precisely the one used by the ancient Egyptians and is commonly referred to by people who study the Great Pyramid as the Pyramid or Primitive inch[20].

The obvious question that springs to mind is: How could a supposedly technologically backward people (compared to our own standards) have been using a unit of measure which is more precise than any currently in use today?

The answer of course, is that they could *not* have been doing so. Which brings us to the inescapable conclusion that if in no other respect, at least as far as distance measuring goes, the ancient Egyptians were superior to us.

Quite a blow to our arrogance in proclaiming technological superiority to any previous age of man, is it not?[21]

The Great Pyramid has a smaller pyramid etched into each of its four sides. This results in a slight hollowing of each side of the pyramid. Furthermore, at some distance from where the original sides of the

---

[20] Lemesurier suggests the name Sacred Digit, which is indeed more elegant, but I shall confine myself to calling it Primitive inch throughout the remainder of this book (I find the irony of the word primitive in this context a fitting one!).
Measurements in Primitive inches will be followed by the symbol  P".

[21] As a matter of interest, another unit of measure known to have been used by the Egyptians, is known as the Royal Cubit.
This distance is slightly shorter than a Sacred Cubit and is equivalent to 20.6284 Imperial inches or 20.60659 Primitive inches.
The Royal cubit was normally divided into 100 *n*. One *n* of course being equal to 0.206284 Imperial inches, or 5.24 millimetres.

pyramid finished, one can find the still existing foundation sockets of the *full-design* pyramid.[22]

The full-design pyramid of course was never built, but this results in the interesting fact that now one has the possibility of measuring three different types of distances for each of the sides, as described below and on the following page.

Figure 5.

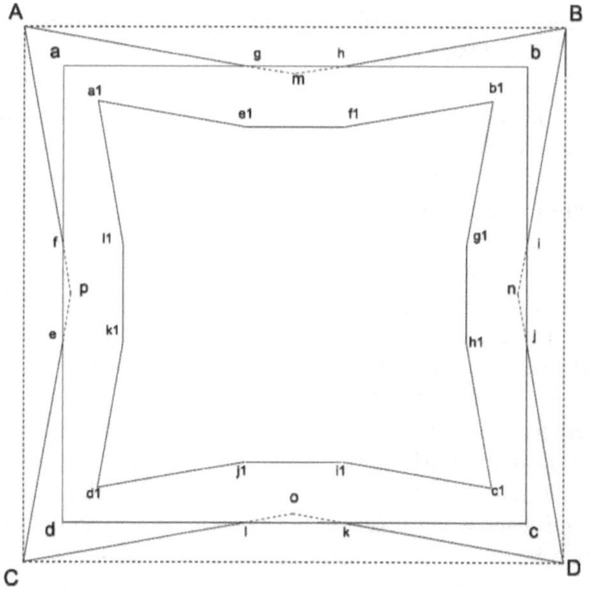

Where:                          a-b-c-d-a = Base as built
      a1-e1-f1-b1-g1-h1-c1-i1-j1-d1-k1-l1-a1 = Base of core masonry
               A-g-h-B-i-j-D-k-l-C-e-f-A = Base of full design pyramid

---

[22] The full-design pyramid, is what the Great Pyramid would be if its sides reached the existing foundation sockets, which it never did, even when it still possessed all its casing stones.

1) A-B = 365.242 Sacred Cubits. 365.242 is the number of days in a Mean Solar Tropical Year

2) A-g-h-B = 365.256 Sacred Cubits. 365.256 is the number of days in the Sidereal Year (Time the Earth takes to complete a Solar revolution)

3) A-m-B = 365.259 Sacred Cubits. 365.259 is the number of days in the Anomalistic Year (Time the Earth takes to return to the same point in its orbital path)

4) A full-circuit (perimeter) measurement in Primitive inches results in the values 36524.2, 36525.6 and 36525.9, once again giving the three types of year possible.

5) The depth of the side concavity is 35.762 Primitive inches which is a unit of measure which appears several times inside the pyramid's passages.

6) Similarly, the difference between the perimeter of the full-design pyramid and the perimeter of the pyramid as built is 286.102 Primitive inches, and this too is a value we find again in several places. It is equal to half the length of the base side of the missing capstone of the Great Pyramid[23] and it is also the exact distance by which the entrance passage to the Great pyramid is offset from the centre-line; certainly this cannot be coincidence.[24]

7) 286.102 divided by 8 is equal to 35.76275. Considering the difference here from 35.762 is *less than one thousandth of a Primitive inch* (0.025 millimetres), for all intents and purposes, it's obvious that 286.102 should be taken as a multiple of 35.762. This is significant when taken in the context of a Pyramid Code (See point 8 below).

8) Lemesurier goes on to do just what the title of his book says. He *decodes* the pyramid. By showing relationships between values and numbers he is able to piece together the meaning of the Great Pyramid, and as it turns out, the Great Pyramid is nothing less than a map for the spiritual paths of mankind written in stone.

---

[23] Lemesurier gives ample evidence to show that the Great Pyramid was originally built either without the top or with a fake top only, which was designed to weather away relatively soon.

[24] Again, Lemesurier gives more proof for this fact than I can hope to muster here.

The incredible thing is that there is no way that Lemesurier could really 'fudge' the numbers.

The Great Pyramid has been measured and re-measured in great detail by many professional surveyors, and it is of note that while Lemesurier makes use of primarily one set of numbers,[25] he goes on to describe the difference between the values he used and those given by another survey, which could also be described as being the one that diverges the most from Rutherford's values.

This difference, when applied to the perimeter of the base of the Great Pyramid amounts to a value of less than one Primitive inch over a total distance of over 900 metres (more than half a mile).

But even more interesting, is that this difference applies to the *full-design* pyramid only and not to the actual pyramid as built.

Lemesurier goes on to explain why the architect could well have meant for both possibilities to be considered, since if Rutherford's measurements are used, the exterior of the Great Pyramid (by using Lemesurier's code) would be interpreted as meaning one thing, while if the other measurements were used , the base of the Great Pyramid would not be exactly square and thus a different *but equally valid meaning*[26] would be suggested. Additionally, the value of 286.102 Primitive inches cannot be a coincidence (see point 6 earlier), and in fact, using Lemesurier's code it not only indicates where the originally hidden entrance to the Great Pyramid could be found, but it shows that the architect of the Great Pyramid wanted to draw attention to the existence of what we now call the *full-design* pyramid, which has never been built.

It is also obvious that the *full-design* pyramid was never *meant* to be built, because otherwise there would not be the correlation between the full-design pyramid and the pyramid as built which the architect obviously wanted to emphasise. In other words, this was just a very clever way of presenting two or more ideas in just one building. Since the full-design pyramid was never built, both ideas must be taken to be possible, and when we investigate these ideas, both *are* indeed valid.

---

[25] Rutherford's.

[26] Both meanings are extensively substantiated by other features of The Great Pyramid.

While, as I have already stated, it is not possible for me to explain in detail how Lemesurier goes about decoding the Great Pyramid[27] I will say that it's purpose concentrates on the predicting (or in the case of past events, confirming) mankind's spiritual pattern of growth.

By measuring distances along the floor in the relevant units (which the Great Pyramid points out by placing a step of appropriate size wherever a change of timescale needs to be applied) one can measure the distance between events in time.

Most common is the use of one Primitive inch to represent one year of time, although units of one $n$ and others are also used.

Occurrences of interest, (such as the birth of Jesus, the second World War etc.) are depicted by particular features of the passages.

It is important to keep in mind that Lemesurier sets out a relatively simple standard to be used for translation of the pyramid right at the beginning of his book and that when viewed in this light, the vast majority of the entire structure suddenly becomes as easy to read as a long, but not particularly difficult, math problem.

Furthermore, it is a very consistent math problem, further indicating, that Lemesurier is most definitely on the right track.

It might be of interest to the reader, to point out certain features mentioned in *The Great Pyramid Decoded* :

A) So accurate is this building, that the death of Christ as forecast by the Great Pyramid[28] turns out to be the 1st day of April of the year 33 AD. Nor can the numbers be juggled, a distance along a stone floor is not open to argument. As the saying goes, "It's written in stone."

B) A period between 1933 and 1944 AD during which an inverted Messiah apparently was to manifest (Hitler as some kind of Anti-Christ?).

C) The establishment of a separate and unique form of human society based on spiritual values rather than material ones in 1999 (21st February is given as the possible date).

---

[27] To explain it in detail would basically require me to reproduce Lemesurier's work, which I find pointless since his book has recently been reprinted, and to explain it in a half-hearted fashion would not do it justice by a long way.

[28] It has to be forecast since the pyramid was already very old when Jesus was born.

D) The return or appearance in the sky of the sign of the Messiah in the year 2034 AD (31st October is given as the possible date).[29]

Some additional points of interest with regard to the Great Pyramid are set out below, which are valid regardless of whether the idea of the Great Pyramid being some kind of oracle is accepted or not.[30]

1.) When compared to the Great Pyramid, the Empire State building comes out a distant second best. While the modern structure is considerably taller, the Great Pyramid weighs some thirty times more and has over twice the volume. In terms of size and complexity, possibly the only present day structure which could be compared to it might be the Chunnel between Britain and France.[31] Considering the theoretical lack of technologically advanced equipment in the building of the Great Pyramid though, its construction cannot be adequately compared to anything we have done at present. Placing the Great Pyramid in the technological context which archaeologists presently do would perhaps be equivalent only to a structure so fantastic that it would have to lie in the realms of deep science fiction. Such a

---

[29] The return of Jesus? Or is the return of some extra-terrestrial group implied? Lemesurier does not seem to subscribe to the theory of alien or extra-terrestrial origins to any great or even palpable degree, yet he goes some way to say that the Great Pyramid appears to actually refer to a physical sign in the sky, and he's forced to mention the possibility of perhaps some sort of space-ship or extra-terrestrial sign which will be visible to the majority of mankind. He also makes it quite clear that for all he knows, the sign will be some sort of spiritual descent of Jesus along with a cohort of angels.

[30] I admit that if I had been told prior to purchasing it, that *The Great Pyramid Decoded* was supposed to show how Jesus will return to Earth to save mankind, I would probably have laughed and never given it a second glance. I feel I must add here, that Lemesurier himself scoffs at the idea of the Great Pyramid being some sort of vindication of Catholic dogma. It *is* though, some kind of an oracle, or prophecy in stone, and as such it would be incomplete if it made no reference to a historical figure which most definitely played a key role in the development of humanity.

[31] And it must be stated, that while the Chunnel is certainly an engineering accomplishment, its purpose is primarily a practical one. Some may argue that its construction, by its very nature, on some level implies that mankind is trying to leave a noticeable mark. Even so, the Chunnel cannot be said to embody any of the spiritual or holistic measurements or symbolic dimensions of the Great Pyramid. It is not aligned with the magnetic poles, nor does it include the length of a year in one or other of its primary dimensions. It does not, in other words, try to convey any sort of message other than: "it was built".

structure (which has been suggested by science fiction writers) might be a cable, complete with turbolift, that connected the Earth to a point high above it, where a space station orbited, or perhaps even to the Moon.

2.) There is also considerable evidence[32] that the Great Pyramid gives the minimum and maximum values for the eccentricity of the Earth's orbit as well as its diameter.

The *phi* ($\phi$) ratio is also incorporated in the basic design of the Great Pyramid. This is especially interesting since this ratio is closely linked to the Fibonacci series, which occurs regularly in natural phenomena.[33]

3.) The Great Pyramid is situated at the exact centre of the geometrical quadrant formed by the Nile Delta.

This feat of engineering would tax even present day builders, and it's difficult to see how this could have been done by a primitive people. This fact was only discovered in 1868 by the United States Coast Survey.

4.) Reference to any equal-area projection of the Earth's surface shows that the Great Pyramid also lies on the longest land-contact meridian.

More importantly, on such a map (on which areas of land are correctly represented) the Great Pyramid is found to lie at the centre of the Earth's land masses. This is said by Lemesurier to be due to coincidence or luck, since the builders could not possibly have known about such things. But in view of the Mars Pyramids, it would seem much more likely to me, that the builders in fact were very much aware of such phenomena.

5.) The ratio of the Great Pyramid's height to its base perimeter is exactly equal to two times *pi* ($\pi$).

The relationship then, is identical to the one you find between the radius and the circumference of a circle.

---

[32] See page 309 of *The Great Pyramid Decoded*.

[33] See pages 316 to 321 of *The Great Pyramid Decoded*.

Using the Sacred cubit as a unit of measure, we find then, that with the exception of the values 35.76 P" and 286.1 P", every one of the basic external and internal measurements of the Great Pyramid is a function of pi and 365.242. When read in the context of the quarter aroura circle, (see point 6 below) this makes it obvious that the Great Pyramid is supposed to embody the planet Earth, in view of its relationship to the length of a year in days. Nor is this the only way in which the Great pyramid represents the Earth by giving precise astronomical measurements.

6.) The Egyptians used a further important unit of measure, the aroura. This is a unit of area equal to a square with sides 100 Royal cubits in length. This would give such a square a side length of 2060.659 Primitive inches.

The area of such a square is also equal to the area of a rectangle measuring 3652.4235 P" by 1162.6025 P".

This in turn is equal to the area of a parallelogram with a height of 1162.6025 P" and base length of 3652.4235 P".

Furthermore the dimensions 1162.6025 P" and 3652.4235 P" are also the diameter and circumference of a circle whose area is exactly equal to a quarter aroura.

See figures 6, 7, 8 and 9 below.

Figure 6.                                    Figure 7.

Aroura square                          Aroura rectangle
Side = 2060.659 P"               Height = 1162.6025 P"
                                            Base  = 3652.4235 P"

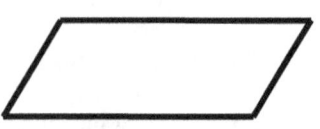

Figure 8.                                    Figure 9.

Aroura parallelogram                Quarter Aroura circle
Height = 1162.6025 P"              Diameter = 1162.6025 P"
Base   = 3652.4235 P"             Circumference = 3652.4235 P"

You will note that the circumference of a quarter Aroura circle in Primitive inches (3652.4235 P") is equal to ten times the number of days in the Earth's mean solar tropical year (365.242 days).

The ancient Egyptians were aware of all these factors, because as Lemesurier reminds us, the writer Horapollon stated: "To represent the current year they depict the fourth part of an aroura."
It is extremely interesting to note then, that the Great Pyramid incorporates the aroura measurement several times in its configuration.
Firstly, two aroura parallelograms are implicit in the vertical cross section of the full-design Pyramid as viewed from any of its four sides, this is shown in the figure below.

Figure 10.

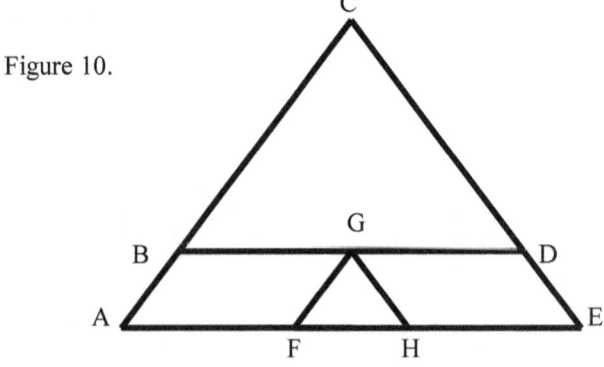

Where:

1) ABCDE is the vertical cross section of the full-design Pyramid.
2) ABGF and GDEH are the aroura parallelograms.
3) BGD is the axis or midline of the 35th course of the Great pyramid, which is exactly 2 Sacred Cubits in height.
4) FGH is the inset triangle evident on each of the four sides of the Great Pyramid, and it is exactly 1/5 the size of triangle ABCDE, which of course represents the entire Great Pyramid.

It is no coincidence then, that the 35th course of the pyramid is so prominent, being not only among the thickest, but also exactly two Sacred cubits in height, making its axis equivalent to the top of the aroura parallelograms. Evidently the architect wanted to draw our attention to this course of masonry (Lemesurier once again gives far more evidence for this than I have).[34]

Furthermore, the quarter aroura circle is also represented in the King's chamber and antechamber in such a way as to define the relationship between the Royal Cubit and the Primitive inch.

The Antechamber is thus in a way the geometrical key to the Great Pyramid. It is therefore interesting to note that it is in this very chamber, that using Lemesurier's code, we find the return of the Messiah. This event in turn (always according to the code) seems to be the key which will finally open the spiritual path for mankind.

It is also worth mentioning, that the circle inscribed by the massive boulders of Stonehenge, has been at least partially understood, and the evidence seems overwhelming that while its true function may never be fully grasped, it most certainly has astronomical connotations so precise and surprising that it has been referred to on several occasions as a 'primitive astronomical computer'. Would it surprise you to know then, that the Stonehenge circle is also a Quarter Aroura circle?[35]

Even if we do not accept Lemesurier's decoding of the Pyramid, it is still impossible for us to deny the fact that the ancient Egyptians had

---

[34] Using Lemesurier's code, it is possible to attribute a number of functions to the 35th course of masonry in addition to the one outlined here, which is perhaps of only passing importance.

[35] See page 303 of *The Great Pyramid Decoded*.

an intimate knowledge of astronomy and that this knowledge is embodied in the Great Pyramid.

In view of these and other facts, it becomes obvious, that the Great Pyramid is not only supposed to represent the Earth, at least in its external aspects, but that its architect intended for some message to be read in its construction. We also discover that it is a far more complex structure than we might have suspected at first glance, and in any event, the idea that the pyramid is merely a burial mound has now become completely absurd.

Nor, uncomfortable as it may be for some, have we (by any means) reached the end of the very factual evidence that suggests the Great pyramid has technology connected to it that is 'out of this world'.

# An Early dawn for the Bronze Age

Robert Bauval, co-author of the best-selling *The Orion Mystery*,[36] was instrumental in an interesting re-discovery.

In September of 1993, he came across a remarkable passage in Charles Piazzi Smyth's book of 1878, *The Great Pyramid*,[37] in which, Smyth was recounting the discovery of the 'air-channels' found in the Queen's chamber by Waynman Dixon and Dr Grant.

The amazing thing, is that Smyth describes how certain items were found in these 'air-channels'.

The items in question were: a little *bronze* grapnel-like item, a portion of cedar-like wood which is described as having possibly been some kind of handle for the grapnel instrument, and a grey-granite, or green-stone ball weighing 8,325 grains.

Bauval gives the weight of 8,325 grains as being roughly equivalent to 0.85 kilograms, but in fact this is wrong, since one grain is equivalent to 0.06481 grams, so 8,325 grains would amount to 539.5 grams, or 0.5395 kilograms, which of course is just a shade over half a kilogram.

It may be of course, that the error is a typographical one, rather than one due to poor workmanship on Bauval's part.[38]

At any rate, interested by the fact that apparently a bronze instrument was discovered in the Great Pyramid, Bauval decided to try and find out what had happened to these artefacts.

Eventually, he was successful, through a mixture of investigative work as well as some measure of luck (an ex-museum employee saw a newspaper article concerning the items and remembered that they were indeed stored at the British Museum where Bauval had already unsuccessfully searched for them).

---

[36] See bibliography for details of this book.

[37] The name Piazzi Smyth is well recognised in Egyptology circles, and so it should be, since he was one of the explorers whom during the 19th Century helped enormously in bringing the awe and mystery of the pyramids, and of the Great Pyramid in particular, to the world at large.

[38] Bauval, after all, later quotes (on pg. 241 of *The Orion Mystery*) a letter from Dixon to Smyth, where the weight of the stone is given as 1lb 3oz, which of course corresponds exactly to the value of 539 grams and some change, showing quite clearly that this must indeed have been the stone's weight.

It was in this rather amazing manner then, that items of considerable importance, which had been 'lost' in the nether regions of the British Museum were re-discovered.[39]

Bauval points out that the finding of a bronze item in the Great Pyramid is a startling discovery, because given the supposed date of its construction (about 2600 to 2400 BC) bronze was not supposed to have been known in Egypt at that time.

There are wide discrepancies as to the length of particular Dynasties, but Bauval gives the values of 3100 to 2686 BC for the period covering the first two Dynasties, followed by the period 2686 to 2181 BC which encompasses Dynasties three through to six.

But according to Bauval (and others) bronze in Egypt was supposed to have been around only since the period given by Bauval as that which spanned the years 2133 to 1786 BC and which is known as the Middle Kingdom.

Remembering then, that Khufu lived in the Fourth Dynasty, the discovery of two vessels dating from the *Second* Dynasty, which had been previously thought to be composed of copper, but in fact turned out to be made of bronze,[40] would seem even more startling than the finding of the bronze grapnel in the Great Pyramid.

Bauval became aware of the two bronze vessels when it was communicated to him by the British Museum on 2 November, 1993.

In other words, if the dates provided by Bauval concerning the various Dynasties are correct, the bronze age in Egypt would have to be moved back by some 550 to 1000 years.

Although at first this sounds like truly astounding news, on deeper investigation, we find it is not all that Earth-shattering after all.

The Bronze Age is generally thought to have occurred between 4000 and 6000 years ago, which would mean that as far back as 4000 BC bronze was being used by certain people in the world.

In fact, the Hittites had supposedly mastered the art of smelting iron

---

[39] One cannot help but wonder; if items as important as these could be misplaced by the British Museum in relatively modern times, how much more knowledge have we lost throughout human history?

[40] See pg. 247 of *The Orion Mystery*.

as far back as the second millennium BC,[41] but when the Hittite empire collapsed in about 1200 BC, the skill was lost for a time.

For these reasons, it is not inconceivable then, that the Egyptians came into contact with bronze as far back as 2600 BC when Khufu reigned.

Nevertheless, as we shall see a little later, Bauval's re-discovery is indeed an important one, and it would be wise to keep it in mind for later reference.

In his book, Bauval also describes in some detail, research carried out by one Rudolf Gantenbrink, whom he befriended in the course of his own studies.

Gantenbrink had developed a small robot for the investigation of the Queen's Chamber 'air-shaft' channels,[42] and on entering the Northern shaft, the tiny robot filmed not only the rod used many years earlier by Dixon to plunge its depths, but also what appeared to be another grapnel-like instrument along with another piece of wood.

Bauval points out that this instrument appears to be made of gold rather than bronze judging by its shine, but since at the close of his book the item is still in the shaft, verification of this point will have to await further research.

More interesting though, is Gantenbrink's discovery of a tiny portcullis some 30 or 40 metres along the Southern shaft of the Queen's chamber. According to Bauval this was supposed to be opened at around February or March of 1994, but in fact, at the time of this book going to press, it remains unopened.

Nevertheless, the prospect of a secret chamber possibly lying behind this tiny stone doorway is an exiting one. Especially as it would seem unlikely that it could have been opened since the time when it was built.

---

[41] No mean feat considering that a temperature of 1100 to 1500 degrees Celsius is required as well as a careful control of atmospheric conditions inside the furnace.

[42] The Robot was called UPUAUT 2, and was the successor of UPUAUT 1, a similar robot, named after the Egyptian Jackal-God, which had been used to travel through the 'air-shafts' in the King's chamber.

# The Age of the Great Pyramid

The people whom have written about the Great Pyramid are numerous indeed, but in the course of the study of this building, we should limit ourselves to those works that present factual information which can be double-checked by actual investigation; which is why I have taken a great liking to *The Great Pyramid Decoded*, but Peter Lemesurier, fortunately, is not the only writer that has made important advances in the understanding of what the Great Pyramid is.

In recent times, the question concerning the true age of the Great Pyramid has come under the scrutiny of a number of serious investigators, and it must be said, that in view of their discoveries, the old belief that its construction began in about 2700 BC can now no longer be held as certain, no matter how dear it is to certain academics.

It is important to point out that the *only* reason that the Great Pyramid is taken to be about 4500 years old is by its association with the Pharaoh known as Khufu.

This association in turn, is based on only *one* piece of evidence:

*The discovery by Col. Richard Howard Vyse in 1837 of the quarry markings painted on the walls of the construction chamber known as Lady Arbuthnot's Chamber, which supposedly identify the rocks used as having been mined for King Khufu.*

Well over 150 years ago, the question as to whether Khufu had really been the builder of the Great Pyramid was one that was already becoming increasingly untenable. Miraculously though, it was at precisely this time, that the first evidence connecting Khufu to Great Pyramid was discovered.

Zecharia Sitchin in his second book of the Earth Chronicles however, has taken great pains to point out that, Col. Howard Vyse and his partner J.R. Hill *forged* that discovery.[43]

Sitchin's work in this regard however, has largely been ignored by Egyptologists and archaeologists alike, mainly for the following reasons:

---

[43] See pg. 253 to 282 of *The Stairway to Heaven*, the second book of the Earth Chronicles by Zecharia Sitchin. See bibliography for further details on this book.

1) Sitchin's work is exhaustively researched and annotated, and while this is an excellent thing from a scholarly or research point of view, it does not make for light reading. In parts of his books, Sitchin can become somewhat tedious even to an interested researcher, due to his endless attention to detail. For this reason then, few people have had the patience to read through his books from cover to cover, resulting in a low level of awareness with respect to his astounding claims.

   It's a bit like reading a book that at the end describes the startling events that led to the Apollo Moon landings, but starts the story by describing how the first rockets were built in ancient China and does not skip any steps in between.

   Very useful, but not necessarily fun.

2) Sitchin (as a result of his research) is an exponent of the theory that we have extra-terrestrial origins, and this of course does not bode well with academics and the so called 'experts'.[44]

3) It would appear that while perhaps well intentioned, Sitchin may have been somewhat too eager in his deductions.

The basic facts (at least as they are found in Sitchin's book) are set out below, although I necessarily gloss over the finer details.

Vyse discovered what he called Wellington's chamber on March 27, 1837, when, after using gunpowder to blast the rock, a small hole was made that reached into the second construction chamber above the King's Chamber (see figs. 11 and 12 below).

The first construction chamber had been discovered much earlier, by Nathaniel Davison in 1765, and it is interesting to note that no hieroglyphs were found in it.

Vyse resorted to using gunpowder for blasting on several occasions in his search for a hidden chamber in the Great Pyramid, and it is

---

[44] I became aware of Sitchin's work only after I had laid out my basic theory, it is interesting to note then, that he arrived at a similar (if infinitely more detailed) conclusion with respect to our origins as I have, although approaching the problem from a completely different starting point. One, I might add, that I would not have had the patience or inclination for, as it entails first learning a lot of ancient Sumerian and other old Semitic languages, including Hebrew.

important to note too, that his attitude towards the Great Pyramid in general, was far from a reverent one.

He viewed the building as a pagan monument and thought it perfectly acceptable to blast his way through the masonry by using gunpowder in generous quantities. So much so, that what today looks like an enormous entrance on the south side of the pyramid is in fact a gash left by Vyse in his attempts at discovering another entrance.

Figure 11.

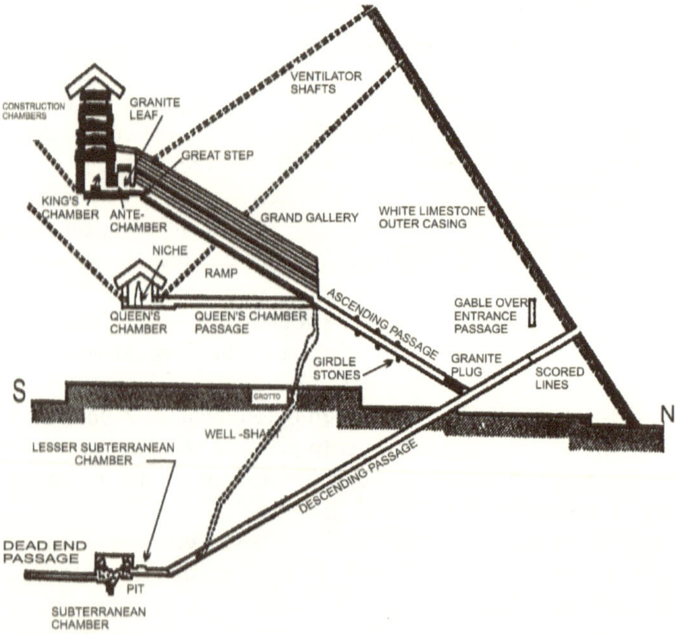

Passage details of the Great Pyramid.

Figure 12.

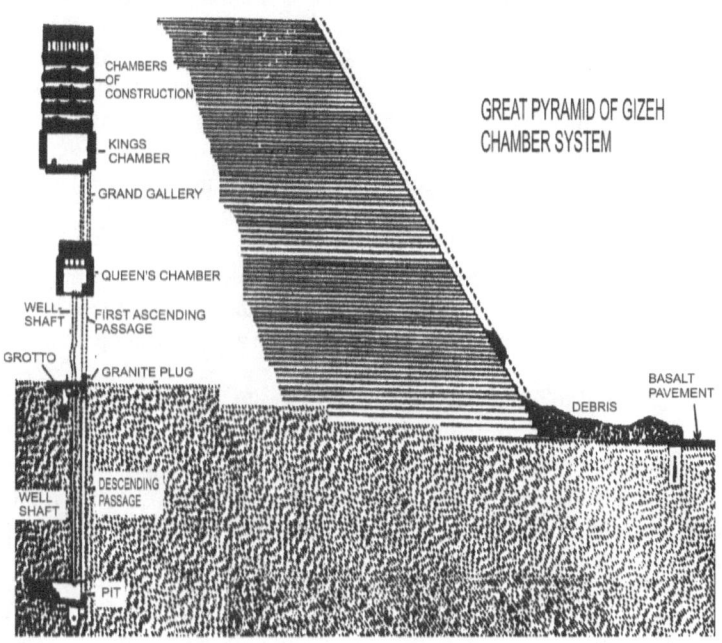

GREAT PYRAMID OF GIZEH
CHAMBER SYSTEM

CHAMBERS —OF CONSTRUCTION

KINGS CHAMBER

GRAND GALLERY

QUEEN'S CHAMBER

WELL SHAFT

FIRST ASCENDING PASSAGE

GROTTO

GRANITE PLUG

BASALT PAVEMENT

DEBRIS

WELL SHAFT

DESCENDING PASSAGE

PIT

Notice that when viewed from the North, the passages of the Great
Pyramid form the figure of a man, symbolising humanity complete with
feet, sex organs, stomach, head and what appears to be a five layered
crown on top of it.

It was only on March 30 however, that after further use of gunpowder
for blasting purposes, the hole was enlarged enough that it would
allow access to people. Vyse and his assistant, J.R. Hill, entered the
chamber and inspected it thoroughly, finding that it had been
hermetically sealed and possessed a black sediment on the floor.[45]

---

[45] Lemesurier points out that this 'dust' was much later found to be the crushed
corpses of thousands of beetles, and that if enough of this substance survived to the
present day, it might be able to be used for conducting radio-carbon tests on it in order
to ascertain its approximate age.

The ceiling was beautifully polished unlike the rough floor, and although it was clear that the chamber had never before been entered, it was bare and empty of any sarcophagus or treasure of any kind.

Later that evening, after having had the hole still further enlarged and having communicated to the British Consul that he'd named the new room Wellington's Chamber, Vyse re-entered the chamber with Hill and two other men, a Mr. Perring and a Mr. Mash (who was there at Perring's invitation) and it was only now, that:

"...we went into Wellington's Chamber and took various measurements, *and in doing so we found the quarry marks*."

As Sitchin himself puts it: What a sudden stroke of luck!

How is it that when Vyse and Hill first explored the chamber so thoroughly they found nothing, but now, in the presence of an additional two witnesses, the amazing discovery was suddenly made?

It's interesting to note too, that immediately after the March 27 discovery, but prior to actually entering the chamber on March 30, Vyse discharged the foreman, a man named Paulo, for no known reason. Similarly, he'd already got rid of Captain Giovanni Battista Caviglia, who'd originally been the one searching for a hidden chamber in the Great Pyramid. Caviglia had been discharged and ordered away from the site on February 13, 1837, directly after his major discovery the previous day in Campell's Tomb of a sarcophagus inscribed with hieroglyphs and masons' red-paint markings on the walls of the chamber. Surely a strange time to get rid of a person that had proved successful in making important discoveries.

Caviglia returned to the site only two days later and for the last time, in order to collect his belongings. For years afterwards, Caviglia continued to make what the Colonel himself describes as "dishonourable accusations" against him. Vyse however neglects to mention the nature of such accusations in his chronicles.[46]

Certainly such behaviour could at the very least be said to be somewhat suspect, but alone, would not constitute any evidence of

---

[46] It is also worth noting that according to Sitchin, on the night of February 12, Vyse entered the Great Pyramid accompanied by John Perring, whom was an engineer with the Egyptian public works department, and pushed a reed through a crack in the ceiling block of Davison's chamber. When it was retrieved unbent, it became clear to Vyse that a chamber lay above. The next morning he fired Caviglia and employed Perring.

forgery, but then, the story of Col. Howard Vyse and J.R. Hill as told by Sitchin has only begun...

Given the similarity of Wellington's Chamber to the one previously discovered by Davison, Vyse concluded that yet another similar chamber must lie above Wellington's.
He promptly fired the remaining foreman (Giachino) on April 4 (again without reason given) and once again through use of liberal amounts of gunpowder, the third construction chamber was finally broken into on April 25, 1837 and named after Lord Nelson.
On entering, along with the black dust, Vyse says he found more quarry marks in the same red paint.
         On May 7 Vyse blasted once again into a further construction chamber (the fourth) and named it after Lady Arbuthnot. His journal entry (according to Sitchin) makes no mention of any quarry marks, although later on, many were found there. Surely another suspicious way of acting.
Most striking of all, several cartouches were found in this chamber, which could only mean that royal names had been discovered.
Cartouches, for those unfamiliar with Egyptian hieroglyphs, are those symbols which are contained by an oval or rectangular enclosure, as shown in figure 13 below, and they were only ever used in the writing of a royal name.

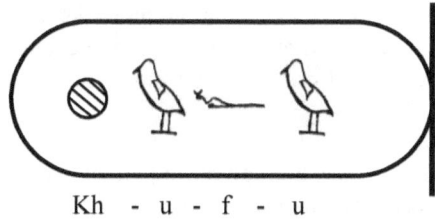

Kh  -  u  -  f  -  u

Figure 13.

When on May 18, an application was made to Vyse by a Dr. Walni for copies of the characters found in the Great Pyramid, in order that they may be examined by Mr. Rosellini, who was an Egyptologist that specialised in the deciphering of royal names, Vyse turned the request down outright.

Instead, the next day Vyse entered the chamber accompanied by Lord Arbuthnot, a Mr. Brethel and a Mr. Raven, where they compared the drawings made by Hill of the markings, with those found on the walls, and later, they "signed an attestation to their accuracy."

Soon afterwards, the fifth and final construction chamber was broken into and here yet more markings, including a cartouche (royal name) were found.

Finally, Vyse's work was done, he'd made an important discovery which (as he allegedly admits in his chronicles even before any of the construction chamber were discovered) was something he'd very much wanted to do.

After having spent large sums of money in the exploration of the Great Pyramid (at great expense to his family) Colonel Vyse had finally joined the ranks of those distinguished, aristocratic officers that had made important discoveries in the field of archaeology, and that had the honour of receiving praise and approval from distinguished societies and the public at large.

Although somewhat lengthy, the British Museum's analysis of Vyse's discovery seemed in general to agree with what the Colonel himself expected; namely, that the cartouches could be read as the royal name *Khufu* or some variation of it. In other words, it appeared quite certain now, that just as had been written by Herodotus many centuries earlier, Khufu must have been the builder of the Great Pyramid.

Zecharia Sitchin however claims that the more detailed aspects of the Museum's analysis escaped attention due to the excitement generated by Vyse's discovery, and presents the following (adapted) account of it.

It was Samuel Birch, the hieroglyph expert of the British Museum who made the original analysis of the symbols, and according to Sitchin, he states right away that the symbols are somewhat illegible due to their having been written in a style that is "...semi-hieratic or linear hieroglyphic..."

Egyptian hieroglyphs are complex symbols, and their carving required a great deal of skill, which is why in olden times in Egypt, the position of scribe was not a lowly one.

As time passed however, with the spread of commerce and the arising need for a simpler way of communicating, eventually a more quickly written form of the hieroglyph evolved. This was a more linear script that is known as semi-hieratic, and it first appears in Egyptian texts only centuries after the time in which Khufu reigned.

In addition, some of the symbols were written in what appeared to Mr. Birch to be hieratic as opposed to semi-hieratic, and these types of hieroglyphs evolved later still. Mr. Birch also pointed out that many of the symbols had certain peculiarities he could not explain.

Of one set in Campell's Chamber, which follows the cartouche with Khufu's name in it, he says that "it would be difficult to find a parallel." And several other symbols were "equally difficult of solution."

In one instance, a sign that means good or gracious, was used as a numeral, a usage that, Sitchin points out, had never been seen before or since this discovery.

Equally frustrating for Birch were a row of symbols (a royal title) immediately following one of the cartouches, they too seemed to be written in a linear style, but additionally, the only similarity found to this set was one found on the coffin of the Queen Amasis. Birch however neglected to point out that this Queen had lived more than 2,000 years after Khufu.

In other words, not only were the scripts themselves written in a way that suggested a later period, but even the titles they represented, appeared only on works which were all from much later times.

As for the main issue at hand (whose name was it that had been found in the Great Pyramid) Birch correctly concluded that not one, *but two royal names had been found within the ancient monument.*

The two cartouches as found in the chambers according to Sitchin are shown below.

Translated as Saufou or Shoufou
(Khufu or HWFW)
Fig. 14

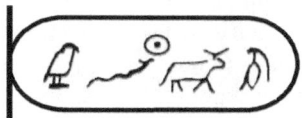

Senekhuf or Seneshoufou
(Khnum-HFW)
Fig.15

This problem of the two pharaohs in one tomb, has in fact been a source of endless embarrassment for the 'experts' that tell us the Great Pyramid is really Khufu's gravestone.

Egyptologists have suggested that Khufu may have simply taken over an earlier pharaohs' pyramid, Sitchin points out that this cannot be the case, because the royal name translated as Khufu, appears only in the topmost chamber, while that of Khnum-HWF appeared in two of the lower chambers. But if Khnum-HFW is taken to be Suphis II, as suggested by the ancient King Lists, and Khufu as Suphis I, how could Khufu (Suphis I) have inscribed his name in a higher chamber than the King that succeeded him?

Another theory that has been presented is that Khufu and Khnum-HFW were one and the same, but this idea is rather inconsistent, since no parallel to it has been brought forward as corroborating evidence. It would be something akin to finding Winston Churchill's name in history books as both Winston Churchill, and say Tonsil Hillchurch. Not a likely story.

But Sitchin's proof (if indeed it is proof) with regard to Vyse's forgery does not end here, in fact, he points out, that the forgers made not one, not two, but at least three tragic mistakes.

Firstly they spelt the name of Khufu as Khnum-HFW.

A mistake that can be attributed to the way the royal name of Khufu was represented in the book on Egyptian hieroglyphs by Sir John Wilkinson's *Materia Hieroglyphica*, which, surprise, surprise, is the only book mentioned (and repeatedly at that) in Col. Vyse's own chronicles.[47]

The second, and even more damning error, (always according to Sitchin) is that Vyse, in the writing of his journal, makes the blunder of tipping us off that he's now become aware of his mistake.

An entry in his diary, states that he'd found a partial cartouche of Suphis (Khufu) on a brown stone fragment outside the pyramid. This event supposedly took place on June 2.

Sitchin says that Vyse probably intends us to believe that he recognised the partial fragment because he'd already discovered the full cartouche on May 27, in the Campell chamber, but he adds that Vyse made one small mistake... the entry in the journal is dated

[47] See The Stairway to Heaven, pg. 271 to 276

May 9, meaning he was perfectly aware of what Khufu's cartouche should *really* look like long before he entered the last chamber.

And finally, the last nail in Vyse's coffin is that, having realised his mistake (ironically as a the result of Wilkinson's new and updated book, called *Manners and Customs of the Ancient Egyptians*) he once again makes another in attempting to correct his fault.

Being advised by Wilkinson's new book that an example of the hieroglyph in question was available in the area of the Sinai and reproduced in the illustrated *Voyage de l'Arabie Petree*, by Leon de Laborde he proceeded (presumably) to acquire this book, which was readily available in Cairo at the time, says Sitchin.

But unluckily for Vyse, Laborde too had made a little error in his book. The first symbol of the Khufu cartouche, (the Kh- part) when correctly written, as has been done in all of the original works of the Egyptians, is represented by a shaded circle thus:

This symbol represents a sieve, but Laborde, in his drawing had shown it as a void circle,[48] thus:   O

Furthermore, Wilkinson sometimes rendered the circle as blank but with a dot in the middle in his own work.

But this was a mistake that would never have been made by an ancient Egyptian scribe, because, the void circle with a dot in the middle, was the symbol of Ra, the supreme God of Egypt.

In effect, Sitchin claims that Vyse and Hill, had not spelt Khufu, but rather Raufu! *An error that would never have been made by a real Egyptian scribe.*

Sitchin further points out in great detail,[49] that Vyse was also the person responsible for the discovery of Menkaure's name in the Pyramid that now bears that name, much like the Great Pyramid is today also known as Khufu's pyramid. But in that instance, all the experts agree, the coffin fragment found in Menkaure's Pyramid, was of a much, much later date, or if you like, a failed attempt by Vyse to forge the creator of this pyramid as well.[50]

---

[48] The rock carvings of course, as in all instances of original hieroglyphs, have always correctly represented it as a shaded circle.

[49] See *The Stairway to Heaven*, pg. 277 to 282.

[50] Sitchin also points out that the Vyse markings are between two and a half to three feet long and about a foot wide as well as being crude and imprecise. Quite a contrast to the fine, precise and delicate artwork of the true Egyptian scribes, no matter the era they lived in. Additionally, except for one tiny exception in the corner of Wellington's

So all is well in the end, the forgery has finally been discovered and we can all get on with our lives. Well not quite, for indeed there is some suspicion as to whether Sitchin has presented the case fairly.

The point with respect to Menkaure's pyramid is certainly valid, although one might not be able to conclusively prove, more than 150 years later, if this was due to dishonesty on the part of Vyse or not, as Sitchin suggests is also evident from the Vyse journals.

In them, it is supposedly described, that the mound of rubbish in which the coffin fragment and bones is found is apparently thoroughly searched without any discovery being made, but only on the second pass are the remains of a mummy and the coffin fragment found. A valid point made by Sitchin is that the bones do not date from the same period as the coffin fragment,[51] meaning that not just one, but two foreign pieces of matter had been introduced into the pyramid and that seems to make the theory of the accademics —that the coffin fragment is part of a restoration attempt performed much later— decidedly suspect.

Nevertheless, it has been said that Sitchin intentionally quotes Samuel Birch out of context and that the markings in the chambers are in fact correctly shaded and not represented by a circle with a dot in the middle.

The problem is that short of actually going to Egypt, gaining permission to enter the construction chambers and photograph the markings, it is far from easy to confirm Sitchin's claims.

The British Museum stores the Vyse chronicles in vaults and access to these documents, while possible, is far from easily achieved. The process is in fact rather time consuming, in addition to which one needs to first familiarise himself with the material in order for a reading of it to either confirm or dispute Sitchin's claims.

Not being able to travel to London myself, I asked my friend Paul Bentley to research this topic for me, which task he took to enthusiastically, yet, not being familiar with the topic, Paul was not really able to tell me conclusively whether the story as told by Sitchin is an accurate one. His first impression is that it may not be wholly

---

Chamber and a partial bird symbol in Campbell's Chamber, no markings were found on the eastern side of the chambers, those used by Vyse to force an entry, but only on the north, south and western walls. Once again, a surprising stroke of 'luck'.

[51] As I.E.S. Edwards points out, the bones are from early Christian times.

faithful to the truth, since in the Vyse journals he came across the cartouches with the circle correctly shaded in. Despite this, due to time constraints and the complex nature of the drawings, he was unable to locate the copies of the markings as found in the construction chambers. In any event, even if he had, photographic evidence would still be best, since if the copies were made by a person not familiar with Egyptian hieroglyphs, (as Sitchin claims is the case with respect to Hill) they may have inadvertently placed a circle with a dot on the copy, where in the original it is in fact a shaded circle.

On contacting a person that seems to have been investigating this very same topic (and whom is of the opinion that Sitchin has been dishonest) via electronic mail, I received the reply that a photograph of the relevant markings might be found in "Stadelmann's *Die Aegyptischen Pyramiden*", but I have been unable to find this work as of two days before going to press.

One piece of hieroglyphic writing that *does* talk about the Great Pyramid *and* Khufu at the same time however exists. It is known as the Inventory Stela[52] and was discovered by Auguste Mariette in the 1850s, in the ruins of the temple of Isis, which lies near the Great Pyramid.

This stela —which is carved in stone, has none of the aberrations found in the Vyse markings and by all accounts seems to definitely be authentic in nature— has nevertheless been called a vile forgery.

The only reason it has been called that though, is because it seems to contradict what the Vyse painted symbols imply; namely that Khufu built the Great Pyramid.

But as we have just seen, the Vyse paintings may be just that: paintings perpetrated either by Vyse, Hill, or both, and as such, they would have no basis for telling us anything about the origin of the Great Pyramid.

In this new light then, we can only conclude that the Inventory Stela, is indeed what it seems to be, an original work dating from the time of Khufu. Especially since to say otherwise, would imply that Mariette had to have created a perfect forgery in all aspects, but one carved in stone at that, and with none of the multiple inconsistencies and errors allegedly made by Vyse and Hill.

---

[52] It presently lies in the Cairo Museum.

A feat so difficult that it would be a kind of miracle in itself.

But as soon as we do this, take the Inventory Stela as a real artefact instead of a forgery,[53] we shatter a myth that has existed at least since 1837.

Because this Stela, which also recounts how Khufu restored the crumbling temple of Isis, *states that both the Sphinx and the Great Pyramid, where already there when Khufu came along*, and in fact the Great Pyramid is talked of as being dedicated to this Goddess rather than to Khufu himself.

Effectively, this means that the Great Pyramid can no longer be attributed to Khufu. Additionally, it raises the question once again, of just who built it. As well of course, as when, and why.

Despite his possible "eagerness" in calling the Vyse markings forgeries, Zecharia Sitchin is not the only one to suggest that Khufu 'stole' the Great Pyramid as his own monument by erecting temples in honour of it all around its base.

Indeed, Lemesurier made the very same suggestion, adding that Khufu may in fact even have changed his name to sound or read like the one of the King whom (perhaps) originally existed long before, and was responsible for the construction of this amazing monument.


In any case, we are now faced with the daunting task of trying to discover *when* the pyramid was actually built.

At first glance, what appears to be a particularly interesting view in this respect, are the findings of the same Robert Bauval, whom we have already encountered in our study of the Great Pyramid.

Bauval spent some ten years generally involving himself in the study of the Egyptian pyramids and myths, and in particular, the study of the Giza pyramids.

His book, *The Orion Mystery*, mainly concerns itself with showing a relationship between the position of stars in the sky and their counterparts on Earth in the form of pyramids.

According to Bauval, the Giza pyramids represent the three stars which are commonly referred to as "Orion's belt" and which are indeed found in the constellation of Orion.

---

[53] As Lemesurier points out on page 109 of *The Great Pyramid, Your Personal Guide*, the Stela may well be a late copy of a Fourth Dynasty original, but why this would mean someone forged it has yet to be explained.

Additionally he tries to show how other pyramids in the Giza area supposedly correspond to stars in the vicinity of this constellation.

Bauval's book is certainly interesting, having been written partly as an autobiographical recounting of his travels and studies.

One can almost feel his frustration and elation at times, as his need for making an important discovery concerning the pyramids fuels him through his journey. If we strip it of such flesh though, we find that his theories are based on rather few and shaky premises at best.

Despite this, Bauval certainly merits some credit, for although not responsible for them, he documents Gantenbrink's discoveries well indeed, and in any event, he was certainly pivotal in the re-discovery of the bronze grapnel and associated instruments.

It is somewhat ironic then, that these two important features of his book are but a small part of it.

Due to the rather complex nature of that facet of astronomy known as precession, a complete critical appraisal of Bauval's work would have to be a lengthy affair. Especially when this is mixed in with the numerous Egyptian myths he quotes and interprets, sometimes, in a way that admittedly fits snugly with his ideas.

In short then, I will have to limit myself to pointing out a few inconsistencies in the broader sense, and ultimately, as always, allow others to decide for themselves.


Precession is that apparent movement which constellations undergo over long periods of time, which results in their changing elevation with respect to the horizon. It is a cyclic phenomenon, and those constellations which seem to move from place to place in the sky eventually return to their original positions. The process is so slow however, that the effect cannot be seen in one human life, as it happens over a timescale of thousands of years.

Bauval's theory with regard to precession and the Giza pyramids then, is basically that the 'air-shafts' found in the Great Pyramid's King's and Queen's Chambers (See figure 16 on the next page) were directed in the past towards certain stars, and noting which stars may have been indicated by them in ancient times, one should be able to discover when the Great Pyramid was built.

By using mathematical calculations to figure out at what elevation the stars were in past eras (the process known as precession) one should be able by trial and error, to show at what time what star was 'visible' from the 'air-shafts'.

Figure 16

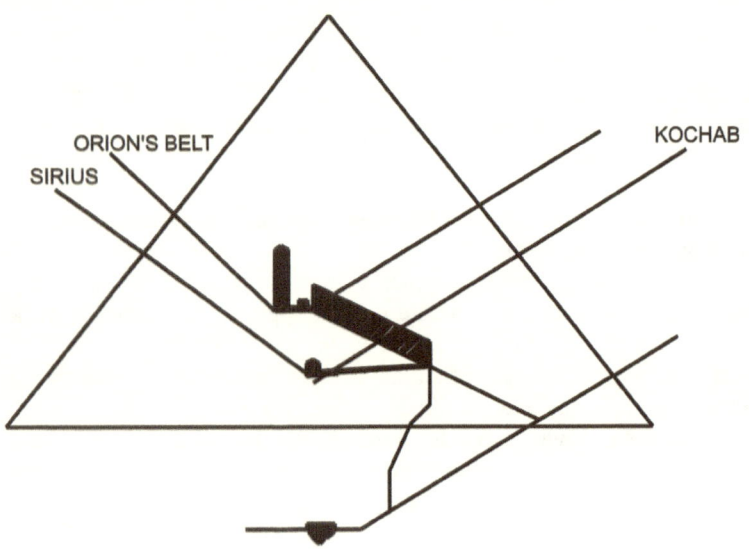

The stars indicated by the various air-shafts according to Bauval's theory.

There are however, a number of premises which need to be validated for such a theory to hold water.

1)  We need to assume that indeed this was the purpose of the so called air-shafts.
2)  We need to agree that even if they were used for the purpose of star gazing (or at least star pointing) their objective was indeed the stars named by Bauval.
3)  It needs to be understood, why a whole pyramid needed to be built in order to sight some stars, since this could have been done quite accurately merely via the more express means of a simple and much smaller ramp.

4)  If indeed Bauval's theory is valid, then we should expect it to
     either predict certain findings or at least to confirm others.

Firstly, the idea that the shafts were used, for the purposes of star-
gazing, cannot be substantiated on any level.
The King's Chamber shafts both have a horizontal section in them,
which allows them access to the King's Chamber. The southern
shaft's lower end also resembles a kind of old bread oven or kiln, as
shown in figure 17 below. The measurements of inches are taken from
pg. 100 of *The Great Pyramid Decoded.*

Figure 17.

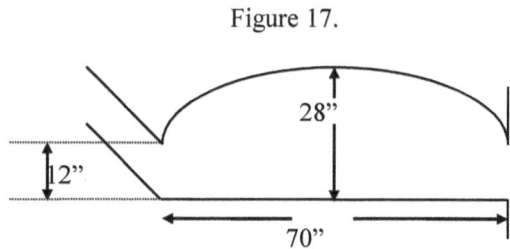

Dimensions of lower end of southern air shaft in King's Chamber.

While the northern shaft, as Lemesurier reminds us that Edgars
reports, takes "...a number of short, sharp bends, each succeeding
bend tending upward and toward the north-west, before it finally
bends northward to proceed directly to the outside of the Pyramid at a
steep angle."[54]
As for the air-shafts in the Queen's chamber, well, they had both been
left with the lower portions uncut. Five inches of stone had to be
removed from the solid block in which the shafts were set before they
could be declared 'open'.
In addition, with Gantenbrink's discovery of a tiny portcullis down
the Southern shaft of the Queen's chamber, we can confidently say
that the Ancient Egyptians would have needed to be able to see
through several inches of stone for them to use these shafts as star-
finders, never mind the fact that in the King's chamber they would
have to be able to literally see around bends.

---

[54] See footnote 61 on pg. 112 of *The Great Pyramid Decoded.*

In view of these facts, we can only say that even if the shafts were intended to point towards the relevant stars, they were not to do so for the benefit of anyone's eyes, but rather for some hitherto unknown and somewhat mystical reason.

While point number (3) makes it clear, that if indeed the shafts were to point at the required stars, then it was for a purely ceremonial reason and not a practical one.

Another peculiar item concerning this theory, is that although Bauval correctly dismisses the Vyse markings as 'graffiti' and further adds that they are the strongest 'evidence' (his inverted commas) that links the Great Pyramid to Khufu,[55] he seems intent on showing that the pyramid was indeed built at this time, and most of his calculations concerning the air-shafts start from the premise that this was so. Clearly an approach that is somewhat confused.

Furthermore, Bauval's seemingly final conclusion concerning the Great Pyramid's air-shafts, is that they were used in religious ceremonies with the mummies of not one, but several Kings.[56]

He adds that the we cannot after all, be exactly sure when the Granite Plug that blocks the upper passages was inserted.[57]

This last statement I find particularly incredible, since it supposedly comes from a person that should be familiar with the Great Pyramid in some depth.

The Granite Plug consist of three massive granite blocks (see figure 11 earlier in this chapter) which closed off any access to the upper chamber for untold aeons. When Caliph Al Mamoun's workmen discovered it in the ninth century, they had to burrow around it, through the surrounding and softer limestone, since the three massive blocks were practically impenetrable being composed of granite, nor could they be taken down to the descending passage, because the ascending passage in which they are located, is *tapered* in the horizontal plane. In other words, the granite plug was built-in with the ascending passage, as a permanent feature.

---

[55] See note 25 on pg. 308 of *The Orion Mystery*, where Bauval even gives a reference for an alternative view of the Vyse markings (W. R. Fix, *Pyramid Odyssey*, pgs. 75-89).

[56] See pgs. 231 to 234 of *The Orion Mystery*.

[57] See pg. 234 and note 32 on pg. 327 of *The Orion Mystery*.

This is further substantiated by the fact that there is no clearance whatever between the granite blocks and the surrounding limestone passage, but, as Lemesurier once again points out, without at least some minimal clearance there can be no sliding of the blocks into place.[58]

But even if we allow Bauval the benefit of a non-existent doubt, and say that the Granite Plug was inserted at some later date, then we have to allow for a very strange set of circumstances.

In order for the plug to be inserted, access to the ascending passage must be allowed for, and even if we allow the most optimistic of situations, we must then be forced to conclude that the plug had to be inserted before completion of the King's Chamber, since otherwise, there would not be any access from the unfinished top portion of the pyramid.

As can be seen from figure 11 however, the Queen's Chamber air-shafts reach a final level that is higher than that of the King's Chamber. So, not only could the King's Chamber air-shafts not even have begun, but also, the ones of the Queen's Chamber could not have been completed. And this of course makes it impossible that they could then be used for any sort of repetitive ceremony, which in any event could not occur, because the King's Chamber was not complete!

This set of circumstances then places quite a large shadow of doubt over Bauval's theory that the Great Pyramid, along with its 'air-shafts' was used as a sort of final ceremonial event by the dead corpses of ancient Pharaohs.

As for his assertion that the Great Pyramid's shafts seem to have pointed towards meaningful stars for the Egyptians at around the time of Khufu, this alone cannot be taken to be evidential of the Great Pyramid having been built at that time.

But in any event, there are two serious queries concerning such a hypothesis, and both come from Bauval himself.

Firstly, as set out by himself on page 180 of his book, and supposedly using Gantenbrink's latest information available concerning the true direction of the 'air-shafts', he gives the following dates for the completion of the relevant shafts:

---

[58] See *The Great Pyramid, Your Personal Guide*, pg. 46 to 49.

| Table 1: Precession dates for air-shafts as given by R. Bauval | | |
|---|---|---|
| **Shaft** | **Angle and Epoch using Gantenbrink's measurements** | **Angle and Epoch using Petrie's measurements** |
| KC south | 45° 00' 00" *c.*2475 BC | 44° 30' 00" *c.*2600 BC |
| KC north | 32° 28' 00" *c.*2425 BC | 31° 00' 00" *c.*2600 BC |
| QC south[59] | 39° 30' 00" *c.*2400 BC | 38° 28' 00" *c.*2750 BC |
| QC north | Not given | Not given |

But if we are to take these dates as correct (using Gantenbrink's measurements) then it would mean that the King's Chamber was completed some 25 to 75 years before the Queen's Chamber, which lies below it!

Clearly an impossible situation, unless of course, as a friend of mine jokingly suggested, we allow that the Great Pyramid was built from the top down.

It would seem then, that Bauval would have better served his cause if he'd used the angle measurements provided by Petrie, which at least are consistent not only with the King's Chamber having been built after the Queen's chamber (by some 150 years apparently!) but also with both sides of the King's chamber having been completed at the same time, which makes quite a lot more sense.

But in reading *The Orion Mystery*,[60] we find that Bauval's preference for Gantenbrink's measurements is also induced to some extent, by the fact that Petrie's angle for the King's Chamber's southern shaft, would mean that the central star of Orion's belt was indicated (Epsilon Orionis) instead of Zeta Orionis, which was thought by Bauval to be the star that represented the Great Pyramid.

---

[59] It is also unclear how Bauval can be sure of this angle, since at the close of his book, the stone door found in this shaft remains unopened, and thus anything hidden behind it remains a mystery.

[60] Pages 136 and 137.

So regardless of which angle we assume to be more correct,[61] there seems to still be a problem with the theory.

Furthermore, we can rule out a margin of 'play' in the calculations as being responsible for the discrepancy between the supposed building of the King's Chamber shafts and Queen's Chamber shafts, because, as Bauval himself points out, he was only willing to allow himself a margin of error of no more than 20 years.[62]

It was the very discrepancy of 150 years[63] that in fact led Bauval to assume that Petrie's angle measurements must have been a little off, and thus to wait with baited breath for Gantenbrink's more accurate data concerning their alignment.

Yet, if we take Bauval's words at face value, how does he explain a discrepancy between the date values (when using the angles provided according to Gantenbrink's measurements, no less) that run from 25 to as much as 75 years in a *negative* direction?!

Effectively this means that the error in Bauval's theory is *at least* 50 years in the best possible case, because even if we take the average of 25 and 75 and assume the mistake to be 'only' 50 years, if any years passed between the building of the Queen's chamber and the King's chamber, these would, of course, have to be added to the value of 50. *An error, that by Bauval's own standards, is two and a half times more than that he professes to allow himself!*

The second doubt which arises in Bauval's assertion that the Great Pyramid was built *c.*2450 BC, comes from his own calculations of precession concerning a much earlier date for its construction.

Although suspiciously vague in his conclusions concerning this topic, Bauval obviously calculated (at least in a preliminary fashion) the position of the Orion constellation during the years 10,000, 10,400 and 10,450 BC and in doing so arrives at the conclusion that indeed the time period is significant with respect to the Great Pyramid.[64]

---

[61] In view of the more advanced technology available to him, I would assume Gantenbrink's to be the superior measurement in terms of accuracy.

[62] Page 138 of *The Orion Mystery*.

[63] Bauval gives a value of 140 years in the text on page 138, as a result of a more refined and detailed account, but 150 years as presented in his Table 1.

[64] See pgs. 198 to 205 of *The Orion Mystery*.

It is interesting to note that Bauval is rather sketchy on just how significant this date is, but to quote him: "Why did the architect who designed this shaft and probably the whole pyramid want to draw attention to this remote First Time date of Osiris in *c*.10450 BC?"

Furthermore, it is obvious that the thought of the Great Pyramid having been built at this much earlier date crossed his mind, because he writes: "We are, of course, aware that 10450 BC is far too remote for archaeologists and Egyptologists to entertain, but these findings challenge them to explain — or dispute — the mounting astronomical evidence."

Too remote for what exactly? Bauval does not say, but the implication is clear.

Additionally, on that same page (203), there is a diagram showing the three Giza pyramids as well as three sketches of the Orion constellation in different positions in the sky and each with a date written next to it, these dates are: *c*.2000 AD, *c*.2450 AD and *c*.10400 BC Admittedly it is a rather unhelpful diagram, as it really conveys nothing of a scientific nature and in fact could be said to exist merely for embellishing reasons, the caption underneath it however, is rather interesting. It reads: "At around 10400 BC the pattern in the sky was mirrored on the ground by the pyramids."

But hold everything a minute! Is it not the very core idea throughout Bauval's book, that *it's because the pattern in the sky around **2450 BC** was mirrored on the ground by the pyramids*, that he concludes that this is when they were in fact built?

So if the same is true of 10400 BC, why could this not be as valid a date for their construction?

Nor can we, in fairness, ascribe the caption to this figure as being one of the typographical errors that one encounters on several occasions throughout *The Orion Mystery*, because it falls rather neatly under a section entitled: "**V The Timaeus: 10450 BC**".

The reference to the Timaeus being the obvious one, namely, the approximate date given by Plato in the Timaeus for the downfall of the 'legendary' Atlantis.

Ultimately, the impression I was left with in reading this section, was distinctly one in which Bauval seemed unwilling to go against the 'public opinion of the experts', but where he also could not completely ignore the evidence that perhaps the Great Pyramid was indeed built as long ago as 10,450 BC.

At any rate, despite the fact that I have not been kind to Bauval in the last few pages, I *do* believe that his theory that the three large pyramids at Giza are meant to represent Orion's belt, has indeed got some merit.

The topic certainly requires further investigation, and in this respect, Bauval has made a very real and important discovery, but in view of the numerous inconsistencies I find with respect to the time the Great Pyramid was supposedly built, as well as the sometimes imprecise nature of Bauval's work,[65] I cannot accept his premise that indeed it was constructed during Khufu's reign.

A 'fact' which he seems to ascribe to even before his 'confirmation' of it, and also in contrast with his rebuttal of the Vyse 'graffiti'.

So far then, it would seem, we have looked at when the Great Pyramid was *not* built, let's see what we can do about discovering when this event may in fact have happened.

◆ ◆ ◆ ◆ ◆

Edgar Cayce was a renowned clairvoyant who died in 1945.

Under a hypnotic trance, he seemed able to make astonishing predictions and have access to unlimited knowledge. While some of the terms used by him under trance were not always understood (or even agreed with) by the conscious Edgar Cayce, his apparent success in making predictions and curing ills, was such that a foundation was set up, as far back as 1932, to examine the stenographic records of his

---

[65] Bauval's knowledge of the Great Pyramid at times seems somewhat superficial, especially for an engineer, viz. with respect to the Granite Plug, or his contention that not just the Great Pyramid, but also the other two main Giza pyramids (known as Khafre's and Menkaure's) as well as the two Dashour pyramids and a lone pyramid at Abu Ruwash, were *all built in a roughly 100 year period!* As I have already shown, in terms of the technology supposedly used, even just building the Great Pyramid alone in a 100 years, would still seem to be an impossible task in terms of labour, never mind of course the other points, already mentioned, with respect to accuracy etc. To suggest that not one, but that all three Giza pyramids were built in merely one hundred years (as well as an additional three others thrown in just for good measure) while still ascribing to the builders the orthodox technology that Egyptologists do, seems to me to take the already farcical theories of the 'experts' to a new level of impossibility. Bauval also makes a few errors with figures, on page 287 for example, he gives the value of a Royal cubit as being 0.5237 millimetres.

trance periods.[66] As these amount to over fourteen thousand, some may still be in the process of being examined.

Cayce predicted that the hall of records (a time capsule containing Atlantean knowledge) would be discovered before the end of the century somewhere between the Sphinx and the Nile. He sometimes described this hall of records as a pyramid, and Lemesurier makes the interesting point that the missing capstone of the Great pyramid is itself, of course, a smaller pyramid.

Cayce, as reported in *The Great Pyramid Decoded*,[67] also predicted the existence of some chambers originating from the right paw of the Sphinx.

At the time of Lemesurier writing *The Great pyramid Decoded* this of course had not yet happened, but by the time his book *The Great Pyramid, Your Personal Guide* had been written ten years later, such chambers had indeed been found.

At any rate, Edgar Cayce made numerous predictions up to about the year 1998, and a great deal of these have been extremely accurate. He also said, while under trance, that the Great Pyramid's construction began in 10490 BC and took 100 years to complete.

I reprint here what Lemesurier has to say about this:[68]

> This dating accords with extraordinary exactness with the date of the last known reversal of the earth's magnetic field, as determined in 1971 by the Swedish scientists N.-A. Mörner, J.P. Lanser and J. Hospers on the basis of geological core-samples (New Scientist, 6th January 1972, p. 7). The end of the reversal period was calculated by them to have occurred some 12,400 years ago—and thus, notionally, in around 10430 BC If, then, we accept that the construction of the Great Pyramid (as dated by Cayce) may have been associated with some cataclysmic event, the possibility arises of a direct link between that event and the magnetic reversal referred to. And one possibility in particular which springs to mind is that the magnetic reversal may have resulted not from changing currents within the earth's core, but from a geologically or astronomically induced 'flip-over' of the earth itself (producing a reversed relationship between the dynamo-'rotor' of the spinning earth and the magnetic 'stator' of its cosmic environment). Such a 'flip-over' would of course have produced tidal

---

[66] By the A.R.E. (Association for Research and Enlightment).

[67] Pg. 207 of *The Great Pyramid Decoded*.

[68] From note 8 on pg. 207 of *The Great Pyramid Decoded*.

waves and geological upheavals of unbelievable magnitude—and certainly more than enough to merit the title cataclysm. It would also accord with the ancient Egyptian priesthood's insistence to Herodotus that such a 'flip-over' had indeed happened at least four times during the period to which their records referred.

Of course, it needs to be pointed out, that a 'flip-over' as Lemesurier refers to it, or at least a severe disturbance (catastrophe, to use a perhaps better term) would indeed be experienced by a planet that was subjected to a collision with a large enough asteroid, perhaps particularly so, if it was largely composed of nickel-iron.

Cayce's prophecies concerning the Great Pyramid and surroundings, have to date been rather shockingly accurate, and if his prediction that a time capsule of Atlantean knowledge will be found before the end of this century holds true, then the next four remaining years to the close of the twentieth century, will indeed be interesting ones.

Another indication that the Great Pyramid is far older than presently accepted comes from the monument itself.[69]

The Great Pyramid gives the values of the Solar Tropical Year, and the Sidereal Year quite precisely (to seven decimal places in fact) as 365.2423499 and 365.2563565.[70] These values vary from the current figures by some 0.0001589 days and 0.00000906 days respectively. Since the length of the Sidereal, Solar and anomalistic year vary slightly over long periods of time, if we work our way back in time to a date when the figures given by the Great Pyramid were correct, we might get a value for its original construction date.

There are some problems with this approach though. The Anomalistic Year is not given explicitly in the exterior design of the pyramid, while the Solar Tropical year variation is arrived at by an extremely complex formula that apparently has not been taken back far enough into time yet, to check against the value given by the pyramid for a date comparison.

Lastly, with values this small, there is bound to be a considerable margin for error. Despite this, the Sidereal year however can be measured against the Great Pyramid's value, since its rate of change is

---

[69] As Lemesurier once again points out in *The Great Pyramid, Your Personal Guide*.

[70] See Pg. 221 of Peter Lemesurier's *The Great Pyramid Your Personal Guide*.

0.00000011 days per century. This would make the Great pyramid over 8200 years old, with a construction date reaching back to 6250 BC or thereabouts.[71]

Considering that this value is almost 4000 years older than the one presently accepted, it certainly puts some serious doubt on the true age of this building.

It is also worth noting that the world's cults and religions, usually chose a sacred form which was directly related to the then ruling sign of the zodiac. Once again, Lemesurier reminds us that during each of the 2,000 or so years for which a zodiacal sign rules, there can be found corresponding cults. Taurus was identified by bull-worship and bull-sacrifice. The age of Pisces began shortly before the birth of Jesus and of course the fish has symbolic significance for the Christians.

Why then, should the Sphinx, which is thought to be a contemporary and guardian of the Great Pyramid, resemble a lion if it was built during the period of 2600 BC or so, when the ruling sign of the zodiac was the bull?

The sign of Leo however ruled from about 10,970 to 8,810 BC which would fit rather well with Cayce's dating of 10,490 BC

While none of the above points can be said to be conclusive of anything, it should be obvious that the Great Pyramid could conceivably be a lot older than we might at first think. Especially since the present 'orthodox' way of having ascribed a date to it, is by no means more scientific than the ones I have briefly outlined above.

It might also be of interest to note, that when giving dates for phenomena buried in the very distant past, scientists have consistently erred on the side of youth. Examples might be the creation of the Universe (thought to have taken place as early as some 5700 years ago by some religiously oriented thinkers not only of the last, but also of the *present* century), the existence of *Homo Sapiens Sapiens* on Earth, the age of the Dinosaurs et cetera, all of which continued to become older and older as technology discovered more ingenious ways of measuring their true age.

---

[71] Lemesurier gives a possible error of ± 400 years. See Pg 221 of *The Great Pyramid Your Personal Guide.*

But in any event, we now reach a penultimate and most interesting point, because presently, we are going to look at a recent scientific discovery, which questions the true age of the Sphinx, companion and guardian of the Great Pyramid, in a most interesting manner.

At the October 1991 Geological Society of America meeting, Robert M. Schoch and John A. West, announced publicly for the first time, that in their opinion, the Great Sphinx of Giza is the result of a much older civilisation than the one which existed in c.2600 BC

It was John West who first began to suspect a much older age for the Great Sphinx, basing his thoughts on the writings of a man named R.A. Schwaller de Lubicz, whom somewhere in one of his books mentions that the Great Sphinx shows geological weathering features that seemed to indicate the Sphinx could be much older than what most present-day Egyptologists would suspect.

But Mr. West, you see, has no formal degrees, no credentials from some emeritus institution, no formal, academic training in Egyptology...in short, as far as the Egyptologists are concerned, even if Mr. West uncovered a 300 foot wide flying saucer from underneath a pyramid in front of a million witnesses, he would still remain, just another crackpot. A pyramidiot no doubt.

Wisely however, West decided to enlist the services of a professional geologist, and asked Robert Eddy, a long-standing friend of his at Boston University, if he might know any open-minded geologists that would be willing to investigate the Sphinx further from a geological point of view. And that, is how West became acquainted with Robert Schoch, a science professor at the same Boston University, that holds doctoral degrees in geology.

Although open-minded, Schoch was initially far from taken in by what he refers to as West's "...outlandish, but interesting." ideas concerning the age of the Sphinx.

Schoch goes on to say that although West mentioned something about getting him to Egypt, he felt West would not fly him over to Cairo "...just to point out what was surely a simple error on his part."

Despite this rather orthodox view, when Schoch started his preliminary study of the Sphinx, he soon began to change his mind; to quote him: "...I became convinced that either the rocks were behaving in very strange ways or West was actually onto something."

Schoch then proceeded to conduct non-invasive studies of the Great Sphinx along with limited seismic investigations.

He also performed a more extensive seismic investigation in April 1991, with the help and instruction of Thomas L. Dobeki, a geophysicist working for a Houston based firm (McBride-Ratcliff).

Schoch then returned to Egypt in June 1991 to get more data and double check the information which they had already collected.

As a result of his investigations, Schoch finally concluded that indeed the Sphinx is considerably older than the value normally stated by Egyptologists. He places the age of the Sphinx at about 7000 years, dating the Sphinx's body (conservatively) to about 5000 BC but admits it could be even older than this.

West, not being constrained by a certain necessary amount of conservatism on the part of scientists, quite unashamedly (rightly so!) admits that in his opinion, the Sphinx was built around 10,000 BC by survivors from Atlantis.

This idea of West's, not surprisingly, finds little favour with the mainstream Egyptologists, but perhaps more disturbing, is the fact that even Schoch's assumptions, which are based on extremely sound scientific investigation techniques, have been attacked in a manner that suggests little evidence of honest scientific argument and much of pigheaded bigotry.

Schoch and West had studied the Great Sphinx from a new angle. By noticing the different types of chemical and mechanical weathering which the Sphinx showed, they concluded not only that it must have been excavated much earlier than presently believed by most orthodox archaeologists, but also, that its construction probably occurred in an early Holocene pluvial period, during which the Sahara was experiencing a wet climate, because only erosion occurring as a result of persistent rainfall and surface runoff would be able to produce the kind of geological weathering which the Sphinx shows.

Schoch is not the first to notice the ancient and anomalous weathering of the Sphinx, but he apparently is the first to have been bold enough to investigate it and then draw conclusions according to the results he got. Nor are the results in question. Schoch's work is not in error. His results are indeed correct, as competent geologists (even those that disagree with him vehemently) are forced to admit. It's the interpretation of those results which is the root of the controversy.

The arguments are lengthy, of a technical nature and require considerable consideration before being understood and hence appreciated. For this reason, it is not possible for me to present them here. Interested person are referred to the bibliography, where I have listed my references with respect to this subject. Even further references are provided by most of the articles I used.

The gist of the matter however is this: Schoch's arguments are as they are not because he is a publicity seeking sensationalist (something West has been accused of, and hence by association Schoch too) but rather, because given the geological evidence, the theory which best sums up all the anomalies present, is the one that the Sphinx is much older than currently believed.

It's Sherlock's version of Occam's razor again; once all the possible alternatives have been analysed and dismissed, whatever is left, however improbable or impossible, must be the truth. Regardless of whether we like it or not!

Obviously, for this statement to be correct our logic must be faultless, which in itself is no easy feat to accomplish, but like I have stated, the quality of Schoch's work is not in question here.

Similarly, Schoch's scientific background and conduct is beyond reproach. In fact, his staunchest 'scientific' critics seem to have associations to weird 'theosophical' or mystical societies much more than Schoch himself, who was trained in a decidedly classical way with respect to geology.

Schoch has also made an enquiry into why his theory should be so viciously attacked, and raises some pertinent, and valid points.

Firstly, it would seem that Egyptologists seem to be more of a 'breed apart' than perhaps any of the more classical departments of academia, like chemists, or physicists.

Secondly, precious few scientists from 'outside' the Egyptology circle have been successful in making an impact on the Egyptologists, since it seems they are staunchly opposed to anything that challenges the limits of their imaginations or their incomplete and sometimes even arbitrary 'scientific proof'.

While an analytical chemist would generally be receptive to the opinion of a physicist, or biologist, the Egyptologist, as a creature, is more insular and insulated. The general personality trait they seem to exhibit most is a forceful indignance for any 'outsider' who tries to delve in the field without proper 'breeding'.

While this is in some ways true of all organisations or groups; and in particular of scientists in general, even among physicists, chemists, or whatever, it seems to be rather more pronounced in Egyptologists; whom in any event, are far less scientific in the main, than a classical scientist like Schoch.

One might wonder at the truthfulness of such statements, which of course are generalisations anyway, but the crux of the matter is that indeed, the trend is a real one, and as Richard Milton in his book *Forbidden Science* points out extremely well, one that is not exclusive to Egyptologists, and I suspect, not exclusive to the sciences either.

It is a sad human trait, that we project our ignorance of the Universe on the world around us, and then we interpret that, rather than the real world which lies hidden by our continuously playing hologram. When this hologram (our personal reality) is challenged, we prefer to ignore the true reality in favour of our own subjective one.

The main arguments against Schoch's theory are one of three kinds. The first we need not concern ourselves with overmuch, because they are more akin to slandering tactics than true criticism. Some person, who may or may not be a trained geologist, voices his disapproval of Schoch's theories before even beginning to analyse the data, then goes on to explain away all of Schoch's *factual evidence* with personal opinions, which have little basis in fact if any, and certainly none in scientific conduct or thought.

Such criticism needs no formal reply, other than being made aware of the fact that unfortunately, such people exist, and they'd rather spend their time criticising anything and everything, without ever doing any actual work themselves. (**Added to original:** *Almost 20 years later, I now think there is actual value in using facts and logic to mock them, if only to expose their illogic and stupidity at best, or wilful dishonesty at worst, and preventing their pseudo-thinking from becoming acceptable, though I fear it may be a battle already nearly lost.*)

Perfect examples of such critics may be the Islamic Fundamentalist who feel that not only the Sphinx, but also the Great Pyramids, are idolatry monuments, and should be destroyed because they predate the prophet Mohammed.

Another classic case of infallible logic ladies and gentlemen:

"If you kill the chicken, then the chicken never existed in the first place."

It is sometimes really frightening and depressing, to think that there are millions of babbling cretins in the world, who think the above statement is valid. Nor am I referring exclusively to any one group of people. Sadly these creatures do not congregate in any one place, but are spread-out throughout the world.

The second type of criticism, is one where, what appears to be an honest critique, is made on what would seem to be scientific grounds. On closer investigation though, it would seem that although possibly the critic is truly in earnest regarding his objections, his reasoning seems to be flawed in some way. Sometimes in ways that are extremely surprising, considering that the individual is usually a soundly trained expert in the relevant field.

This type of objection is of the same kind that might have been voiced with respect to the Wright brothers by Simon Newcomb, (and the vast majority of the scientific community of the time too) professor of mathematics and astronomy at John Hopkins University.

Just weeks before the Wright brother flew at Kitty Hawk North Carolina, the professor had written an article in *The Independent* that showed 'scientifically', that powered human flight was "utterly impossible."[72]

The criticism was surely well intentioned, but decidedly in error nevertheless. And these sort of criticisms are also the ones which have plagued *all* the major inventors, philosophers-scientists and great thinkers of history.

The term *genius*, does not apply to persons that have abilities so far removed from normal men that they are above the rest of humanity. In fact, while certainly geniuses do seem to have a kind of 'spark' for reaching certain conclusions by intuition, the fact is that, prior to reaching such a spark, a great deal of thought and effort went into the project. No one just woke up and invented the light-bulb.

Edison had more than 10,000 failures before he finally succeeded in achieving what, once again had been thought impossible.

The crux of the problem, is that few men (or women) have that kind of tenacity of purpose. That urge to get to the bottom of something even if it kills you. And lacking this burning need for complete understanding, the mediocre minds of the majority, cannot at first grasp the lone genius who's taken the time to find things out for himself.

---

[72] See *Forbidden Science* pg.11. Details of this book are in the bibliography.

The fact that one can have several post-doctorate degrees in a field, is however, no inoculation against this kind of mental lethargy.

And we should also keep in mind the fact that while a person can be a genius in one field, he can also, simultaneously, be a complete fool in another.

In this respect then, it would seem that although earnest, the second type of Schoch critic, is just —quite simply— wrong; as is evidenced by a truly logical and unbiased analysis of the material at hand.

The third type of critics, and the most useful in a sense, are those that scientifically attack Schoch's theory, but seem to focus only on the weak points and gloss over the more relevant ones, or in other instances, make an error similar to the second-type critic with respect to an important point of Schoch's theory.

As anyone that has done even just a cursory reading of the history of science will attest, theories are seldom ever finite and correct.

In fact, I can't think of a single example of such a theory.

Even Einstein's theory of relativity, for all its merit, is now beginning to crumble around the edges.

It's nothing to be ashamed or afraid of. It's a natural process of learning and growing, which everything in the Universe must follow.

So certainly, while Schoch may not necessarily be correct about all his assumptions, it is the general improvement on our current ideas that we should be focusing on, and not those points which bring little merit, or which although perhaps completely wrong, do not invalidate the basic premise of a theory.

In short then, Robert M. Schoch's analysis of the Sphinx —initiated by the enquiring mind of John Anthony West— seems to me to have all the traits of one of those theories that has first got to go through the long corridor of stone-throwers before it becomes generally accepted as in fact having been wronged and being an advancement on current thought.

As a last note on the Sphinx, it is worth mentioning that Schoch's theory is not universally rejected.

A group of scientists from Waseda University in Tokyo, using different criteria and methods, arrived to similar conclusions in 1987, (before Schoch) stating that the Sphinx is older than presently said to be by mainstream Egyptologists.[73]

But then, perhaps the Japanese —whose basic religious beliefs are presumably not threatened by the Sphinx, and for whom the age of this monument could be said to be more of a scientific curiosity than an assault on one's personal beliefs regarding the history of mankind— might just be trying to find out the truth, without regard for the consequences, unlike many biased persons, among which, a large number of Egyptologists.

Lastly, a peculiar trend for Egyptologist to follow when they are low on ammunition, is to state that since there is no real evidence of an advanced civilisation having existed before the ancient Egyptians, the Sphinx and Giza Pyramids could not have been built before the time of these people.

Apart from the fact that such a statement just flies completely against the Sherlock Holmes[74] theory on what is and isn't possible, the more obvious point one should note, is that *just because no evidence has been found, does not mean none exists!*

In fact, if the Sphinx and Giza Pyramids *have* been built by people who came before the Egyptians, then *they are themselves evidence.* And pretty startling and obvious evidence at that!

Yet it's been ignored as such, even in the face of obvious, logical, and factual evidence such as the one I have touched on in this chapter. So what chance would a 10,000 year old piece of technologically advanced equipment —like say an electric potato-peeler, for argument's sake— have, when more than 6 million tons of rock both at Giza and in other locations around the world, have been completely ignored!?

Somehow I don't think my money would be on the side of the potato-peeler, which no doubt, in the hands of a garden variety Egyptologist, would become a ritual instrument for the cleaning out of sand grains from camels' toes.

In a way, this type of logic, concerning a lost civilisation is just another version of the idea that:

---

[73] See footnote 2 in the Fortean Times article by Schoch for references to these investigations. Details of the article in question are to be found in the bibliography.

[74] As given to Sherlock by his creator, Sir Arthur Conan Doyle.

"If you kill the chicken, then the chicken never existed in the first place." The only difference being that it may go something like this: "If we can't find the chicken, then it never existed."
The idea of the advocates of the first statement and those of the second both working towards their own goals at the same time, in the same place, is decidedly frightening for those of us that would like to know the truth about chickens; whatever that may be!

The implications of an advanced civilisation in the distant past, are that humanity may well have had more than just one or two 'ups and downs'. It begins to give a lot more weight to the idea that this planet is periodically visited by catastrophes (of whatever nature or origin) that almost completely destroy the human race, who then builds itself up again, but loses most of its knowledge and history in doing so, which perhaps only survive as myths.
It may be hard for us to accept that we may be completely at the mercy of an earthquake, or a wandering rock from space, or even extra-terrestrial beings from some distant star, but the truth of the matter is that if such is the case, then denying it will certainly not work in our favour. In fact quite the opposite. In beginning to allow for such possibilities, we would then be starting to take appropriate steps; for such is human nature.
We would build more earthquake-proof buildings, place more telescopes in space to give us more advanced warning of things like wild asteroids, and we would also begin to investigate the UFO phenomenon with more science and less hype.
And if along the way we also happen to discover that we have to rewrite the history books, then so be it.

♦♦♦♦♦

Finally, as a last question on the Great Pyramid, how would we reconcile ourselves with the fact that, while every other building or object built by the ancient Egyptians, including the other smaller and far less accurate pyramids, is literally covered in hieroglyphs,[75] the three massive Pyramids at Giza, which are without a doubt the most impressive of all the pyramids found in Egypt, did not yield so much

---

[75] Pyramids were not inscribed with hieroglyphs on the outside as a general practice, but their insides, in the case of pyramids other than the Giza ones, did indeed produce hieroglyphs.

as a single Egyptian hieroglyphic inscription other than the ones 'found' by Col. Howard Vyse?

There is however, *one inscription*, found in the stonework beneath the point of the entrance gable, which for reasons best known to themselves, Egyptologists tend to ignore, and that as far as anyone can tell, would seem to be the only original piece of carving which may have survived to the present day.

It appears to be of a much older origin than the inscriptions carved near it in the nineteenth century by a German Egyptologist in honour of the Emperor Friedrich-Wilhelm IV of Prussia, and it is in a code or language which remains unknown to this day.

Peter Lemesurier gives the inscription as looking like the one shown below.[76] And I think you'll agree, if I say that there is something a little eerie about these symbols.

Figure 18.

The only presumably original carvings
found anywhere on the Great Pyramid.

The more we study this building the more obvious it becomes that it could not possibly have been built by the supposedly 'technological-ly challenged' ancient Egyptians. This fact on its own may confuse us or lead us to reconsider ancient history, but perhaps, by itself would not lead us immediately to the conclusion that there is an extra-terrestrial influence in our origins or history.

Lemesurier highlights this, because he obviously was not aware of the Face on Mars when he wrote *The Great Pyramid Decoded*.

---

[76] See page 32 of *The Great Pyramid Your personal Guide*.

As early as page 13 of that book, he suggests that in view of the features of this incredible structure, it must have been built by a technologically advanced people. He suggests that such people may have been the few survivors of the supposedly mythical Atlantis, which, in legend at least, has always been associated with a high level of technological achievement.

In other words, the implication is that civilisation may have reached peaks previously unsuspected, only to collapse again to the point of barbarism, as a result of some major catastrophic event. Lemesurier adds, that fantastic as this idea may sound at first, *the only other alternative* would seem to be what he calls "Von Däniken's controversial postulation of men from another world" and he goes on to say that this idea seems to him to be much more unlikely than the one he puts forward.

For my part, I tend to agree with Lemesurier with regard to the building of the Great Pyramid. I feel that this structure probably *was* built by people of a more Atlantean descent than Martian or extra-solar, yet it becomes obvious to me, in view of the pyramids on Mars that there most definitely *is* an extra-terrestrial connection somewhere along the line.

# Summary

The main objective and summary of this chapter is a simple one. The Great Pyramid could not be built with the technology attributed to the ancient Egyptians. The conclusion is that it was built by people who had knowledge of advanced methods of construction, and if Lemesurier's decoding method is applied, then these people could apparently also predict or 'read' the future of the human race on Earth.

Regardless of whether Lemesurier's code is accepted or not, the fact remains, that an advanced race was involved in the construction of the Great Pyramid. This of course is consistent with the original theory of survivors of an Atlantis. There is, in my opinion, as well as that of numerous geologists, enough evidence to substantiate the possible existence of Atlantis at some point in the past.[77]

Furthermore, it seems consistent to assume that Atlantis was an outpost of the more advanced humans, and that the few survivors of this quasi-magical land, eventually mixed with the more Earthly humans and that their technology gradually vanished over the millennia, but not before they left some evidence of their passing.

Regardless of the Atlantis/Outpost connection, it has at least been ascertained in this chapter, that indeed, the Great Pyramid of Giza on Earth exhibits traits of a very advanced technology.

Furthermore, such technology could definitely be considered to be of an extra-terrestrial nature, and in consideration of the Pyramids on Mars, this makes perfect sense. In fact it verifies something that the reasonable investigator was bound to have already concluded or at the very least suspected, namely that there is some sort of connection between the Martian pyramids and the ones on Earth.

If you turn back to the summary of Chapter 2 you will note that point 10.d was taken to be one of the most important for the verification of the general theory of alien origins for the human race.

---

[77] Interested readers should refer to the bibliography for books which deal with Atlantis in some detail.

Since this point has now been proved to fit with the general skeleton, we have taken the first step towards fleshing out the theory in more detail.

In view of the discovery of what appears to be a bronze instrument in the Great Pyramid however, it would seem that we have also validated point 10.c. to some extent, especially since the only impartial and truly scientific evidence, seems to indicate that the Great Pyramid is indeed much older than has previously been suspected.[78]

In conclusion then, upon closer investigation, the Great Pyramid shows signs not only of being a structure that exhibits signs of a hitherto unsuspected high level of technology, but also of having yielded up objects, that seem to be 'displaced in time', for according to all our orthodox archaeological evidence, bronze should really not have been found in the Great Pyramid, and particularly so if it originates from as far back as a distant 10,000 BC or thereabouts.

It is also worth pointing out, that the technology required for the completion of the Great Pyramid, is in fact *not* with us even today, not only from a purely engineering point of view, but as we shall learn in later chapters, also (and perhaps more significantly) from what may be called a 'mental' point of view.

In view of these facts, then, the Great Pyramid fits extremely well with the skeleton-theory outlined in chapter two.

---

[78] An iron plate was also found in the southern 'air-shaft' of the King's Chamber in the Great Pyramid by J.R. Hill, in 1837. Supposedly the plate was found between a join in the masonry, suggesting an origin contemporary with that of the Great Pyramid, but in view of Hill's involvement with the Vyse markings, one cannot help but wonder at its true pedigree. It's an ironic twist of fate, that what may have turned out to be perhaps one of the most important discoveries made in the Great Pyramid, has instead been tarnished by association with what may have been a less than honourable individual. On the other hand though, for those that feel that neither Vyse nor Hill were forgers of any kind and that the Khufu markings are authentic, this plate would seem to be a bit of a problem since it would mean having to explain how the Egyptians could forge iron. Either way then, whether original or not, this plate does serve some purpose, and that is, to embarrass still further, the orthodox version of the story concerning the Great Pyramid

# 4      Mars

| Table 2:    Earth and Mars Data Compared | | |
|---|---|---|
| **FEATURE** | **Earth** | **Mars** |
| Surface Gravity (Earth = 1) | 1.000 | 0.38 |
| Equatorial Diameter (Km) | 12,756 | 6,794 |
| Sidereal Period of Axial Rotation (Length of one day & night) | 23h 56m 04s | 24h 37m 23s |
| Length of one orbit around the Sun | 365 days | 687 days |
| Escape Velocity (Km per second) | 11.2 | 5.03 |
| Inclination to Orbit | 23°27' | 24°46' |
| Density (Kg per m$^3$ ) | 5,517 | 3,933 |
| Mass (Earth = 1) | 1.000 | 0.107 |
| Main Atmospheric Composition | Nitrogen 78.08% Oxygen 20.94% Argon 0. 93% | $CO_2$ ± 96% Nitrogen 2.5% Argon 1.5% Oxygen 0.1 % |
| Average Surface Pressure      (Earth = 1 Bar) | 1 | 0.0074 (Less than 1% of Earth's) |
| Average Surface Temperature (Kelvin, Centigrade, Fahrenheit) | 298 K 25° C 77° F | 230 K -43° C -45° F |
| Min-Max Temperatures (Kelvin) | 200 K to 333 K | 150 K to 303 K |
| Approximate Magnetosphere depth on Sunward side | 10 Earth Radii | Negligible |
| Albedo | 0.36 | 0.16 |
| Mean Distance from Sun (Km) | 1.496 x 10$^8$ | 2.279 x 10$^8$ |

The table on the previous page shows the main characteristics of the planet Mars as we find it today, and compares them with the ones we have here on Earth.

We note that in some important aspects, Mars is peculiarly similar to Earth. The layman might be drawn to notice the fact that Mars has a day and night very similar in length to Earth's. In fact, so closely do these figures resemble each other, that astronomers have gone so far as to give the Martian day the name of Sol. On Earth we have days, on Mars they have Sols. I however will be referring to them as Martian Days or as MAR.[1]

The likelihood of two planets in the same solar system naturally having a length of rotation which differs by less than 3% is of course small. To suggest that Mars or Earth (or both) were terraformed by extraterrestrials simply on the merit of this one factor would not only be hasty, but downright alarmist and unscientific. By taking another glance at the table however, we also notice that Mars has an inclination to its orbit which is just a little over 5% more than the Earth's tilt of 23 degrees and 27 minutes.

Once again, taken in isolation, this fact would no doubt be far more likely to be the result of coincidence than the will of extraterrestrials (whom, if this *was* the only evidence, would no doubt be referred to as little green men.) Taken in conjunction with the length of the Martian Day (MAR) though, this set of 'coincidences' begins to look at least a little suspect.[2]

---

[1] Sol of course is also the name of our sun. Personally I feel the name Sols for Martian days is an unfortunate one. *Martian Day* may sound uninspired, but certainly it's neater. For purposes of expediency and in order to avoid confusing a Martian day with our primary star, I have chosen to refer to a Martian Day as a MAR, borrowing the first three letters of the planet's name for this purpose. The plural remains MAR and no S is added, in a similar manner to words such as SHEEP, which is identical in its plural or singular form. The word mar will be used in a similar fashion to the word day and, depending on context, can represent either just a single period of sunlight (half a rotation), or a period equaling one complete rotation of the planet (which encompasses both one day and one night). When referring specifically to a Martian night, the prefix NI can be added to the word mar, forming the new word NIMAR. Once again no S would be added to the plural form. This methodology could of course also be used for other planets equally well.

[2] Simply because we happen to have the pyramids on Mars, should not mean we get the easy way out, especially since the scientific community seems pretty intent on claiming they're a figment of overactive imaginations. Remember that in order to be logical, we should not look for evidence which fits in with the Martian Pyramids and then say "Ha, I told you so!" We should analyse each piece of the puzzle separately at first, noting any points of interest. Once all the pieces have been analysed, we can then

Thanks to its inclination to its orbit and the length of a mar, Mars has seasons just like Earth, complete with Spring, Summer, Autumn and Winter[3]. The length of each of these is given in the table below.

| Table 3: Duration of Mars' Seasons | | | |
|---|---|---|---|
| **SEASON** | | **DURATION** | |
| Northern Hemisphere | Southern Hemisphere | Mar | Earth Days |
| SPRING | AUTUMN | 194 | 199 |
| SUMMER | WINTER | 178 | 183 |
| AUTUMN | SPRING | 143 | 147 |
| WINTER | SUMMER | 154 | 158 |

Despite this, the sceptical observer is bound to notice that today, life for unprotected humans on the surface of Mars is impossible.

With an atmosphere so rarefied that it would not sustain life even if it was composed of pure oxygen, this comes as no surprise. Similarly, the temperature on Mars is anything but friendly.

While in some equatorial regions the temperature may get as high as 30°C (85°F), in the main, the average temperature on Mars is well below the freezing point of water at some -43°C (-45°F), but it gets much colder than this closer to the poles and at night. The Viking landers operated for two full Martian years, and in that time recorded

---

try to fit them together. The neater the fit, the more likely that the overall theory is correct, especially if we have to make little or no 'adjusting' to any of the pieces in order to 'make' it fit. If there were no pyramids or Face on Mars, any astronomer or scientist that were to suggest that Mars was terraformed by aliens, basing such a theory exclusively on the duration of a mar and the inclination of Mars to its orbit being similar to Earth's, would no doubt be made into a laughing stock. While my adventuring spirit and sense of wonder might prevent me to call such a hypothetical scientist an outright cretin, I too would no doubt think he was sadly mistaken. If however, such a scientist was to supply additional evidence that proved Mars once had intelligent life on it, we might reconsider what at first appeared to be a ludicrous proposition.

[3] These can still be observed today since the polar caps expand and contract as the Martian year proceeds from winter to summer.

temperatures during the Martian winter that registered well below -100°C, with a minimum value close to -123°C (-148°F). Nor would it be surprising if values even lower than this were found at the poles. There is also no free standing water on Mars,[4] and the only water present is in the form of clouds, fogs and trapped as ice at the poles and possibly, as some astronomers theorise, as permafrost under the top layer of Martian soil.

Another important factor which would make life on Mars rather troublesome is its apparent absence of a magnetosphere.

The Earth is enveloped in a magnetic cloud which is known as the magnetosphere. This magnetic field is not actually spherical in shape. It is thought that if it were, the magnetosphere would extend at least 100 Earth radii all around the planet. This does not happen though, because the sun has the effect of squashing the magnetosphere. The particles and energy radiated by the sun, are usually referred to collectively as solar wind, and this is a good descriptive name.

Imagine that we have a snowball floating in mid-air in a hot climate. As the water ice became warmer, and began to evaporate, you would see a thin film of water vapour enveloping this snowball. Since the snowball is floating in mid-air, the evaporating water would emanate from every part of the snowball and create a kind of gaseous envelope around the more solid snowball. Now imagine that a light wind came along. The gaseous envelope would then be thinner on the windward side and stretch out behind the snowball.

The Earth's magnetosphere behaves in much the same way and while on the side facing the sun it is only about 10 Earth radii deep on average, on the night side (the one shielded from the sun) the magnetosphere is thought to extend up to a 1000 Earth radii or more, and in any event it extents out well beyond the moon's orbit which lies at about 60 Earth radii in distance.

The Earth's magnetosphere is important because it protects us from the harmful radiation of the Sun. People complain about the evaporating ozone layer, and this is indeed a tragic thing, since the atmosphere also plays a very important part in protecting us from

---

[4] Possibly some small puddles or dew may form in the depths of the deepest canyons where atmospheric clouds can be observed, but these would last only a few seconds and probably no more than a minute since it would go from ice to liquid for a very short period of time as the temperature rose during the daytime and would then almost immediately evaporate due to the thin atmosphere. There are no lakes or rivers on Mars today.

radiation, but if the magnetosphere were to suddenly 'switch off', humans would not continue existing for very long at all on this planet. On average, a person on Earth can expect to be exposed to about one fifth of a rem[5] per year due to background radiation and artificial causes. The more rem a person is subjected to in the shortest time, the more likely that he or she will develop cancers, leukaemia, or in severe cases, radiation sickness and eventually a painful death as the body's genetic structure begins to break down and malfunction in all areas.

Astronauts orbiting Earth, have been known to receive up to a full rem in a ten-day flight period, and one should remember that these people were still protected by the Earth's magnetic field. Beyond the Earth's magnetosphere, the chance of receiving a lethal dose of radiation of several hundred rems is certain if given enough time.

Solar radiation is given out in dangerous amounts during the occurrence of solar flares. These gigantic explosions on the Sun's surface are known to occur more frequently during sunspot activity and it is now generally accepted that sunspot activity occurs in a roughly eleven year cycle. Solar flares and sunspots are also know to squash the Earth's magnetic field considerably and in doing this affect the weather systems on Earth in a generally adverse manner. Since 1992/93 was considered to be a sunspot period, perhaps then, we can begin to appreciate why the recent spree of floods and freak weather occurred, although I feel blaming it all on the Sun would be irresponsible, because surely, mankind is more at fault here.

Now that we know this, and considering that in this chapter we're primarily looking at Mars with a view to determining if life on it may have been possible at some date in its past, we must obviously ask the question, how could humans (or at least humanoid Martians) survive on a planet that has no appreciable magnetic field?

Lastly, as a matter of interest, we might notice that while not a barrier to long term occupation, the gravity on Mars is only 38% of Earth's. A man that weighed one hundred kilograms on Earth would only weigh 38 kilograms on Mars.[6]

---

[5] Rem stands for Roentgen equivalent man, and is a measure of radiation.

[6] On a further point of interest, if we assume that any Martians that may have existed on Mars were closely related to *Homo Sapiens Sapiens* in origin, then as a result of the lower gravity, the average Martian would probably have been quite a bit taller than the average Earthling and also muscularly and skeletally weaker.

At this point then, we have learned a little of how we find Mars today, and examined all the main reasons why under such conditions, humanoid life on Mars could not exist for any appreciable length of time. There are other minor problems, such as the lack of free-standing water and so on, but as you will see later, if we could address the main problems of atmosphere, temperature, and particularly the lack of a planetary magnetic field, the other problems would rapidly disappear without us having to worry too much about them at all.

This is so because, as we should know by now in view of the pollution problems we are creating here on Earth, a planet by its very nature is an ecosystem, where each factor affects every other factor, and fixing one thing would generally bring about the betterment of several others.[7] For example, the adding of an atmosphere to Mars would not only make the air breathable if that atmosphere was composed of at least some 20% or so of oxygen, but it would also stabilise the temperature ranges, so that the variation between maximum and minimum would be far less than it is now. Similarly, by the same process of trapping solar radiation, the planetary temperature would be raised. This in turn would facilitate the existence of free standing water for at least some period of the year, which would probably be confined to lying mainly in the warmer equatorial regions. Having an atmosphere would also provide some measure of protection from solar radiation. So merely by adding an atmosphere we would have gone a long way towards fixing the other problems.

The main obstacles facing us with regard to life on Mars then, can be summed up as being:

1) Lack of a breathable atmosphere.

2) Extreme temperature ranges and lack of free standing water.

3) The lack of adequate protection from solar radiation in view of the absence of a magnetosphere.

---

[7] The reverse of this can indeed be seen here on Earth, where the spoiling of one aspect of the ecology results in a worsening of many other factors.

We have already seen that Mars has some uncanny characteristics with respect to its inclination to its orbit and its length of the day; and the existence of seasons as a result of these factors, but let us now look at Mars a little more closely to see if there is any evidence to suggest that it did indeed support life in a distant past.

The first and perhaps most important point in this regard that we should take note of, is that Mars shows clear signs that free standing water did indeed exist on its surface in the past. This is not some hare brained idea which originated in my skull alone, but rather one with which astronomers that study Mars all over the world agree with. It may not be a much publicised fact (perhaps for reasons I touch upon in Part II) but a fact it is. Mars once had free standing water in large amounts. This is obvious when looking at the ancient and now dried up river beds and channels which are particularly predominant in the equatorial regions and southern hemisphere. The fact that these channels are in the very same region of the planet that we would expect them to be if Mars once had an atmosphere which permitted it to have free standing water, is of course reassuring. (See plate 9 for a Viking Orbiter image of a section of dried up Martian river valley).

It is particularly important to understand that if Mars once had free standing water, it also *must* have had an atmosphere which was much thicker than the one we find today. I say *must* because today on Mars, with an atmospheric pressure of only 6 or 7 millibar, free standing water would evaporate explosively.
If water was present in large enough quantities to create rivers and lakes and possibly oceans too, then the atmospheric pressure must have been a lot higher in the past than it is at present.

If we had flowing water though (as we must to create channels) it's obvious that for at least part of the Martian year the surface temperature was above the freezing point of water. And knowing this, we can assume that in the equatorial regions, even the Martian winters must have had a temperature well within the one we would require for human habitation. The people of Icelandic and Eskimo origin no doubt live in a colder climate than the one we could expect to have existed long ago on Mars in the equatorial regions.

So far then, we have water, an atmosphere and a temperature range which would permit humans to survive. Quite suddenly then, we have gone from thinking of Mars as a sterile and dead planet, which it no doubt is now, to one which had flowing water, an atmosphere which could be breathed by humans if it contained enough oxygen and no other toxic gasses, and a temperature range over the Martian year which once again could support human existence.

This is a fact. At one point in its history Mars had these things. This is no hazy photograph which can be conveniently labelled as a 'trick of the light' because it does not fit with our preconceived ideas.

This is solid, incontrovertible fact. Written on the face of Mars in a way to be as indelible and uncomfortable for some of us as the Great Pyramid of Giza is for others of our kind.

Of course there is still the problem of the magnetosphere. How did the Martians survive the increased radiation they were exposed to? In trying to answer this question I have formulated my own theory with regard to the magnetosphere, which however may not sit squarely with a lot of people, be they laymen or scientists. Before expounding on this topic though, I would like to point out a few things.

1) Regardless of whether my personal explanation of the magnetosphere is accepted or not, the fact remains that the atmosphere on Mars already helped to lower the incoming radiation considerably. Depending on the density and composition of it, it is possible that the Martian atmosphere may well have been more successful at screening the Sun's radiation than our own. Some evidence for this may be found in the planet's present day albedo factor. The albedo factor is the amount of radiation reflected by a planet. The atmospheric clouds of Earth have an albedo of between 44 and 80 percent, while the land surfaces exhibit an albedo of between 8 and 40 percent. Mars has an albedo of 16% on average when compared with Earth's 36%. Considering that Mars has a tenuous atmosphere (to put it mildly) when compared to Earth, it's impressive just how much radiation it reflects back. Significantly, Mars' atmosphere seems particularly adept at reflecting light closer to the blue wavelengths. Ultraviolet radiation of course lies just beyond the blue and violet, hence the name. Of all the radiations which would make life difficult if not screened, ultraviolet is thought to be perhaps the most harmful. If the way present-day Mars reflects radiation is any indication of what the Martian system was capable of when it had a much

denser atmosphere, then it's not unlikely that the Martian albedo factor was once much higher than the one of Earth. This of course would go a considerable length to ensure a more radiation free existence for the Martians even if their magnetosphere was much weaker.

2) Once again, regardless of my views on the magnetosphere of Mars, the existence of an atmosphere and free standing water remains undisputed. Some of course, will say that the atmosphere need not have been breathable. And if they do not accept the Face and pyramids as being artificial constructs such persons may then be loathe to accept that Martians ever existed.

I would like to draw attention to the fact that this is not really a logical approach. Apart from the fact that we have already drawn attention to the unlikelihood of the great Pyramid of Giza and in so doing added to the already independently examined likelihood that the Face and Martian pyramids *are* indeed artificial in origin, we have also examined Mars to some extent and found that it had many factors that, while unsuspected at first perhaps, turn out to meet us directly head on. We ask the question:

*"Mars, could you have supported life?"*
And it replies: *"I have a day and a night very similar to your own. I have seasons like yours if almost double in length. I had running water, a thicker atmosphere and a stable temperature in the required range."*

That sounds suspiciously like a *YES* to me. The more optimistic among us may even go so far as to say that I've shown conclusive proof that Mars was terraformed to behave like Earth. While I do not expect the majority to agree with this without further evidence, it must be said that by now our views should have changed quite a bit from the more orthodox ones we had until recently. At the very least, one should see that we need to review a lot of what we presently accept on faith, and once again, I am inclined to say to the non-believers: silence is equivalent to assent.

3) If we accept the Face and Martian pyramids as artificial, then of course, immediately the question of whether the atmosphere was breathable and the protection from solar rays sufficient becomes idiotic. If the pyramids are there, the Martians must have been there. If the Martians were there, then the atmosphere must have been breathable and the protection from radiation sufficient.

4) In view of points 1 to 3 above, it seems obvious to me that we should at least allow the possibility that Mars could once support life. If we do not and say: "NO!" right here, we may close the door to further evidence because we'll have stopped looking. The logical thing to do, even if we want to remain very sceptical, is to allow the possibility of life, so that we may enquire a little further and perhaps find more evidence to back this theory up.

In order to address the magnetosphere question, it seems to me that we should first know a little more about the nature of this phenomenon. What is it exactly, and were does it come from or how does it originate?

The commonly accepted theory is that a magnetic field arises as a result of a planet's metal core. It is thought (as a result of studying seismic waves) that Earth has a core that is composed primarily of iron. It is also theorised that the outer core is liquid iron while the inner core is solid.[8] The magnetosphere is thought to originate as a result of this supposedly iron core, although no one has yet (to my knowledge at least) advanced an adequate theory to explain exactly why or how this should be so. In fact, until relatively recently, the magnetosphere was not even known to exist, and as far as mankind is concerned, the true study of electromagnetic phenomena can comfortably be said to be in its infancy. Nor does it take a genius to realise this, when we consider that to the best of our knowledge, this is the first time electricity is an accepted commodity for large sectors of the world's population. And despite the fact that many use electricity hundreds of times a day every day, very few of us really

---

[8] I find this assumption unlikely personally, and would think the inner core more likely to be in liquid form than the outer core, but I have not researched the matter in depth, and I suppose that given the much higher pressures to be found at the core, it may well be that the inner core is indeed solid. Having said that though, I should also point out that the temperature is much higher too at this depth.

think about how it really works. We can take some comfort in the knowledge that very few people actually can really explain electricity well, and even they are limited to explaining *how* it works, not *why* it works, which of course would be much more interesting. Electricity and magnetism are inextricably linked, but of the two, the far more mysterious phenomena is magnetism. It is no accident then, that when we try to discover more about the magnetosphere we meet with a very thick fog indeed.

The human aura has recently been accepted by science as being a true and verifiable phenomena. In fact, although Kirilian photography[9] has shown conclusively the existence of this phenomena for several years, orthodox science has only recently begun to take notice of it.

As it turns out, all living creatures and objects are surrounded by an electromagnetic field which is commonly referred to as the aura.

Mystics for centuries have talked about how this field is connected to the *chakras*[10] of the human being and how the human aura is in fact

---

[9] Kirilian photography, named after its discoverer, is the photographing of humans, plants, animals and objects with special equipment. In these photographs, the aura of living things can be seen. Some people are able to see the aura with the naked eye, and Barbara Ann Brennan in her book *Hands of Light* describes a method for doing so. I have followed the instructions in that book in this regard and must admit to indeed having seen the aura of both a plant and my hands. Subsequently I have been able to make considerable progress in this field, but that, as they say, is another story.

[10] Chakras are centres of power or vortexes through which the human aura seems to draw in the surrounding energy. There are said to be a number of smaller Chakras and twelve major ones which operate in pairs. Their location is approximately: **Pair 1:** The perineum and the top of the head. This pair is usually referred to as two individual Chakras, the one located between the anus and front genitals being number one and the one at the top of the head being number seven. The remaining pairs are perpendicular to (and meet in) the spine, and are referred to as Chakra two to Chakra six.
**Pair 2:** The sacral area at the back and the area below the navel at the front.
**Pair 3:** The area known as the solar plexus at the front and the corresponding point in the spine at the back.
**Pair 4:** The heart area at the front and corresponding point in the spine at the back (roughly the middle of the shoulder blades).
**Pair 5:** The throat at the front and neck at the back.
**Pair 6:** The place between the eyes at the front and corresponding point at the back of the head.

The soul is also said to reside at the meeting point of the three Chakras found in the head. This point corresponds to the location of the pineal gland and it is interesting to note that Plato, in ancient times, had knowledge of this. Similarly, it is interesting to note that while the function of the pineal gland is unknown, recent studies indicate that composed of different layers. These layers are usually referred to as bodies, because each layer encloses the one preceding it.

The Etheric body for example is said to encompass the physical body which we are all familiar with. The Astral body is said to encompass the etheric and the physical and so on. Like a set of Russian dolls that all fit into each other, the human aura is said to be composed of several bodies.

These things have been written and spoken of by mystics, monks, and Indian gurus for centuries (and anyone interested in the verification of this statement can do so by making a visit to their nearest well-stocked library and enlisting the help of the librarian in locating the esoteric section). It is very interesting to note then, that indeed we do have a human aura, that indeed we do have Chakras and indeed it does seem that the human aura is divided into fields of ever subtler intensity.[11]

Living creatures continue to have this aura as long as they are alive. Unfortunately I have not come across any studies or even opinions as to what exactly happens to the aura at the moment of death, but I would suggest that it fades. Whether this is instantaneous or happens over a period of time is arguable, but I hold the human aura to be intrinsically connected to that quality we call life, and this has several levels of manifestation. A human corpse may well be devoid of what, for want of a better word we may refer to as the soul, the mind or the life force characteristic of humans, but it is by no means devoid of all life.

The individual cells continue to survive for a considerable amount of time before they too die as the body's systems break down and make a continued existence impossible even for the individual cells. My theory then, in regard to a human body, and I would welcome some research into this matter, is that on the point of death we would see a release of a majority of the magnetic phenomenon we refer to as the aura in a relatively short period of time, whether this period is a few seconds or a few minutes I cannot guess, but I would estimate for it to be no longer than a half hour and possibly as short as a second or two. The method by which death came about, I feel, would also play a part

---

it may serve primarily a regulating function for the entire body, as it is thought to be responsible for informing the body as to conditions of light and dark, making the pineal gland a light sensitive organ. Quite fitting when one thinks of the association of light, enlightenment and the human soul.

[11] See the bibliography under the Human Aura heading for references on this and other related subjects.

in this. There would however, be left a residual magnetism in the cells of the body, and this would gradually fade as the individual cells perished. Perhaps during decay there may be a momentary renewal of magnetic phenomena as bacteria and other foreign lifeforms are propagated in the corpse, but upon completion of putrefaction, very little or no magnetic field would be left beyond the one intrinsic to all matter.

It is important to note that if we reduce a human being to his constituent atoms (a lot of hydrogen and oxygen, an appreciable amount of carbon and a myriad other elements in varying and declining quantities) the combined magnetic field produced by these atoms would be very, very, much less than the one we find in the exact same atoms when are arranged in such a way as to produce a living human being.

In a way, the human aura could be said to be the 'proof' of the existence of the soul. Whether this aura remains 'coherent' after leaving the body or not and as to the nature of an afterlife such a coherent aura may or may not proceed to, I dare not speculate on here. I feel I'm already stepping on many toes as it is without taking on this task as well.

In any event, I hold the aura to be intimately connected to the quality we refer to as 'life'.


Now then, the Earth has a magnetic field, and indeed so does Mercury, but Mercury has a field strength of less than 1 percent that of Earth's, while having a composition that is thought to be 80% Iron-Nickel, and a total mass of about 5.5% that of Earth (Mercury is much smaller than Earth).

Venus has no appreciable magnetic field and Jupiter, which is composed of gasses in their liquid state, has a very strong and complex magnetic field.

From this it should be obvious that the theory of a magnetosphere existing as a result of a metal core is suspect indeed. If this were the case then Mercury should have a magnetic field equal to about 10% of the Earth's and Jupiter shouldn't really have one at all[12] while Venus

---

[12] Jupiter's magnetic field is attributed to the hydrogen it is composed of in large quantity being in the 'metallic' state. Due to the intense heat and pressure, the interior of Jupiter is thought to be composed at least in part of individual hydrogen atoms which not being bonded in the molecular state ($H_2$) become electrically conductive. Why this should create such a strong and varied magnetic field is still very unclear to me though.

should have one similar to Earth since it is thought to have a similar core.

My conclusion then is a simple one, and while I do not hold that it is necessarily correct, it is the one I choose to believe until a better one comes along. Mars used to have a magnetosphere. Perhaps not a very powerful one, but one that combined with the atmosphere shielded the Martians from excessive solar radiation and made life on its surface possible. Then something happened to effectively 'kill' the planet. I shall discuss why and how I think this event came about in more detail later, but not to put too fine a point on it, a massive asteroid impact occurred —or more accurately, several— and in the resulting crash, Mars died. Literally.

I'm not going to expound on the theory of whether a planet is conscious or not and whether it classifies as a living organism or not, such speculation would be quite pointless for now and must be left either for later or for others to pursue. What I am saying is that I believe that if Earth was subjected to a massive asteroid impact which nearly tore it apart, which burned off the atmosphere and killed all life on it by boiling off all its oceans, its magnetic field would not only be disrupted, it would diminish considerably and if the planet and its ecosystem did not recover in time from the impact (as it would not if it was of so severe a nature) then it may disappear altogether. In other words, in a very similar way that the aura of a person leaves the body once he or she dies, so I hold does the aura (or magnetosphere) of a planet disappear when it too dies.

# The Death of Mars

I believe Mars used to harbour life, until a catastrophic event that very nearly tore the planet asunder occurred. The evidence for this, once we begin to look for it, is obvious.

If we look at Mars as a whole, we find that it is a planet of incredible contrasts. While being just a little over half the size of our Earth, it has features on it that put our highest mountains and deepest oceans to shame. Valles Marineris is a canyon that measures about 4000 kilometres in length while being 700 kilometres wide at its widest point. Even when not at its widest yet, the canyon walls can be as distant as 120 kilometres, and it reaches a depth of about 6.5 kilometres. At the other end of the scale we have Olympus Mons. The largest volcano of our entire solar system. It reaches into the thin sky of Mars for a distance of 27 kilometres. On Earth, our tallest peak is Mount Everest. At a height of 8848 metres it's a little less than one third the size of Olympus Mons. While the Marianas Trench is a shade over 11 kilometres deep. The total range from deepest to highest point on Earth is not quite 20 kilometres, while on Mars it's about 33.5.

How did Mars come to have such drastic geographical features?

The answer may be found in a feature called Hellas Planitia. This is a crater so vast that at first it was thought to be a plain. In fact it is the result of a massive asteroid impact. This crater is about 1600 kilometres across (1000 miles) and some five kilometres deep (three miles). Similarly, Argyre Planitia is another huge impact crater, it lies to the west of Hellas and is about 600 kilometres across (360 miles). Both these enormous craters are in the southern hemisphere of the red planet. (Refer to plate 11 for a sketch map of the main Martian features).

In all, Mars has 16 impact craters with diameters greater than 250 kilometres (150 miles), but the interesting thing is that almost exactly on the opposite side of Mars from where the centre of the Hellas crater is, we find the imperious Olympus Mons, and not him alone, for three other huge volcanoes, Arsia, Pavonis and Ascraeus are also found in the area. All of them lie on a raised bulge known as the Tharsis bulge. So pronounced is this bulge (about 9 kilometres in height) that each of these three volcanoes, despite being smaller, reaches a height comparable to Olympus Mons (27 kilometres).

It is thought that it is not a coincidence that all four volcanoes have a very similar height. It could well be that this was the maximum height a volcano on Mars could reach before the underlying crust became unstable. Adding to the interest of the area is the Valles Marineris, which ends in a chaotic, delta-like, expanse of broken terrain. This area is criss-crossed by canyons and known as Noctis Labyrinthus. It lies only some 200 km south of Pavonis Mons. The conclusion seems obvious. The asteroid that created the Hellas basin, nearly split Mars in half. The Valles Marineris is not a canyon in the true sense of the word. No river carved it out of the Martian surface. Instead it is a fault line. A massive crack of the crust that is testimony to the catastrophic impact which Mars underwent.

So is the Tharsis bulge, along with its massive volcanoes, as is indeed Olympus Mons, all lying opposite the Hellas basin and to a lesser extent, also opposite the Argyre crater.

Perhaps the most interesting feature concerning these vast craters is that they are thought to have happened when Mars still had an atmosphere. The craters should be much more pronounced than they are if they had occurred when Mars had already lost its atmosphere. This is because a crater that occurs on an airless world as a result of an asteroid collision would spray ejecta[13] all around the impact point. And this would of course be much in evidence, but the ejecta on Mars is distinctly less than the amount expected.

If the impact occurred when Mars still had an atmosphere though, this helps explain the phenomenon. Firstly, the ejecta would not travel as far from the impact point, and secondly, the materials which were capable of existing only in the presence of an atmosphere would have evaporated, liquefied and flowed further away, melted or frozen; effectively being 'absorbed' by the surrounding landscape a lot more than if there had been no atmosphere, since in such a case, erosion would be minimal.

This process of 'absorption' is normal on Earth, where also thanks in large part to vegetation and animal action, a crater of considerable size can blend into the landscape in a period of a few hundred years to such an extent as to be practically unrecognisable from the surrounding landscape. Every airless planet of our solar system which has a solid surface shows extensive asteroid impacts.

---

[13] Ejecta is the material an asteroid carves out of the planet when crashing into it.

The Earth was not spared visitations by numerous asteroids either, it's just that our planet is very much more dynamic than those airless worlds. For a start, a lot of the smaller asteroids burn up in the atmosphere, and only the bigger ones get to reach the surface. Of those, 70% end up in the oceans or lakes of the world, and the few craters which are actually formed on land are quickly covered up again by the weather, vegetation and the generally changing and evolving landscape.

It is also important to note that while scientists speculate about the age of the Martian craters and volcanoes, until a manned study is undertaken, it is not possible to say with any sort of certainty what the age of these features is, nor is it possible to say how long ago Mars had an atmosphere. All that is known for certain is that at some point it did, and now it does not. The only dating possible on Mars is relative.[14] That is, we can say that some feature is probably older or younger than another feature, but we cannot say by how much or what the age of either feature is. I find it particularly relevant then, that the Hellas and Argyre craters are younger than most of the others and that the Tharsis bulge volcanoes are also youngest of the volcanoes, making both set of events the most 'recent' events.

Interestingly, the southern hemisphere shows from *ten to a hundred times* more cratering than the northern hemisphere, which is particularly less marked by the larger sized craters. The significance of this uneven bombardment is unknown by scientists for the moment, and no satisfactory explanation can be given.

The result of being bombarded by at least two large asteroids and several smaller, but still very damaging ones, is the difference between the Mars of long ago with an atmosphere, oceans, and life (in my opinion at least), and the Mars we see today. An airless desert with freezing temperatures. More evidence of this can be found in the distinctly red tint of the unfortunate world.

The Martian soil is thought to be rich in silicates,[15] which indicates a high quantity of oxygen trapped in the soil. In fact, the presence of a high oxygen content in the Martian soil is pretty much

---

[14] This type of dating is taken from the amount of asteroid impacts a feature shows. Since asteroid impacts are thought to be relatively constant in nature, a feature covered in many small craters is considered much older than one that has very few craters on it.

[15] A mixture of oxygen and silicon or silicon compounds with metals.

agreed upon by the various scientists that have studied Mars, since oxygen is in fact what gives Mars its red-brown colour. In a sense, Mars could be said to be covered by a layer of rusty soil. How did all this oxygen get trapped in the soil though?

The answer to the question is of course unclear, but the best theory that astronomers have come up with so far (to the best of my knowledge anyway) is that the water on Mars was disassociated by ultraviolet rays.

When subjected to ultraviolet radiation, the water molecule splits to its constituent gasses hydrogen and oxygen. In the absence of a high gravity and substantial atmosphere, the hydrogen, being a light gas, would escape into space, while oxygen being a reactive gas, would undoubtedly react with whatever was at hand to form more stable compounds. The only thing at hand of course would be the Martian surface, and this explains the presumed high oxygen content of the soil.

But how did the ultraviolet rays get such easy access to the water on the surface of Mars? This could only happen if whatever kept them at bay was either destroyed or removed or somehow 'switched off'. This something of course would have to be any previously existing magnetosphere, but ultimately the Martian atmosphere itself would have to be neutralised. Perhaps though, with the magnetosphere removed, the ultraviolet rays would have eventually broken down the Martian atmosphere, and once this was done, it would be a short step for the ultraviolet radiation to disassociate the water molecules.

It is also generally accepted that Mars had a much higher nitrogen content, further placing its ancient atmosphere a step closer to our own. The presence of oxygen in the Martian soil in the form of peroxides[16] is also theorised.

It would seem then, at least based on the evidence we have, that Mars most certainly could have had a breathable atmosphere, composed of oxygen and nitrogen, much like our own.[17] It is important to note that

---

[16] Peroxides contain both hydrogen and oxygen, although associated in a different configuration from the water molecule. The existence of peroxides in the Martian soil would further support the theory of water disassociation due to ultraviolet radiation.

[17] I personally feel Mars had a higher oxygen content than we have on Earth today, with a correspondingly lower content of Nitrogen and other gasses. I base this reasoning on the seemingly high amount of oxygen that is presumably trapped in the

the basis for this idea with respect to the atmosphere of Mars, is one that originates from the minds of people that in all probability would laugh at the idea of Mars once having supported humanoid life.

The reason they would laugh, as I have stated before, is not because Mars could not have supported such lifeforms, but rather because most of us have been trained to think in certain ways, which often stifle our imagination and curiosity, sadly masking the truth of our origins in the process.

There is one more point which may be crucial in determining what happened to Mars. If we assume that the planet was indeed 'killed' by a meteoric bombardment, we should first examine what we would expect to happen in this theoretical situation and then compare it with what we have at present, and see how well or otherwise it fits.

There would be good cause to suppose that at least some of the meteorite impacts may well have been caused by carbonaceous chondrites.[18] I say 'good cause', because we have two excellent examples of the types of asteroids Mars was bombarded by. They are called Phobos and Deimos. The two tiny moons of Mars.

That these two satellites did not come with the original planet 'as built' is obvious from their very irregular behaviour, and there is no dispute on this point from any camp I know of. I will describe the two attendants of the War God[19] in more detail later, for the present however, I must ask the reader to be patient and take it on faith, even if just for a few moments.

Both Phobos and Deimos are carbonaceous chondrites, and if Mars was bombarded by this type of asteroid, we can expect several things to have happened.

---

surface elements of Mars, but the lower amount of nitrogen that scientists generally seem to expect. I am not basing this on any scientific method, nor do I claim to be correct in this regard, but my guess as to the oxygen to nitrogen ratio would be something like maybe 30 or even 40% of the atmosphere being oxygen and the rest nitrogen and other mainly inert gasses, with perhaps a higher percentage of carbon dioxide (3 or 4% ?) than we have on Earth.

[18] A carbonaceous chondrite, also known as a C-type asteroid, is thought to be composed mainly of carbon based compounds.

[19] Phobos and Deimos (Fear and Dread) in ancient Greek mythology, were known as the attendants of Ares (Mars) who of course was the God of War.

Firstly, since these types of asteroids have a rather low density and are thought to be composed of frozen carbon compounds and possibly volatile gasses, their entry into an atmosphere at high speed would result in a large part of their mass being burned up. This burning up of course would result in large quantities of carbon dioxide being given off. On another note, if a pocket of volatile gas (hydrogen for example) was present, it might well cause a heated explosion which would (in a high oxygen content atmosphere especially) burn the very air the asteroid was travelling through. One such asteroid or two, even if rather large would not necessarily destroy the entire atmosphere, but a multitude of them, most of which would burn up completely before reaching the surface, could well result in a massive atmospheric flare-up. The very air would catch fire, burning any vegetation or people as well as the asteroid itself as it travelled to it destructive conclusion. In addition, it has been calculated that the huge Martian volcanoes alone would have been quite capable to produce a lot more than the total carbon dioxide found on Mars today. Since I consider the volcanoes to be a direct result of the asteroid impacts, the loss of a breathable atmosphere on Mars would coincide with the replacement of it by vast quantities of carbon dioxide.

Secondly, the burning up of the Martian atmosphere of course would give ultraviolet rays free reign to then 'boil' off whatever water survived the initial onslaught of the asteroid impacts, and in any event, all life on the planet would perish in a very short period of time. The only possible survivors would have to be bacteria of some particularly resilient type. Even then, their survival, if they are in any way similar to their Earth cousins, would be doubtful. If Mars retains any of its original life forms, they're not going to be very evolved ones in terms of intelligence.

Thirdly, if my theory with regard to a planet's magnetosphere disappearing when that planet 'dies' is accepted, then, after having been struck by asteroids large enough to create the Hellas and Argyre Planitia, along with the other less large but still enormous craters, Mars would have been adequately 'killed', further ensuring that with the magnetosphere gone, the ultraviolet radiation had free access to the planet's atmosphere and water.

# Phobos and Deimos

Phobos measures about 27 by 22.5 by 19 kilometres and lies only 9377 Kilometres (5827 miles) from the centre of Mars, meaning it's only 5980 kilometres above the Martian surface. It completes one orbit of Mars in only 7 hours and 39 minutes. This makes it the only moon in the entire solar system that circles its parent planet in less time than the planet takes to revolve on its own axis. Because Phobos is able to run around Mars quicker than Mars rotates, it is seen to rise in the west instead of the east, as observed from the Martian surface. Phobos is so close to the surface that its irregular shape could be seen by the naked eye of an observer situated on Mars.

At latitudes above 70° though, (closer to the poles of Mars) such an observer would not be able to see Phobos at all, since it would lie below the horizon.

The surface of Phobos is particularly irregular on the side of it which faces Mars; it is here that we find crater Stickney.

Stickney measures 9.6 kilometres in diameter (6 miles) and is a deep depression in the potato-shaped body of Phobos. Nearby we find another two craters, also rather large, being about 5 kilometres across. On Phobos we also find a lot of rubble. It is thought that this material is ejecta that was originally thrown outwards by the impacts which formed the craters, but if so, we then have the problem of explaining why so much of it fell back to the surface of Phobos since its gravity is extremely low, with an escape velocity[20] of only 15 metres per second (54 kilometres per hour or about 34 mph).

Given the size of Stickney, it would be safe to assume that the impact that created it occurred at high speed. If so, the rubble that was excavated in that event, would also have been moving quite fast. When you consider that a baseball player can pitch a ball at speeds of up to 140 kilometres per hour (90 mph), it becomes difficult to explain how material kicked up by a meteoric impact which formed a crater nearly 10 kilometres in diameter could be travelling at just a

---

[20] Escape velocity is the speed an object has to move at in order to escape a body's gravitational pull. On Earth this is 11.2 kilometres per second, meaning that an object that leaves the surface of Earth at an initial speed of 11.1 kilometres per second would eventually fall back to Earth, but one that set off with a speed of 11.3 kilometres per second, would get away.

little over a third of this speed. The rubble found on Phobos must also have fallen back very slowly, since it did not create secondary craters. Of even greater interest however are the striations or 'lines' found on the surface of Phobos. These markings seem to be cracks radiating from the crater Stickney and converging on the opposite side of the moon. The grooves are about 700 metres wide and 90 metres deep near the crater, but only about 100 metres wide on the opposite side of Phobos. Up to now scientists have assumed that the grooves are the result of the impact which created Stickney, a collision that nearly split Phobos into a myriad of smaller fragments.

If this were so, we would expect crater densities on the surface of Phobos to be lower in the grooves than in the surrounding terrain, since the soil found in the cracks would be younger and therefore have been exposed to micrometeorite impacts for less time. This however is not the case, and Phobos exhibits a constant crater density that is equal both inside and outside the grooves. Partly as a result of this, some scientists have gone on to speculate that the tremendous tidal pull that Phobos is subjected to as a result of its vicinity to Mars must have played a part in the creation of the striations.

Deimos on the other hand, orbits Mars at a distance of 23,459 Kilometres (14,577 miles) from the centre of Mars and measures 16 by 11 by 9.6 kilometres. It orbits Mars in 30 hours and 18 minutes, meaning it's travelling around Mars with a speed about thirty-seven percent slower than Phobos.[21] Its surface is heavily cratered, but the largest crater is only 2.2 kilometres (1.4 miles) in diameter, and most craters are partially filled in, being observable mainly as a result of the relative brightness of their rims. The depth of the fill material found in the craters has been estimated as being between 5 and 10 metres. It is not clear why the surface geology of the two moons of Mars should be so different, but they do share certain peculiarities. Both are among the darkest objects found in the solar system. They reflect only 6% of the light they receive, making them darker than coal. Both also have incredibly low densities, which the Viking craft estimated as being about twice the density of water.

---

[21] Assuming a perfectly circular orbit around Mars, Phobos is travelling at some 2.13 Km per second, while Deimos is moving at about 1.35 Km per second.

As a result of this low density, Iosif Shklovskii, a Soviet theorist, jokingly commented in 1960, that Phobos and Deimos may be hollow. This comment caused quite a stir, especially because it came from a reputable source, and it was only later that Iosif explained he'd said it in jest.

I for one am not sure that Iosif was wrong in saying Phobos and Deimos may be hollow. In fact I'm not even sure that Iosif was joking when he said it. He would not be the first person that on impulse expressed a thought he held to be true and met with an unexpected and ferocious attack from the 'experts'. The majority of people in such a situation would probably either retract the statement or say it was not meant seriously. A few, more stubborn persons may even be nudged by interested parties to retract or discredit the irritating comment. Considering the history of the Soviet Union in regard to 'thought policing' the idea cannot be entirely dismissed, but it's probably more likely that simple peer pressure was at work here. In any event, I am not suggesting that Phobos and Deimos *are* hollow; I am merely saying that the possibility should not be dismissed entirely, especially in view of the facts we have uncovered up to now and the ones which will be uncovered later as we progress in this fascinating subject.

Another point of interest with respect to both Phobos and Deimos is that each satellite is in an almost critical orbit. Phobos is so close to Mars that it is close to the Roche limit; a point below which a moon breaks up and more or less slowly,[22] falls towards the parent planet. Deimos instead is close to the point at which the gravitational pull of Mars would become too weak to keep it as a satellite. Furthermore, both satellites keep the same side always facing the surface of Mars. In other words, the period of their rotation is equal to the time they take to complete one orbit. A feature we find in our own moon, who also keeps the same side facing Earth. This feature is attributed to the tidal forces which pull on the two small bodies.

But perhaps most interesting of all, is the fact that *both satellites have circular orbits which lie directly above the equator of Mars and this is something which according to our best scientific knowledge should not happen naturally.*

---

[22] If the planet has an atmosphere, the degeneration of the orbit is greatly accelerated, but in the case of Mars as it stands at present, it is estimated that Phobos would take 100 million years to break up and crash to the Martian surface.

Of course the Universe is vast, and we should expect it to occasionally be in conflict with our most trusted theories. If there is any chance that an asteroid could end up in a circular equatorial orbit around a planet by natural means, then it most certainly must have already happened *somewhere* in the Universe. Why not in our very own solar system? There are two very good reasons why not.

Firstly, our solar system is already pretty unique in consideration of the fact that Earth holds a variety of lifeforms, additionally, we have already seen how Mars in all probability could also harbour life at one time[23] and may well have had a similar atmosphere. These things in themselves are already 'longshots' from a mathematical point of view. They become more acceptable only if we assume that intelligent life is far more prevalent in our stellar vicinity than we have supposed up to now. If we do this, then of course it becomes obvious that Phobos and Deimos have the type of orbit that they do because it was manipulated by intelligent beings many years ago. If we do not accept this assumption, then we must explain why so many 'longshots' of probability seem to happen in our solar system and in particular in the vicinity of Mars.

One longshot is a possibility, two becomes improbable at the very least, but three or more become a definite trend that proves the events are in fact not longshots at all; for whatever reasons, be they natural or otherwise.

In this case, the fundamentals of astral physics are pretty well proven so we would have to conclude the reasons are 'unnatural', in other words, artificial in origin. Created by the hand (or appendage) of some intelligent beings.

Secondly, even if we accept that an unlikely set of events came to produce a circular and equatorial orbit for one of the satellites (Phobos' encounter with the meteorite that created Stickney for example) how would we explain the fact that the other also has the same characteristically circular and equatorial orbit?

This second point in fact ties in with the first argument again. As someone before me said : "One miracle is an improbability. Two is a trend."

---

[23] If we accept the Face as being artificial in origin, then this is not a probability but a fact.

Because of their obvious difference in structure from Mars, their strange orbits and C-type asteroid configurations, Phobos and Deimos would appear to be captured asteroids. This theory is one which finds much favour in the scientific community, and one I tend to agree with. In fact, on this point at least, there seems to be little doubt. If Phobos and Deimos had been asteroids that wandered a little too close to Mars and naturally became captured by that planet's gravity however, their orbits would not be circular, but elliptical, and they would be tilted to the equator. Nor is it probable that their orbits would resemble each other since they would have been randomly related to Mars and the possibility of two originally extraneous objects falling in a similar orbital pattern around a planet is very, very remote in the best of cases. Similarly, if their capture occurred by natural means, it would be very unlikely indeed, that Phobos' orbital period would be so short.

For these reasons then, I propose that while Phobos and Deimos are 'captured' asteroids, their capture by the Martian gravity was aided by the involvement of extraterrestrials beings with a high level of technology.

So unconventional is the motion of Phobos and Deimos that when first discovered, they were thought to be artificial satellites. All I am saying is that this view was not entirely wrong. Their surface and composition may be natural in origin, but their location was achieved by artificial means. If I am correct in this assumption, then it is not unthinkable that we may find evidence of alien technology on or even *in* one or both of these small moons.

Once examined by a manned expedition, the crater Stickney and related grooves may prove to be less natural in origin than what we may have at first supposed.

If Phobos and Deimos were placed in their present day orbits by intelligent beings, there must have been a reason for it. If that reason is the one I propose in this and the next two chapters, then we may even find some sort of guidance or tracking mechanisms on Phobos, or at least some evidence that such devices may at one time have been placed inside or on the surface of the moon.

Until further exploration of Phobos is made though, this possibility cannot be said to be very likely, and in fact, even mentioning it may hurt my case, since no doubt, narrow minded experts are bound to pick on it as a 'typical example' of my 'flights of fancy'. Once again, I must stress that it is not my aim to prove correct in every detail of this particular field of enquiry. I am merely interested in showing the correctness of the broader view, and in this respect I have encouraged the reader on a number of occasions to check up on the key points of my claims in order to satisfy him or herself that they are indeed correct.

In the course of the fascinating journey of discovery I undertook while researching the material for this book I could not, cannot, and probably (and hopefully) never will be able to completely restrain my imagination, so it is inevitable that occasionally I make a claim that sounds as fantastic as : "There may be alien technology buried inside Phobos". All I am saying when making such 'ridiculous' claims is that I would not be overly surprised if they turned out to be correct. On the other hand, nor would I be crushed if they turned out to be completely erroneous. The points which hold my interest (and hopefully the reader's as well) are the main ones. The fact that Mars once had flowing rivers, lakes and an atmosphere and temperature that would probably have allowed humans to live there with little trouble. The Face and Martian pyramids. The impossibilities of the Great Pyramid at Giza. These are the main points on which the entire book hinges and not whether or not there is alien technology on Phobos or how tall the average Martian was. In short, it is hoped that the critics will focus on the main points and not the little and mainly irrelevant ones.

For some reason though, human nature seems to get stuck on the details and miss the whole point. This is clearly evidenced by the sad fact that most of us are more interested in keeping up with the latest gossip on some famous actor than in using the same amount of time and energy to develop a new skill and thereby improving our own lot. At least in part, this book is an attempt to try and change that.

# Summary

## 1. What We Know

Mars once had running water which implies a denser atmosphere and a temperature that permitted the water to flow. Judging by the oxides present on Mars today, the atmosphere may have had a high oxygen content as well as a considerable amount of nitrogen. In addition, Mars had seasons, which it retains to this day, and a day and night almost identical to Earth's. The planet's tilt is also very close to the one exhibited by our own planet.

Mars was nearly ripped apart by at least two very large asteroid impacts which formed the Argyre and Hellas basins. A number of smaller asteroid impacts seem to have preceded these two large impacts as well.

The majority of bombardment by meteorites seems to have occurred in the southern hemisphere.

Phobos and Deimos are almost without a doubt captured asteroids, this is evidenced by their composition being similar to the one of similar bodies found in the asteroid belt and their alien nature when compared alongside the composition of their parent planet. The problem with their being captured asteroids is that they have orbits which are circular and which lie directly above the Martian equator; this phenomenon cannot be adequately explained using conventional astronomical models. Their surfaces also exhibit an unusually high amount of rubble and Phobos at least has curious markings on it which might be the result of the meteoric collision which formed the crater Stickney. The surface features of the two moons are quite different and in both cases their orbits are extreme in nature if at opposite ends of the spectrum.

## 2. What I Propose

It is my belief, that the entire Martian ecosystem was destroyed as a result of massive meteoric bombardment.

The evidence for this is plentiful, as has already been mentioned several times, and indeed hard to ignore or deny. It's difficult to turn a blind eye on a crack 4000 kilometres in length after all, to say nothing of the huge shield volcanoes.

The fact that the southern hemisphere suffered far more than the northern one in terms of meteoric impacts is intriguing. At first it would seem to suggest that the 'lower' part of Mars passed through a meteor shower while the northern areas got away relatively unscathed. This idea though has some basic problems with it. First of all it would indicate that the meteor shower was relatively small since only half the planet was attacked by it. If it was not small, then it means that Mars only 'dipped' its lower half into the meteoric 'cloud'. If this is so however, we would expect this group of meteors to periodically return to Mars and we have no real evidence of this. To the best of my knowledge, there is no mysterious asteroid cloud out there in space which periodically pelts Mars. I suppose some are bound to say that over the millennia the asteroid cloud in question spent itself by repeatedly coming into contact with Mars, but once again this does not stand up to scrutiny. If this was the case Mars would exhibit signs of a far more evenly distributed meteoric bombardment. Secondly, if Mars only dipped its lower end into the cloud, where is the other half of it? Presumably half of the meteor shower got away.

Lastly, it might be suggested that the meteor shower was a wandering one, that just passed through our solar system and Mars just happened to be in the way, but once again, there is no evidence to support the theory of 'wandering' meteor clouds. If Mars came into contact with a meteoric cloud which partially survived the clash, then it should still be out there, orbiting our Sun.

All of this seems to lead back to the same point again and again, and that is: The meteoric bombardment which initially destroyed Mars was the result of a small and probably concentrated cloud of asteroids, most of which spent themselves on Mars in the meeting. In this case then, the destruction of Mars would be quite sudden, and that is why one side of it shows so many more impacts than the other.

Over the succeeding millennia the various remnants of the original meteoric cloud along with the regular number of ordinary meteors that every planet encounters on a regular basis, helped to form numerous craters on Mars, since after the initial and dramatic contact, Mars now had very little atmosphere left, and even relatively small meteorites could reach the surface without burning up completely.

And yet, despite the evidence, there are still some serious problems with the theory of a small and concentrated meteoric cloud coming into contact with Mars. First of all, such an unusual "cloud" that is so densely packed to have such an effect on a planet the size of Mars has never been observed and no models for it exist. Secondly, given the enormity of emptiness that our solar system is composed of, it seems unlikely that a small group of asteroids, was in precisely the right place to come into contact with Mars in such a way as to very nearly tear it apart.

Thirdly, the cloud would have to be small enough to leave no trace of its passing other than the corpse of Mars and yet large enough to destroy the planet's ecosystem. Where would such an unusual cloud of meteorites originate from? In nature, it would be a strange beast indeed.

If we were to add no more to this story, the dilemma would remain a perplexing one indeed. But if you recall, what started us on this peculiar journey was the fact that there seemed to be an artificial relic on the surface of Mars. If the problem is analysed in this context, then suddenly it becomes a lot easier to find some answers. The vital clue to this riddle is to be found in Phobos and Deimos. If we assume that these satellites of Mars have a natural origin, it becomes almost impossible to explain why they have the orbits which they do. If however we assume that Phobos and Deimos have their peculiar orbits because they were *placed* in their present day locations by intelligent beings, then the mist of uncertainty suddenly lifts and a much clearer picture can be seen.

As we have already seen, if there was intelligent life on Mars, it must have been connected to Earth (the pyramid link). This in turn would show a high degree of technology. And here, it is necessary to re-examine the theory of terraforming. If both Mars and Earth had a similar atmosphere, similar tilt, similar length of day, similar types of seasons, similar climate and humanoids of similar origins living on them, then the conclusion that at least one of the planets (and possibly both) was terraformed becomes inescapable.

As I hope I have already made clear, the chances of all these factors being the same just by chance are so infinitely small that for all practical purposes it might as well be zero.[24] When we add the humanoids with a common origin however, even that remote chance is

---

[24] The mathematician in me cannot help wanting to give this probability a number in the form of a percentage. Based on a purely 'gut instinct' guess I would place the chances of two planets in the same solar system sharing all the characteristics mentioned at somewhere like $1 \times 10^{-150}$ percent. This number is really tiny indeed, but given the vastness of the universe, it is still large enough that it would ensure that *somewhere* it held true, as I suppose it must to some degree.

eliminated, and we must surely look for a different type of explanation other than pure luck or chance. As it turns out, that explanation is terraforming.

At this point then, it becomes necessary to examine the question of terraforming in more detail before we can proceed.

## Terraforming

The process of shaping a planet to one's needs is a long one and bound to take several thousand years even with advanced equipment. The technology used would require as a minimum:

1) Complete knowledge of genetic manipulation processes for both flora and fauna.
2) Access to incredibly vast resources.
3) Advanced means of space travel:
   a) If interplanetary only, this would mean large bases and a continued peaceful existence on the home planet. This peaceful period must last for several millennia for the terraforming of a nearby planet, since surely, the task would tax the inhabitants of the home world to their limits.
   b) If interstellar, then this in turn implies a correspondingly higher level of technology in all other fields would also have been reached, since the gap between merely interplanetary space travel and interstellar space travel is an enormous one. Consequently, a star spanning race as opposed to a planetary spanning one would probably be able to terraform a planet in much less time. Even so the period would probably still number at least a few hundred years.

Let's first examine the idea that only interplanetary space travel is involved.

Since in the ancient history of our planet we have no evidence whatever, not even by way of legend, that we had the type of set up mentioned in point (3. a) above, it seems fair to conclude that the base of operations would have to have been on Mars. This however proves problematic. Mars is about half the size of Earth. How would such a small planet manage to completely terraform one as large as Earth? And if it did so, why would it let giant monsters (dinosaurs) run around for millions of years? Nor is it plausible that Mars was destroyed before the advent of the dinosaurs, because then it would mean the pyramids on Mars would have to be over 300 million years old, something that surely is highly unlikely.

The evidence then, seems to be against the possibility that only interplanetary travel was involved. We must now allow for a people that with the capacity of interstellar travel, began terraforming worlds around stars other than the one/s in their own solar system.

The very concept of an interstellar race of beings adds a new level to the whole problem before us. While up to now we have been concerning ourselves merely with our tiny shred of the Cosmos, suddenly we are confronted with the possibility of a star-spanning origin for the human race. Where did the first humanoids really originate? From which star did we first come from? Just how big is the Empire of these humanoids? Do they inhabit many worlds or only few? Do they span across the Galaxy or are they limited to a few stars relatively close to each other? Do other worlds still harbour our distant cousins or are we the only ones left? If they are out there, why have we lost touch? Why do we not seem to have a mythology and history rich with the idea of our interstellar origins? Has this sort of mythology/history really never existed or have we merely lost it? If so why and how?

These and other countless questions which the idea of interstellar travel raises must go unanswered for now at least, since at present we are tied up with the question of terraforming, and given the vastness of the subject at hand it seems wise to proceed a step at a time, especially since as we have seen, sometimes taking one step opens a multitude of paths before us. We best choose our way carefully, lest we get lost.

Despite the additional questions it raises, the theory of an interstellar origin for the human race at least eliminates the immediate problems before us. A race with interstellar capacity would certainly have the required technology to terraform planets.[25]

---

[25] It might be relevant to note here that if one day in the distant future Earthlings were to terraform a distant planet for the purposes of developing an intelligent race of beings there and they had the ability to do so, it's very likely they would decide to give such a planet seasons similar to the ones we have on Earth. The importance of seasons for the development of farming cannot be stressed enough. Without seasons there would not be any farming although crops might grow all year round, and farming of course is of paramount importance for the development of an intelligent race of creatures.

Here then we seem to have resolved the first part of the puzzle, and yet another question looms before us. Surely a race this advanced would have been able to prevent the death of Mars? Given some advance warning, even we today, have the technology at hand to destroy an unwelcome asteroid that was headed for a collision course with Earth. Granted, the number of asteroids which collided with Mars could have been far greater in number than one or two, perhaps they numbered in the hundreds, but given the far superior level of technology the Ancients[26] must have had in view of their interstellar and terraforming capacities, surely they could have prevented the cataclysm? Especially since no trace of this meteor cloud remains today, meaning it could not have been all that large a cloud to begin with (presumably the survivors of the cataclysm, if any, did not bother to chase the offending meteor cloud and then methodically wiping it out *after* it killed Mars. What would be the point? Some insane sort of revenge !?!?)

Perhaps the asteroid cloud struck when most of the aliens were still only setting up base, so to speak, but then how come we also have humanoids on Earth? And why did the Ancients not send others from their home world to investigate the disappearance of their fellows? The answer to the asteroid catastrophe must lie elsewhere.


The clue, once again, is to be found in Phobos and Deimos. Their paths suggest they were *placed* there by intelligent beings, but to what end? Surely if some sort of satellites were needed then artificial ones would have been built?

It is my theory, that Phobos and Deimos are the remnants of a war so brutal and horrific, that our own world wars look like childhood brawls in a schoolyard by comparison.

If asked what the most terrible weapon of war is, most people would reply : "The atom bomb". Certainly, atomic energy, when unleashed in this manner, is devastating. What happened at Hiroshima and Nagasaki at the close of the second world war cannot be forgotten by those who were alive at the time and have survived to this day, nor (it

---

[26] From here on I shall refer to the original terraformers of Mars and/or Earth as "The Ancients". Their nature is at this stage unknown, they may be human or they may not, they may have created us or we may be their descendants. These questions cannot be answered now and perhaps will not be answered satisfactorily for aeons to come, but I need a way to differentiate between the Ancients, the Martians, the Earthlings and any other bunch of humanoids or creatures that turn up, so for this reason, Ancients it is.

is hoped) should later generations forget this tragic event. Even so, an atom bomb, or even several, going off on a planet are far from the worst thing that can happen to the people living on that globe.

If one really wanted to wipe out all the inhabitants of an entire world, atomic bombs would not be the best or even most cost efficient way to go about it. Nor would one need to build a Death Star, like in the famous Star Wars movie created by George Lucas, although some pretty fancy spaceships would certainly be required.

The best way (if one can use such a term in this context) to destroy an entire planet is quite simple really; simply take an asteroid of decent size (about 25 kilometres across will do fine) and change its direction so that it's on a collision course with the offending planet.

Nickel-Iron asteroids travelling at some 20 kilometres per second would be the weapon of choice, but if you're running low on the heavy metals just about any type will do.

If you are concerned that the defenders might blow it up before it makes contact, there are a number of ways to ensure success.

You could send a whole bunch of asteroids, but this would probably require a lot of ships. It would be simpler to just send a few really massive asteroids on a collision course. This way, even if a couple of nuclear missiles succeed in impacting on them, they'll just break up into smaller pieces. Smaller, but no less deadly. And in any event, your ships can be on standby to both neutralise surface based defences as well as shoot down any missiles travelling towards your deadly asteroids.

Also, if we use the analogy of four bullets from a high powered rifle travelling towards a watermelon, by attacking the bullets with tiny nuclear missiles, all that's changed is that instead of the water melon being ripped open by the four slugs, now it gets peppered with the equivalent of tiny buckshot pellets.

In the case of Mars, the watermelon was not completely ripped asunder, but it came close enough, and lost all it atmosphere, water and resident life forms to boot (as well as any magnetosphere it may have had previously). In other words, what happened to Mars was no accident. It was the result of a terrible conflict. And Phobos and Deimos are simply unused bullets. In fact, crater Stickney on Phobos, may not have been created by a meteorite at all, but by a missile that was designed to break it up into tinier pieces,[27] which would also explain the cracks on it.[28]

Fear and Dread then, are good names not only for Phobos and Deimos, but also for the emotions their relatives must have struck into the heart of any living creature on Mars long ago, as they plummeted towards the surface. We can only hope, untold years later, that their suffering was brief, as indeed it must have been.

After the planet was pretty much wiped clean of resistance, perhaps a couple more bullets were shot at it just for good measure. These last asteroids of course would probably not have been attacked at all by missiles as they went in, because by this point, most or even all of the defences of the planet would have been destroyed already.

I believe at least two asteroids arrived on the surface of Mars practically unscathed by any planetary defences. They formed Hellas basin and Argyre Planitia.

If this theory is correct, these last two asteroids would also have been amongst the last to land and also the ones that caused the most damage in terms of topography. Geologically speaking of course, we have already seen that both Argyre Planitia and the Hellas basin, as well as the shield volcanoes (whose creation I attribute directly to these two larger asteroid impacts) are amongst the freshest features on Mars.

Finally, this would also explain why the southern hemisphere of the planet received so much more attention from asteroid impacts.[29]

The obvious place to start bombarding a planet with asteroids must be the areas around which the planetary defences lie. These in turn would probably be near or even in the same location to the population centres with the highest densities. As we have already seen, the

---

[27] In this context then, it's very interesting to note that crater Stickney faces the Martian surface, from where the defender's missiles would presumably have originated. Similarly, although not impossible, it becomes hard to see how a meteorite impact which supposedly created Stickney and sent Phobos closer to Mars could possibly be facing the planet rather than away from it, which is the direction from which the impact would have occurred.

[28] I will explain why the cracks have the same number of micrometeorite impacts as the rest of the soil on Phobos in the next chapter.

[29] As already mentioned, from ten to a hundred times more impacts are found on the southern hemisphere of Mars, a number of magnitude that is difficult to explain considering the supposedly random nature of meteoroid impacts.

ancient rivers of Mars are to be found predominantly in the southern hemisphere. Population centres of course would have been near water originally, and even though over time the technology probably increased, the biggest population concentrations would still lie in the original locations, in other words, near rivers and freshwater. The same is true on our own planet of course.

At this point then, we seem to have answered most of the questions that we began with. Certainly we also have opened up a vast number of new ones, but before we get to those, let's first make sure that we have been competent and fair in our assumptions.

My proposal certainly seems to be almost too fantastic for words. In fact it seems so dramatic one is at first more likely to associate it with the plot of some special effects filled science fiction movie than with any reality we know of. And yet, the only barrier to its being a real possibility appears to be the one we find in our own minds.

In fact, as I hope to have shown up to now, despite its seemingly unlikely nature, it is the only set of circumstances which satisfactorily explains all of the mysteries we have encountered up to now.

I cannot stress enough, that what at first seems like the work of a deranged mind, occasionally is much later accepted as being based in truth.

Galileo Galilei was placed under house arrest for the last years of his life and risked burning at the stake for stating the Earth moved around the Sun, because less than 300 years ago we still burned people at the stake for such 'heresy'.

Leonardo Da Vinci conceptualised the helicopter long before it was built in reality.

Einstein's theory of relativity, which he wrote when he was 26 years old, was disputed for years and only today are we approaching a more complete understanding of its subtler yet far-reaching implications.

One could go on *ad nauseam* with such examples, but ultimately it is for each individual reader to decide whether my statements are based in truth or are the work of an overactive mind. All I can add is that I once again remind the sceptics to do their own research on the matter. For myself, I am happy enough in the belief that bar tragic accidents and if the human race does not wipe itself out, I should live long enough to see whether I proved to be correct or not.

Before we move on, there remains the problem of explaining why the cracks found on Phobos should have a similar crater density to the uncracked parts of the satellite. The reason is simple if one is willing to stretch the boundaries of imagination still further. Phobos was created only shortly before being nuked by the missile which caused Stickney and the corresponding cracks. What do I mean 'created'? I mean just that. Phobos did not exist until a short time before the impact that formed the cracks. To find out where exactly I think Phobos (and Deimos) originate from, we have to turn to the next chapter.

# 5     Titius-Bode, Chaos Theory & Phaethon

## Prelude to the Chapter

In this and the next two chapters, I allow myself a wider degree of speculation; readers of a sceptical nature are thus forewarned.

In any case, the ideas presented here, even if they should prove to be wrong in their entirety, do not affect the overall validity of the main concepts presented in the last four chapters.

◆ ◆ ◆ ◆ ◆

The attempt to predict the distances between planets was first made by Johann Titius in 1772. This theory however was mainly publicised by Johann Bode, with whose name it was primarily associated for a considerable time. Later, in an attempt to restore rightful credit, it became generally referred to as the Titius-Bode law.

In devising this theory, Johann Titius was trying to find a mathematical method for determining where planets would be found in relation to their primary star. If successful, such a method would have been (and still would be) very useful. In 1772, not all of the presently known members of our solar system had been found.

If however one could use a mathematical theorem that stated a planetary body should be found at a given distance from the central star, this would limit the field of search considerably and consequently aid in the discovery of new planets.

When Uranus was discovered in 1781, the Titius-Bode law seemed to have been confirmed, as its distance from the Sun was close to the one predicted.

This discovery was instrumental in encouraging the scientific community to search for the apparently 'missing planet' at 2.8 AU[1] from the Sun (see table below). When in 1801 Giuseppe Piazzi discovered Ceres and later other small bodies were found in what today is known as the asteroid belt, which lies at approximately 2.8 AU from the Sun, the Titius-Bode theory seemed to be on even firmer footing.

Despite this apparent success however, Neptune's distance from the Sun does not agree well with the one predicted by the Titius-Bode law, and the distance of Pluto is far closer than the predicted value. In addition, it is felt by some astronomers that Johann Titius arrived at his figures by what is occasionally described as "arbitrary mathematical manipulation".[2]

| Table 4: Actual orbital distances from the Sun compared with those predicted by the Titius-Bode rule | | |
|---|---|---|
| **PLANET** | **Actual distance (AU)** | **Predicted distance (AU)** |
| Mercury | 0.39 | 0.4 |
| Venus | 0.72 | 0.7 |
| Earth | 1.00 | 1.0 |
| Mars | 1.52 | 1.6 |
| ——— | ——— | 2.8 |
| Jupiter | 5.20 | 5.2 |
| Saturn | 9.54 | 10.0 |
| Uranus | 19.18 | 19.6 |
| Neptune | 30.06 | 38.8 |
| Pluto | 39.4 | 77.2 |

[1] See Conventions Used at the beginning of the book for a definition of AU.

[2] The Titius-Bode 'law' is given under the section on Conventions Used at the beginning of the book.

For these reasons, today, the Titius-Bode law is generally viewed as not being accurate enough to deserve the name law and is known as the Titius-Bode 'law', the use of inverted commas signifying it's to be taken none too seriously. It is mainly ignored by astronomers as an outdated, and inaccurate system for the predicting of planetary distances from its primary star. Normally, if the Titius-Bode 'law' is given any credit at all it is only to say that it proved useful in its time by encouraging the search for other objects.

Considering that the Titius-Bode 'law' has been pretty much rejected, it is surprising to find that as of yet, no satisfactory replacement theory has come about. In fact, it is for this very reason, that it is still with us. Had a better theory been devised, surely the double-barrelled name Titius-Bode would only be a footnote in what could be called the history of astronomy.

Personally however, I feel that the Titius-Bode 'law' (which I shall call the Titius-Bode rule from now on)[3] deserves a lot more credit than it has received. Before I explain why I feel this way, allow me to introduce the mathematical concept commonly referred to as.....

# Chaos Theory

If memory serves me correctly, chaos theory originated as a result of weather prediction.

In our age, computers are fed with all the available meteorological data on a daily basis, to allow them to 'see' how the weather evolves from day to day. The way in which the weather is predicted, is by allowing the computer to 'run ahead' of the data. In other words, the computer is supplied with the weather data for several days and then is asked: "Based on what you have seen so far, what's your best guess for the next few days on how the wind will move, where these low pressure areas will shift to, what the temperature and humidity in this particular spot will be..." and so on.

The question obviously is not asked literally but in terms the computer will understand. In effect, what the computer does is to estimate the changes that will take place in each of hundreds of thousands of

---

[3] If astronomers feel the word *law* is too strong for a theory that is only partly correct, then perhaps instead of placing it in inverted commas they could simply refer to the theory as the Titius-Bode rule, since the word *rule* generally has a looser meaning attached to it than *law* in scientific circles.

individual bits of data. A human could do it too, perhaps better than a computer, but not at the same speed. While a person would still be busy guessing how the temperature in one place will change for the next day, a computer can have estimated how the temperatures in all the cities of the world will have changed in the next few months without even breaking out in the electronic version of a sweat.

A computer theoretically can predict the weather indefinitely.

You could ask it what the weather in Greenland will be like on October 7th of the year 3456 AD and you would get an answer. It would be a wrong answer (unless an incredible luck factor played a part!) but you would get one.

The reality of the matter is that with each guess the computer makes, it introduces a tiny error, and since its next guess is based on that slightly wrong first guess, its second guess is going to be out by a similar amount, *plus* the original error introduced in the first guess. Very shortly then, your theoretical model for the weather pattern is going to disintegrate. Despite this, making the machine go through the exercise is not a waste of time, because the computer guesses are good enough that they can predict the weather reasonably accurately for the next few days. I don't remember what the actual figure is, but let's assume that the computer can predict the weather for a month with reasonable accuracy. In this case then, day 1 of the prediction would be most likely to be correct, with perhaps 97% or higher chance of the weather turning out like the computer said. Day 30 of our prediction would be least likely to have weather that is close to the one reported by the computer. For the purposes of our example, let's say that day 30 has a 45% chance of being correct.

Now comes the interesting bit. Some clever guy whose name I forget,[4] noticed that the figures supplied to the computer where accurate to something like twelve or more places after the decimal for data that required only to be measured to the nearest unit.

Being somewhat of a computer expert he realised that by reducing the figures to only say eight places after the comma he would save the computer an appreciable amount of time, and surely, he thought, the results would not be adversely affected. So he proceeded to reduce the

---

[4] My sincere apologies to all concerned. I did try to look it up, but chaos theory is a pretty difficult subject to get any information on in this part of the world. The original information came to me via a documentary which unfortunately I could not tape since I had no VCR at the time.

number of values given after the decimal place from twelve to eight.[5] Over the next few days, it became evident that from being able to predict the weather for a month, the computer was now barely able to predict it for a day or two, and certainly, long before a week was up, the weather model of the computer had little to do with actual meteorological conditions of the real world.

The figures were placed back to show twelve numerals after the decimal point, and once again the computer could do a decent job of predicting the weather for a month.

They were dropped to eight places after the comma once more and again, the computer could scarcely predict the weather for the next couple of days. Obviously, those four digits right at the end made a big difference despite the fact that they were insignificantly small by human standards.

The effect of those numbers was considered so small that chaos theory became associated with the phrase: "A butterfly beating its wings in Tokyo can create a hurricane in New York".

This discovery is far more important than is at first apparent, because it implies that the whole of existence as we know it follows a definite pattern that is discernible by mathematical means. The universe around us may look chaotic and unpredictable to us, but in fact it seems to follow very precise and definite laws. The problem is that we are only just becoming able to measure those very fine tolerances which the universe follows. In a sense, we see the universe as chaotic, not because it is, but because we are too "imperfect" to see the very, very, *very*, strict order that every atom of matter follows.

The universe in fact is perfect; and being perfect it cannot help but follow order in the strictest of senses.

The question that springs to mind of course is:

"If the universe is perfect, and we are part of the universe, then why are we not perfect?"

The answer could well be that indeed we *are* perfect, and so is *everyone* around us. We just can't see it yet (or maybe ever) because the scale we operate on is too small.

Let me make an analogy. Pretend that each atom of matter was conscious, and had thoughts just like you or I. Do you think that an

---

[5] These figures may not be exact, as I'm telling the story from memory, but the principle remains unchanged.

atom of oxygen floating around in the atmosphere would know that it is moving according to very precise formula, as was every other atom around him? Of course not. Even we, who are supposedly far more conscious than an atom, cannot predict the movement of a single atom. At best we can predict the movement of several zillion atoms taken as a whole for a short time, which is all that weather prediction is. So, like the atom, we may well be blissfully ignorant of the perfect way in which we move. Even the apparently senseless violence and wars may be part of a plan so large that we cannot begin to comprehend its ultimate significance.[6]

If this answer is correct, then the question of an immense power or intelligence lying quietly behind the existence of the entire universe is no longer a question, but a fact. And furthermore, since it would seem that this power chooses to express perfect order, it would also be safe to describe it as benevolent. Each person of course can interpret this as they see fit, but now you know why a lot of physicists are starting to find it difficult to deny the existence of what is normally referred to as 'God'.

Chaos theory is also closely linked to the fractal shapes which we find repeated again and again in nature. As I have pointed out before, a fractal shape is one which looks chaotic and random, but which follows a very definite mathematical formula. Leaves of trees are now known to be fractal in shape, as are snowflakes and a host of other seemingly random shapes we find in nature.

Can you begin to see why maths is truly a universal language, and thus the importance which the Great Pyramid has, since its message is one left not in hieroglyphs or runes, but in the form of a massive mathematical riddle?

---

[6] Remember that even in the world of the atom 'senseless violence' is rife. An electrical discharge in the form of lightning ionises billions upon billions of atoms, literally tearing molecules apart.

# Titius-Bode and Chaos

If you did not know about chaos theory, the last couple of pages may have been entertaining, but what does all this have to do with Titius-Bode and in turn what does Titius-Bode have to do with the Face on Mars or anything else?

If you have begun to wonder that, then you may be starting to appreciate what an incredibly vast subject it is that we have set for ourselves to tackle here. Consider that by this point you have had to learn a little about:

1) The general layout of the solar system.
2) The general concept of interstellar travel using Einstein's theory of relativity.
3) Nuclear rocket engines.
4) Computer enhancement and analysis of photographs.
5) The dimensions and implications of the Great Pyramid.
6) The terrain, features and history of Mars in some detail.
7) The Titius-Bode rule.
8) The general principles of chaos theory and fractal shapes.

Nor are we even at the halfway mark, and I'm afraid it gets worse before it gets better. Nevertheless, while the situation remains manageable we should persevere. If nothing else we're at least broadening our general knowledge.

Let us consider the relationship between chaos theory and the Titius-Bode rule then. It should be obvious by now, that what chaos theory is saying is that the further you move away from the source, the less likely you are to hit the right spot. This is true of target shooting as well as anything else. We can predict the bounce of a billiard ball when playing pool quite accurately for a bounce or two, a professional player might be able to do it for six or seven bounces, but no one can predict where the ball would be one or two hundred bounces down the line, that's because tiny imperfections in the surface of the pool table, the ball and even the cue stick, while almost insignificant on their own, taken together, enlarge the error not by addition, but by multiplication, ensuring that in a short while of travel along the pool table's surface, the direction of the ball is anyone's guess.

Similarly, it is my opinion, that the Titius-Bode rule is relatively accurate for planets which are closer to the Sun, but becomes inaccurate as we move further and further away. You will note from Table 2, that the Titius-Bode rule is really quite useful at least up to Jupiter. In fact even Saturn and Uranus are pretty close to the predicted spot. Admittedly, the value for Neptune is so far out to be practically useless, and the one for Pluto is even more so. But in any event, a theory which can reasonably predict at what distance from the Sun Jupiter would lie must have some merit. Remember the scale model of the Solar System we constructed back in chapter one?

Using the Titius-Bode rule on our model is equivalent to predicting roughly were a ping-pong sized object would lie from a beach ball that is 183 metres away. Not too shabby.

I think to also expect it to come up with a right answer for an object the size of a grain of sand that lies 1.4 kilometres away from the same beach ball is a bit much. Also, the fact that the Titius-Bode rule predicted a planet should exist at 2.8 AU as well as at 19.6 AU *before any planets were known to exist at those distances,* proves that the Titius-Bode rule, however inaccurate at long distances, is a real rule. If Johann Titius had merely looked at where the planets were located and then made up some mathematical equation that predicted these distances, he would surely not have made an equation that pointed at 2.8 AU, since no known planet existed at this location.

My purpose in going to some length to vindicate the Titius-Bode rule however goes beyond trying to give credit where it's long overdue. Certainly if the Titius-Bode rule is re-examined in detail by scientists, as is my hope, and found to be useful, then it may be fitting to remember both the Johanns a little better in our history books, but the main reason for the presentation of the above arguments is, as always, a practical one.

If the Titius-Bode rule is at least partly correct, then it can be instrumental in our search for new planets around our nearby stars. The Hubble Space Telescope would be instrumental in discovering planets around one of our neighbours, but the Titius-Bode rule may come in handy when we begin to search for smaller planets closer to the parent star or when we look at stars further away and the telescope's magnifying power is at its limit for the purposes of planet discovery.

After all, even if only half of my theories concerning contact with extra-terrestrials in human history are correct, then finding out where the nearest hospitable planet outside of our Solar System is may prove to be quite useful, and any refinement of the Titius-Bode rule will of course further improve our ability to do this.

# The Missing Planet

It is of course a fact that at 2.8 AU from our sun there are no planets, but what we do have, in a belt that stretches approximately from 2.2 to 3.3 AU is a group of small bodies which are generally referred to as the minor planets or asteroids. The entire group as a whole is known as the asteroid belt, and it lies between the orbits of Mars and Jupiter. It is interesting to note that the average of 2.2 and 3.3 is of course 2.8 AU, as predicted by the Titius-Bode rule, and indeed this is normally taken as the value which represents the 'middle' of the belt.

If you recall, what first prompted us to look at the asteroid belt were the Martian satellites Phobos and Deimos, which are thought to have originated from this location.

And if you recall, we still have the problem of the cracks associated with crater Stickney on Phobos.

Since these cracks appear to have the same kind of micrometeorite impact density as the rest of Phobos it would seem that they must have been created at the same time, or shortly after Phobos itself was created. This very notion of course would imply that Phobos was originally part of a larger whole. Now this, is an idea which I found particularly intriguing, and I began to form a picture in my mind of where Phobos may have originally come from. I wondered if perhaps where we now have an asteroid belt we once had a whole planet. At first I thought this could surely not be the case, and the idea must be one I had left over from badly explained astronomy lessons given to me long ago by some relative or other. I also supposed that the idea of a planet having once existed between Mars and Jupiter was one more likely to have its basis in science fiction than in fact.

I must explain here that this thought was based exclusively on intellectual grounds. Intuitively, I had no problem with the asteroid belt once having been a planet. It was my mind that rebelled at the thought, not my gut. That should already have told me something, for it has been my experience in life that when the intellect and the intuition are in conflict it is always the intuition that holds the core of truth, although the intellect may certainly play a part later to define and refine that truth. Imagine the surprise my intellect received then, when researching the matter a little further, I discovered that the theory of a planet existing between Mars and Jupiter is one which conventional science has shared with my intuition for a long time.

It is a theory by the way, which held more favour in the past than it does today. The main argument against it seems to be:

If there was a planet between Mars and Jupiter, what did it crash into in order to be reduced to so much rubble?

This is a good question, especially if one has no knowledge of the things I have talked about in previous chapters, and without any supporting evidence, the idea that some crazy asteroid, comet or other 'lost body' came along and obliterated both itself and this theoretical planet, does indeed seem questionable. Remember the arguments we made against a natural asteroid strike for Mars. Same thing applies here, if to a lesser extent. The big question remains: With all the empty space available, why should the offending object/s find it necessary to be in the precise spot that would result in a drastic collision with the theoretical planet? Isn't it much more likely that there never was a planet here and instead the asteroids just formed individually in much the same way that planets do? A boring theory perhaps, but surely one that is more likely to be correct, no? As it happens, **NO** is precisely what the answer to that last question is.

For a start, many asteroids are composed of nickel-iron. The South African museum in Cape Town has at least three good examples of this type of asteroid. These are scarred and pitted pieces of meteorite weighing 550, 650 and 1172 Kg respectively. Each of them has had a small piece cut off, and it is at this polished surface that one can see the metallic nature of the meteorites. The polished surface is so strong that it cannot be scratched with a fingernail. In order to leave a small mark of negligible depth one would require perhaps a penknife. Nor is there any stony intrusion of any kind that can be readily discerned in the objects.

If you were to take just over a tonne of iron and nickel mixed together and you applied a blowtorch to it liberally for some time, then buried the whole messy lump in a field for some years, letting it rust enough to give it that characteristic reddish-brown colour, you would have a pretty good approximation of what the largest of these meteorites looks like.

It's one thing to say that in the creation of a planet metals and other heavy elements congregate and become part of the planet, sometimes in very large amounts; but it's quite another to say that enough of these elements come together to form a small body almost exclusively composed of them. I am not a qualified astronomer, having learned what I know about this and related subjects primarily by being self-educated in them trough much reading, but it seems to me that any body which might be naturally formed in space would be composed of a multitude of elements.

Depending on the available matter, one body may have more or less of a particular element than another, and two different bodies may thus differ considerably in composition, but surely not to the extent that a nickel-iron asteroid differs from a C-type asteroid[7] (carbonaceous chondrite). Especially when both types of asteroid lie in the same area of space, and thus would have had to have been created out of the same gaseous elements present in the area at about the same time.

If this had indeed been the case, then surely the asteroids would be more uniform in composition.

Also, why should the asteroids in the belt have been formed as individual pebbles while the other planets formed as one main body with occasionally smaller satellites? Once again, the fact that the asteroids share a more or less common orbital path, would seem to suggest that when the elements were still in the form of gas they would have gravitated towards one common point to form a single planet, in much the same way as the other planets were formed.

It is feasible of course that perhaps a number of planetoids were originally formed instead of one, but the asteroid belt is composed of well over 3000 objects, and as far back as 1979, about 2125 of these were known to have regular and permanent orbits around the Sun. What could possibly prompt this area of space to produce over 3000 individual objects instead of a single planet or at most a few objects? Scientists who ascribe to the theory that the asteroids were originally formed in a manner similar to the way we find them today, attribute this unusual occurrence to Jupiter's gravitational effects on the still forming planet. This theory has some merit to it no doubt, and for this reason a single planet may never have existed in the area. Instead a few objects may have inhabited this region of space, and indeed such a theory has been proposed by astronomers. Over the millennia the few objects may have collided and thus formed the present day

---

[7] It is also interesting to note that the asteroids which give spectral measurements consistent with carbonaceous chondrites exist primarily on the outer edge of the asteroid belt, while those that are thought to be heavier (silicate or stony-iron) are generally found on the inside edge, although any significance which this fact may have is not readily apparent.

multitude of asteroids. This may indeed have been the case, but I still find it suspect that a relatively low number of planetoids would come into contact with such drastic consequences, especially since Jupiter's presence, a further 2.5 AU away at its closest, would surely not have been all that drastic. In any event, surely the theory that the asteroids were formed individually cannot be wholly correct in view of these facts.

The estimated total mass of the asteroids amounts to some $3.1 \times 10^{21}$ Kg ($6.614 \times 10^{21}$ lb.)[8] The Earth, by comparison is estimated to be almost 2000 times more massive than this, weighing some $5.97 \times 10^{24}$ Kg.[9] Since the average density of the asteroids is said to be about 3.5 g/cm$^3$ the diameter of a planet which comprised all of the known asteroids would amount to a distance of about 1200 kilometres.[10] Considering that our own Moon has a diameter of some 3476 kilometres, this theoretical planet could not be said to be very large.

Let's however be a little more fair in determining the original size of this theoretical planet, which for simplicity I shall refer to as Phaethon[11] from now on.

---

[8] The value is taken from pg. 84 of the book *Asteroids* (see bibliography) which was written in 1979. However, the total mass given in 1962 on pg. 152 of *Astrophysical Quantities* (again, see bibliography) was $1.7 \times 10^{21}$ Kg. This means that in a period of about 17 years, the estimated total mass of the asteroids has nearly doubled. Admittedly, in this period considerable advances in planetary astronomy were made, but then, advances have also been made in the last 16 years, covering the period from 1979 to 1995. I was however unable to find more recent data that covered this topic in some detail. In any event, the point does not make as much difference as one might think, since the relationship of mass to diameter is not a direct one. Even if we triple the total mass of the asteroids, the diameter of the proposed original planet they came from would not similarly triple, but would increase by only some 30%.

[9] This number by the way is said by some to be a direct multiple of the weight of the Great Pyramid, which in this case would have to weigh $5.97 \times 10^9$ kg, which it may well have done when it still had casing stones.

[10] Those inclined to do so can check for themselves using the formula for the volume of a sphere, ($V = \frac{4}{3} \prod r^3$) and $D = M \div V$, where D=density, M=mass and V=volume.

[11] This name for the 'missing planet', of which the asteroids are supposedly a remnant, was chosen (rather aptly) by Professor S.V. Orlov of the U.S.S.R. Academy of Sciences in 1950. The names of planets in our solar system derive in most cases from the ancient names of the gods of Greek mythology. See appendix D to see how the legend of Phaethon may be an analogy for some of the events which resulted in its destruction.

The first thing to be taken into account, is that if Phaethon ever existed, and it was indeed smashed by a collision, then the pieces of it that remain are bound to be those with a high relative density, since these would be the most difficult to dissipate in the form of gas or tiny pebbles. In other words, Phaethon's original density could be expected to be considerably lower than that exhibited by the surviving asteroids. The density of Phobos is about $2g/cm^3$, which is probably too low, but may be closer to the original value.

In addition, a great deal of the original material would have been lost in a number of ways. Most of the water-ice and other volatile substances could well have been reduced to their gaseous forms and thus have dissipated into the vacuum of space over the millennia, while a considerable amount of the denser material could have been placed in such eccentric orbits as a result of the explosive collision that it either perished by falling into the Sun and other planets or is still in evidence today as cometary bodies. It is difficult to place a value on these losses, but let us be generous, and assume that 70% of the original mass was lost and no longer resides in the asteroid belt.

Finally, there is still one way in which the total mass of Phaethon could be underestimated. Much of the planet could have been reduced to bodies too small to be seen by our telescopes, especially since most of the asteroids are very dark. Remember that an asteroid of 300 Km in diameter will still be merely a point source for even our best telescopes. Small asteroids and fine dust could conceivably amount to another generous 50% of the remaining matter.

Let's now add all of these factor together and see what new value we get for the diameter of Phaethon. First of all we have to double the existing mass to account for matter too small to be seen. This brings the total existing mass of Phaethon to $6.2 \times 10^{21}$ Kg. According to our generous allowances though, this amounts to only 30% of the original mass of Phaethon. Multiplying by 3.33 periodical will give us Phaethon's theoretical original mass. This comes to 20.66 (periodical) $\times 10^{21}$ Kg. Now when we work out the original radius, and hence the diameter, we must remember that we have given Phaethon a new density of about 2 $g/cm^3$.

The final answer is that using these figures, Phaethon would have had a diameter of about 2700 Kilometres, still considerably smaller than our own Moon.

In view of these calculations, it must be said that even if Phaethon existed, it probably was not a very large body at all. Its low gravity and distance from the Sun would in all probability have made it a hostile place, with little or no atmosphere and any water present probably being in the form of ice.

For these reasons then, one may think it unlikely that Phaethon was ever colonised by the people that had populated Mars and to some extent Earth. I would tend to agree, and I feel that if Phaethon did exist, at best it had only some sort of remote outpost, similar to one we may have on the Moon today if we decided to build a permanent lunar base. It is also possible though, that Phaethon was never a single planet, but a collection of small planetoids, perhaps Ceres, Pallas and Vesta, the three largest known bodies of the asteroid belt are the remnants of three of the planetoids that originally existed here, although I feel they probably have lost a considerable amount of mass through several collisions. In such a case, I would estimate that there may have been up to four or five bodies, instead of just one, but I would guess no more than this.[12]

With the exception of the three already mentioned, these where destroyed to smaller chunks of rubble, and it is from here that we get our metallic asteroids.

Metal is thought to generally be found in its most concentrated form at the core of planets, so if we smashed a planet up, the globules of hot metal that form its core would eventually cool and become the floating boulders of nickel-iron that we find in the asteroid belt today. It is furthermore my opinion, that metals will only form in largish quantities in the cores of bodies that are large enough to provide the necessary accumulation of matter. It is for this reason then, that I assume the original number of planetoids in the asteroid belt probably numbered no more than five, in which case, the smaller ones would also probably have had a very low, or even negligible amount of metal.

I would think that the most likely outcome however, is that Phaethon was not a single entity, but rather a combination of larger parent body (a small sized planet), and a few accompanying satellites or moons. Such moons of Phaethon as may have existed would of course be

---

[12] It is always possible though that Phaethon was originally just a single planet with perhaps a small satellite or two.

small and composed of lighter substances, and Phobos and Deimos may then —perhaps more accurately— be assumed to be parts of one of these moons rather than of Phaethon itself.

In view of its size and shape though, it may also be that Ceres was not ever a part of Phaethon, in which case the theory of Phaethon being composed of several bodies of similar size would be more correct. This view however is in my opinion inferior to the one of Phaethon having been composed of a single planet with accompanying moons because I would think that Ceres has a more regular shape due to the fact that after having been broken up, the molten liquid core of Phaethon would naturally settle into a spherical shape as it rapidly cooled.

And in any event, Pallas and Vesta, being considerably smaller than Ceres —and Pallas in particular being irregular in shape— could conceivably have been part of the same original core body.

Table three below gives the principal characteristics of these three bodies, although it should be kept in mind that these values are not necessarily precise. A certain amount of error is expected, since these figures were arrived at by observing relatively small objects which are one and a half times further away from us, at the best of times, than the Sun.

| Table 5: Characteristics of three largest asteroids | | | |
|---|---|---|---|
| FEATURE | Ceres | Pallas | Vesta |
| Polar diameter (Km) | $909 \pm 3.2$ | $580 \pm 20$ | 573 |
| Equatorial diameter (Km) | $975 \pm 3.6$ | $555 \pm 18$ | $525 \pm 25$ |
| Mass (Kg) | $9.5 \times 10^{20}$ | $2.25 \times 10^{20}$ | $2.5 \times 10^{20}$ |
| Density (g/cm$^{-3}$) | $2.07 \pm 0.04$ | $2.7 \pm 0.9$ | $3.46 \pm 0.5$ |

If I am right about Phobos originally being a piece of the Phaethon object/s, then it may well be that Phaethon was destroyed in a similar manner (if more drastic) to the one used to destroy Mars.[13]

---

[13] In which case, the fact that the denser bodies lie towards the inside edge of the belt may be a hint that Phaethon was originally a single entity. When 'shot' with a fast moving asteroid that may well have travelled right through it, the core material

Even if this idea is rejected completely though, it still does not invalidate the conclusions we have arrived at in the previous chapter. In fact, it remains true that the cracks on Phobos are evidence that Stickney was formed shortly after the formation of Phobos itself. Basically, all that this means, is that Phobos was formed by the destruction of a larger body (hence the theory that Phaethon may truly have existed)[14] and crater Stickney is the result of some sort of collision which took place shortly after the formation of Phobos.

If we accept this basic precept, then the case is one of the following:

1) In the event that the original parent object was destroyed by natural means, then Stickney would similarly have to have been formed by an early impact with another body shortly after formation.

2) In the event that the parent body was destroyed by artificial means, then Stickney could have been formed either as a result of natural impact with another body shortly after having been formed, or as a result of further artificial means (e.g. the proposed impact with a missile from Mars designed to destroy it.)[15] In either case the impact which created Stickney would have still occurred relatively soon after the formation of Phobos.

It must be said here, that if Phobos was artificially created, I would find it highly unlikely that Stickney was formed by a natural collision, because since Phobos would have to be transported to Mars orbit, it's probable that it was relatively clear of any debris which may have

---

(especially if molten) would naturally travel in the direction of the exit wound, leaving most of the crust material 'behind', and thus nominally on the outer edge of the belt.

[14] Even if the theory of Phaethon is rejected, I still believe that Phobos is a piece of an originally larger parent body, all it would mean is that the asteroid belt was once composed of far less objects than it is today. A theory which is held to be true by science regardless of whether the idea of Phaethon as a single entity is accepted or not.

[15] A wild idea for which there is absolutely no evidence whatsoever, but that nevertheless captures the imagination, is that crater Stickney on Phobos may have been excavated on purpose by the aggressors in order to perhaps mount an engine that would drive it to its final destination. When Mars was finally destroyed this was no longer necessary, the engine was thus removed and Phobos left with Stickney. I must stress that this really is the science fiction writer in me speaking, (especially since as will become evident in part II, I believe there would have been no need for such an engine, see also footnote 16 below), still.....

endangered the ship doing the transportation. Then again, a ship capable of moving a Phobos sized object through some 180 million kilometres of space may well be impervious to harm from meteoric impact. It may well protect itself from asteroid collisions by virtue of the same technology which allows it to transport Phobos in the first place,[16] so even if Phobos had been artificially created, Stickney could still have been a natural occurrence after all.

If we assume that the Phaethon object/s was/were destroyed by artificial means by the same aggressors that killed what must have been a relatively successful Martian population though, then we must ask ourselves why. There are basically only two main reasons:

1) Phaethon in some way proved to be a threat or at least was inhabited to some extent by the Martians.

2) To have a readily available supply of ammunition with which to bombard Mars and/or Earth. While this may appear to be a relatively minor reason, it may have been quite important, especially if objects of the desired size where few in number before Phaethon was destroyed. If only one or two asteroids or tiny moons existed for example, it would have been wise to use them to break up Phaethon first, in order that more would have been created in the blast, and a plentiful supply of deadly asteroids could be ensured.

If point 2 was the main reason, then Phaethon was destroyed merely as a matter of strategic necessity, but if point number one was the reason, then we open up the possibility that Phaethon (whether as a single body or as a small group) may have had some sort of outpost on it.

---

[16] In this context, see the discussion on anti-gravity machines in Part II.

# Summary

What originally brought me to investigate the asteroid belt is the origin of Phobos and Deimos, since they appeared to have come from this area. Furthermore, the intriguing striations of Phobos seemed to indicate that crater Stickney was formed shortly after the event which created Phobos itself. In fact, these striations, in my opinion are solid evidence that Phobos was once part of a larger body, otherwise, they would not show the same amount of crater impacts as the rest of the moon.

This knowledge seemed to point to the fact that the asteroid belt may originally have been composed of far fewer objects, and that the thousands we observe today are the result of two or more of these bodies having collided.

As we have seen in chapter four, there is some evidence that seems to substantiate the idea that a catastrophic war took place, which destroyed Mars. For this reason, any process which caused any planetoid to break up as a result of a collision should come under closer scrutiny.

Now, it is entirely possible that Phobos, along with Stickney were created by natural collisions. In this case, the creatures that attacked Mars with asteroids were simply lucky and had at hand a ready supply of ammunition for meteoric bombardment of Mars, but certainly, we should at least allow the possibility that Phobos was created by the aggressors for this very purpose, and in this case, the question then arises as to just what the composition of the asteroid belt was before such aggressors came along.


I have already pointed out that it is my belief that originally, where we now have an asteroid belt composed of thousands of boulders, there may have been as few as one or two bodies and probably no more than five. It is also the view of astronomers today, that while the single planet theory has been almost completely discarded, most would be agreeable to the idea that originally an inferior number of planetoids existed.

Since I have already proposed that the events which reduced Mars to a dead planet were artificial in origin, it should come as no great surprise, that the destruction of the Phaethon object/s may also have occurred as the result of intelligent manipulation instead of random chance. The motive after all exists.

In view of all this then, it is more than likely, that the original objects found at approximately 2.8 AU from the Sun were a lot fewer in number in the distant past, than has previously been suspected.

In this case then, the theory of only one or two bodies having existed between Mars and Jupiter becomes more likely, and this in turn further strengthens the position of the Titius-Bode rule.

# 6  Advancing the Theory of War & Interstellar Origin

Up to now, we have been primarily concerned with discovering some truth about facts that lie in the distant past, but we have made no real attempt at contemplating what the implications of those facts are. We have seen how there is some very real evidence that supports the idea that the human race may well have very different origins from the ones we have supposed until now. In fact, if the Face on Mars and nearby Pyramids are shown conclusively to be artificial constructs, then this is no longer merely a theory, but an indisputable fact.

As it turns out though, NASA and others may have vested interests (as well as some good reasons) for seeing to it that the Face on Mars is never proven to be anything other than a 'trick of the light'.[1]
I will touch on some of the reasons for this in part two, but for the moment I would like to stick with the matter at hand, which is primarily to try and begin to make some sense of all the facts we have accumulated up to this point.

For a start, it is about time that we attempt, even if only in a very rough way, to give some sort of timescale to the events we have so far only described.
We might want to start with what we have here on Earth, namely the Great Pyramid at Giza, the Sphinx, and any other relevant information that may give us a clue as to how long ago some of these events took place.
Also, regardless of whether we find any conclusive evidence in this regard, it would seem wise to think that whatever the age of the

---

[1] Never may be a strong word, perhaps they're planning to keep it a secret only for five or six more decades, but for you or I, fifty or sixty years may just as well be an infinity since it's unlikely we'll still be around then.

Atlanteans,[2] the Martians were older still. This is especially true if we assume that the Atlanteans were the survivors of the war which destroyed Mars. On this basis then, we can begin to give an approximate date to certain events.

The Great Pyramid is held by most archaeologists to be about 5,000 years old or so, but as I have already mentioned in chapter three, there is some evidence to suggest that it may be up to twice that age. If we accept Plato's account of Atlantis as being at least partially correct, we find that the war that supposedly took place between the Atlanteans and those who dwelt within the pillars of Herakles (the strait of Gibraltar), is said to have occurred 9,000 years before the time of Plato. This would mean that the war between the Atlanteans and the Europeans of the time took place about 11,400 years BC, give or take a couple of hundred years either way. If we assume that the Great Pyramid was built by the surviving Atlanteans sometime after this war, then this coincides rather well with a dating for the Great Pyramid of at least 10,000 years.

In any event, it seems highly unlikely that the Face on Mars could possibly have an age of less than 11,000 years, because it would mean that as recently as 11,000 years ago, Mars had an atmosphere and running water.

---

[2] For purposes of clarification, the Martians, Atlanteans, Hominids, and Unknown Aggressors are defined below:

**The Martians** : The people who inhabited Mars and whom were attacked by an unknown aggressor.

**The Atlanteans** : These would be the technologically advanced people that lived on Earth in the past, and whom supposedly made their home on Atlantis. Technically speaking, the Atlanteans could be Martians that have formed a colony on Earth.

**The Hominids** : The technologically backward peoples of Earth, be they Neanderthal men or relatively evolved *Homo Sapiens Sapiens*. These people know of space travel only because of the Atlanteans if at all.

**The Unknown Aggressors** : These are the people that destroyed Phaethon and Mars and perhaps also bombarded the Atlanteans. Apparently though, some Atlanteans survived the attack as did Earth and a good portion of the Hominids. It could well be that the unknown aggressors were actually Martians or Atlanteans that for some reason or other went to war with one another, and in this case, the unknown aggressors would have made their home in our same system, the great War having been some sort of Solar System Civil War. At any rate, even if the unknown aggressors were based outside the Solar System, it seems logical to assume that they were at least in some way related to the Martians and/or Atlanteans, and were humanoid in origin, rather than totally alien.

Currently, astronomers seem to think that the features found on the Tharsis bulge are about 200 million years old. This dating however is arrived at by crater counting techniques.

As we have seen, it is my opinion that Mars was destroyed by artificial means, by using large asteroids as 'bullets' with which to shoot Mars. The use of such weapons would result in a far higher number of craters being formed in a very short time than would otherwise be the case if these were to occur naturally. This in turn would lead scientists to assume the age of the Martian terrain is a lot greater than it is in reality, because the high number of craters present would lead them to thinking that they must have occurred over a very long period of time.

It is approximately known how many meteorites enter the Earth's atmosphere in a period of a year. Since most of these burn up in the atmosphere, it is also approximately known that the majority of them are not very large at all. Given these values, astronomers are bound to say that the larger craters found on Mars occurred over a period of many, many years, especially since they are numerous. This would make perfect sense if the asteroids that formed those large craters were of natural origin, since (more or less) meteorites bombard each body in the Solar System at a relatively constant rate. If however the bombardment of Mars was the result of a war, as I have outlined in chapter four, then the age of these craters is far less than previously thought. Furthermore, in view of such events, there may be a lot more debris in Mars orbit than normal, and this too would result in more craters being formed than normal.[3]

Even given all of these variables however, it would seem to me completely unrealistic to think that the Face on Mars is only 6,000 years old, which would be the case, if we assume the age of the Great Pyramid is only 5,000 years and that perhaps only 500 to a 1,000 years passed between the destruction of Mars and the building of the Great Pyramid.

Unrealistic as this value is, it gives us at least some starting point. We can now say with some certainty that the war which destroyed Mars could not have taken place any closer to our own time than 6,000 years ago, as a bare minimum.

---

[3] Particularly if the Mars based defence systems included asteroid-busting missiles in their arsenal of weaponry.

Similarly, the notion that the Face on Mars is as old as 200 million years is quite improbable, nevertheless, despite the huge range, we can at least begin to say that the Martian pyramids are between 200 million and 6,000 years old.

A statement so vague may at first seem to be meaningless, yet, as we'll see shortly, it is not without its uses.

If Mars was destroyed by the means I suggested, it is quite acceptable to say that the age of the craters may have been overestimated by a factor of at least 200 times.[4] In this case then, the features found on the Tharsis bulge would at most be one million years old. Even this value seems to be far too old for the Face on Mars, but we might take it as an upper limit, in a similar fashion to the way in which we took six thousand years as a lower limit. Personally I feel that even ten or eleven thousand years ago is too close to our own time, and feel that the Martian War took place long before this; perhaps as recently as 30,000 years ago, but probably even earlier than this, maybe 100,000 years ago. This of course would mean that The Face on Mars may be as old as 100,000 years, or more.

Some of Richard Hoagland's adverts for his videos claim the Face may be as old as 400,000 years. Nearly a half million years. An impressive age for any structure, especially an artificial one, but then thinking about it, after Mars has been destroyed, the only thing which could wreck any surviving structures would be naturally occurring meteoric strikes, and while certainly these occur from time to time, destructive ones are not as common as one might think, and even if a large enough meteorite were to fall on Mars every day, it would still take a very, very, long time before one landed precisely in the spot necessary for the destruction of the Face. Besides, for all we know countless other monuments similar to the Face have already been wiped out by such meteorites, and in all probability some structures indeed have been destroyed in just such a manner; the Face may in fact be one of the few remaining artefacts of a very ancient human civilisation.

---

[4] I would be very grateful if a qualified astronomer took it upon himself to estimate the age of the craters on Mars if he starts from the premise that they were artificially created. I would estimate that the results would be at least 200 times less in terms of age, but I only possess enough know-how to perform the crudest of calculations in this regard, and would welcome a more scientific means of arriving at the error factor that is involved in the estimation of the age of the craters on Mars.

As of going to press, it has come to my attention that Hoagland's claim is based on the principle that the last time during which the Face could be seen under optimum conditions from the nearby City (so that the summer solstice sun rises directly over the eyes of the Face) was approximately some 400,000 years ago.

In any event though, I would presume that the Face could not be much older than one million years at the very most, or younger than six or seven thousand years.

A clue to the time when the Martian War occurred may be found in the Hindu Vedas, which are amongst the oldest scripts known to survive to the present day.

It is said that they describe events which took place many thousands of years ago. The Vedas which survive in written form however are but a part of the 'myths' which have been handed down for thousands of years through word of mouth.

Some of the astronomical information found in these scripts, seems to substantiate the claim that they are truly ancient, and could be over 8000 years old, although the events they describe are said to have occurred as long ago as thirty or even sixty thousand years ago. The gods described in the Vedas have some very interesting attributes. Perhaps most interesting, is that they consistently travel by means of 'flying chariots'.

Similarly intriguing are the weapons these gods use in battle. Some are described in ways that give them properties which are eerily reminiscent of atomic bombs going off, while others seem to describe some sort of giant laser system that is used to incinerate enemy armies. The effects of these weapons are described as being similar to a supernova, which is correctly identified as being what happens to a sun at the end of its life.

How did people that lived thousands of years ago know about supernovae when we have theoretically only recently discovered this facet of solar death?

The Vedas are a complex set of writings, incorporating many names for gods, heroes, phenomena and objects which twist the Western tongue to its limits. Despite this, when reading Vedic works, even a person with little knowledge of their background or of Hindu language (like myself) is bound to come to the conclusion that either the ancient Hindus had an astounding imagination along with pretty

accurate astronomical knowledge, or that their legends are related to
something that is of a definite extra-terrestrial nature.

The most interesting types of descriptions concerning flying saucers
and weapons of war that should have no place in our ancient history,
are to be found in literal translations of the *Ramayana* and *Maha-
Bharata,* with a description of how to build some of these flying ships
being found in the *Samarangana Sutradhara.* Apparently, also the
Tibetan books, specifically, the *Tantjua* and the *Kantijua* contain
references to prehistoric flying machines.


I have undertaken only the most superficial of studies with respect to
the Hindu Vedas however, in part this has been because translations
of the required texts are not all that prominent in Cape Town's central
library, but more importantly, it is advisable (although not necessarily
imperative) that persons undertaking such an endeavour have either
some knowledge of the exact meanings of some Sanskrit words, or at
least access to people or books that can translate them accurately,
since many words used remain in their original form and do not
convey their true meaning easily.

This I did not have, and while the general meaning can still be
understood, what is a 'fire arrow' to one person could easily be
misinterpreted as being an 'incendiary bomb' by another, and having
no knowledge of Sanskrit, I thought it best to leave the correct
interpretation of these ancient hymns to others.[5]

The Vedas (to my knowledge) seem to concentrate on wars which
took place mainly on Earth, or at least in the skies above Earth, and
could therefore be interpreted as being a much later conflict than that
which destroyed Mars. On the other hand, the destruction of Mars
may have happened shortly before the events described in the Vedas,
if this is so, then the Martian War may have taken place as recently as
some sixty to thirty millennia ago.

My personal feeling however is that the destruction of Mars probably
took place several hundred millennia ago, and while Hoagland's value
of some 400,000 years may be somewhat arbitrary, I find it has a
certain appeal, and considering the flimsy evidence present to give us
a more precise date, we may as well adopt this value; particularly
since it also falls close to the mean of the upper value of one million
years and the lower end value of six millenniums.

---

[5] See appendix B however for more information on the Vedic Hymns in question.

# The Neanderthal Problem

Whatever the value of the Vedas in determining when the Martian war may have taken place, the age of the Face cannot realistically speaking, be said to be any younger than 11,000 years or so; and in all probability even this value is far too low.

This in itself presents some problems even if the Face on Mars is accepted point blank as an ancient artefact, because if we accept the idea that an advanced race of hominids existed on Mars in our distant past, we then have to explain what a rather *primitive* race of similar *hominids* was doing on Earth at around the same time.

I for one, do not believe that Neanderthal man was really a spaceman. Neanderthal man was Neanderthal man, with all that that implies: caves, sticks, stones, fur clothing, et cetera, et cetera.

Now, I believe I have already outlined to some extent the limits of coincidence, and it would be preposterous to say that Neanderthal man was not *somehow* related to *Homo Sapiens Sapiens*. In other words, if the Martians and Atlanteans were of the more evolved human variety (as they surely were) then what the hell was a primitive hominid doing running around in the same area at around the same time?

It just doesn't make sense, does it.

The conclusion seems to be that a more evolved species of human existed alongside a less evolved member of the same species. But this is just not naturally possible. A species evolves at a pretty steady rate, and no amount of local variation could account for one group evolving to the point of interstellar travel while another group of the same species remained trapped in the stone age.

In fact, such a situation would be highly unnatural, and can only make sense if we begin to assume that the human race did not develop along a steady evolutionary path but one which took great leaps and bounds. Again though, this sort of thing does not normally happen. Heightened radiation levels may produce severe mutations in offspring in a relatively short time, but this is generally of a detrimental nature to the race, since broadly speaking, high radiation levels destroy or 'misalign' the various genes of the DNA strand which holds the very code of life. Besides, there is no evidence that we had such high levels of radiation in our distant past, and even if that *were* the case, we would all be pretty much immune to high levels of radiation today, since our forebears would have had to stabilise at some point in order to grow as a vastly more advanced but homogeneous group, and of course, we are all far from immune to high dosages of radiation.

It would appear then, that while the Hominids of Earth and the Martians/Atlanteans were certainly related to one another from a genetic point of view, they did not evolve at the same time. I am tempted to say that they did not have a common origin, but as you will note later, I *do hold* that they had a common origin.

The only answer which seems to make sense to me is outlined below.

# A Tentative Reconstruction of Human Origins

I would maintain that the human race, both the Martians and Hominids like Neanderthal man, were genetically engineered by yet another, even more evolved set of extra-terrestrials. It is this more evolved group (which I shall refer to as the Ancients[6]) which are also responsible for any major terraforming of Mars and/or Earth.

It is the Ancients too which genetically manipulated life on perhaps several planets, apart from the ones in our own Solar System, in order to create intelligent life.

Here then, briefly, is the scenario:

A group of very advanced aliens travel through the galaxy genetically engineering creatures and terraforming planets so that conditions on them will become favourable for the evolution of intelligent life. At first the Ancients content themselves with ensuring a planet has a favourable temperature range, suitable atmosphere and so on. As the aeons pass, a planet may in fact develop life-forms of its own, which, while perhaps being quite advanced from a biological point of view, may not be intelligent enough for the purposes of the aliens. (We'll get back to what these purposes might be a little later, for now, let's just continue with our story.)

On Earth for example, conditions were so favourable that much diverse life developed, and as it turns out, this particular type of life, while being very successful, had reached an impasse.

The dinosaurs were very capable as a species, but in over 300 million years, they did not seem to have developed anything other than the most rudimentary type of intelligence, as is evidenced from the existing survivors of that era, namely the crocodile, the tortoise, snakes, and a few other mainly reptilian or aquatic creatures.

Primates and mammals, with more potential for evolving into intelligent beings, had little chance of 'Coming of Age' while the dinosaurs ruled, so the Ancients decided to get rid of them.

Exactly how they did this is not clear, but the radioactive iridium layer that dates back to 70 million years ago, might be a hint. It is interesting to know, that even if the Earth underwent a full scale nuclear war, not all life would perish. Certainly, without intelligent

---

[6] See footnote number 26 in chapter four for a definition of the Ancients.

intervention it is doubtful that any intelligent life would rise out of the ashes of such a conflict even over a period of millions of years, but if the nuclear holocaust was a little better orchestrated, and genetic manipulation by a third party arranged after its completion, within a relatively short time we might see the Earth become populated by a myriad of funny creatures. Most of these of course would rapidly become extinct as the more successful ones (perhaps as a result of some genetic advantage, but possibly also from the guiding hand of the geneticists in charge of re-population) got a foothold on this remade planet.

In this context, it might be useful to remind ourselves of what was said back in chapter two regarding the rapid and disparate evolution of hominids and primates some fourteen million years ago.

The fact that an amazing amount of different primate/hominid species suddenly surfaced out of nowhere at that time in our prehistory, and furthermore that many of them vanished rather quickly from a geological point of view, is suspicious.

Such a set of events would seem to give some support to the theory of genetic manipulation by intelligent beings.[7]

Now, if this process is occurring separately on a number of worlds, some of which may even be in the same solar system, it is only logical that some planets, for whatever reasons, would produce humanoids that evolved quicker than others.

Partly this may be due to more favourable conditions, the more attentive manipulation of particular genes, or just a lucky combination of many factors; but also, the planets that were terraformed first would probably evolve intelligent beings before the ones that were terraformed at a later date.

Let us now suppose that Earth and Mars were relatively new projects, but that orbiting a nearby star, was a planet that possessed humans which had reached a far higher state of development than the poor Neanderthals to be found on Earth.

---

[7] Readers may find this difficult to believe, but the problem of vast differentiation amongst hominids fourteen million years ago, came to my attention only after I had made the first speculation with respect to genetic engineering by extra-terrestrial entities. In other words, I came across this supporting evidence only after having laid out the basic theory, having based it on other, perhaps less reliable 'evidence' from various myths and legends. Generally speaking, this is a promising sign that a theory may be correct.

In fact, these humans may have been so advanced that their technology perhaps approached the very one that was used by the Ancients to create them, and it would be this same technology which would have enabled them to travel from their star to ours.

Before I can go any further with this idea though, it becomes necessary to begin to ask why the Ancients would go to all the trouble of trying to create intelligent life on so many planets.

## The Plan of the Ancients

What reason, could possibly be important enough to drive a race of advanced beings (the Ancients) to try and create intelligent life wherever they could? The question is a large one, and a number of answers come to mind. Cosmic loneliness might be one of them. Scientific enquiry might be another, but these alone are unlikely to have the Ancients pursuing this task endlessly, and in view of the fact that I hold at least two races of humans were created by the Ancients, (the Neanderthals and the Martians) it would seem that some purpose more than science drives them.

Although, if we assume their technology is so advanced to render them practically infinitely rich, (and this must be true to some extent) it may well be that they would not be satisfied with merely one 'experiment' and in fact may want to perform several similar ones; just like we build newer and bigger particle accelerators, despite their cost, in order to expand our knowledge of subatomic phenomena.

I maintain however, that there is a principle at work here, which amounts to more than mere loneliness or scientific curiosity.

It is the concept of truth. I do not want to further burden the reader with an autobiographical description of my beliefs with regard to theology, but suffice it to say, that even when I was a convinced Atheist, which I have been for most of my life, I believed in truth.

Ultimate truth.

I used to believe that everything was either true, in its purest, ideal form, somewhat true (in a less than perfect form) or false.

Little did I realise, that by believing in truth, what I was actually saying was that I believed in some higher power.

My problem was with religion, not with this higher power.

As I age, I have come to realise that truth is relative, just like everything else. Einstein in a sense said that the entire Universe is relative, and his theories have been proved right again and again. The truth of a thing, depends not so much on reality, but on where the observer happens to be in relation to the phenomena being observed.

For example, if you are in a spaceship travelling at close to the speed of light and carrying a perfect watch that says you have been in that ship for a year, it can be said that you would be truthful if you said : "I have been on this ship for a year". Yet a person on Earth, with a similarly perfect watch would say that a lot more than a year has passed since you entered that spaceship. He too would be telling the truth. Which truth is 'truer'?

Neither. Both statements are true and valid, from the point of view of each observer. If however we step further back, if we could hop to some higher dimension where time is a meaningless factor, then we would see that neither party has it completely correct. Each merely has a piece of the truth, and perhaps, we would also realise that although we may be able to see more from our higher vantage point, our own view may well be merely a small piece of a still higher truth. It could well be that the Universe is in fact just that; a series of truer and truer 'levels', that continue to be refined into infinity.

It is my opinion, that even God (or whatever one may chose to call it) continues to strive and reach for a higher level of truth. Atoms try to become molecules, molecules try to become mono-cellular organisms, they in turn evolve into multi-celled organisms, who in turn try to evolve into higher and higher forms of life. Each of us strives to be more than he or she is, and when we reach there, we try to go higher still. This is the nature of the Universe, to evolve, to grow. To come closer to 'truth'.

Would it surprise you to know that what you are is a small piece of the Universe that has begun to become aware of itself? Because regardless of whether you believe in God or not, regardless of anything at all, that is precisely what each one of us is.

When we use the term 'Universe', most of us get this hazy picture of vast expanses of space with stars and planets, and galaxies scattered across it, but few of us remember to include ourselves in the picture.

'Universe', does not just mean stars and galaxies and planets.

It means everything. Absolutely everything that has ever been, that is now, or that will be in the future. It means all and any planes of existence. In short, the term Universe, encompasses literally all that can exist; and that of course, includes you.

That you are a part of the Universe is undeniable, but that in itself is not the really interesting thing. The interesting thing is that you happen to be amongst the most sentient of creatures on this planet. It may be helpful here if I define sentient.

A plant in some way is sentient. It feels pain and is dimly aware of its environment. An animal, while being far more sentient than a plant is not quite as sentient as a human being. Depending on the animal, it may be far less sentient (say a grasshopper) or very nearly as sentient (or possibly even more so) than a human, in the case of say a dolphin or blue whale.

Now, sentient beings are a very special part of the Universe. They represent the tiniest fraction of the Universe. A number so small it would seem to be insignificant, except it's far from it.

Do you remember when back in chapter four, I stated that the atoms that compose a living human being, taken as a whole in the form of individual elements, do not display anywhere near the same electromagnetic phenomena as they do when connected in such a way as to form a living person? In a way this is the same. Although sentient beings amount only to the very tiniest of fractions of the Universe, their impact is far more important than their relative mass. This is easily seen.

Take one hundred boulders, each weighing as much as a person and place them in a field. Come back one thousand years later, and very, very little will have changed. But place one hundred persons in a field, and come back just *one hundred* years later, and you might not even recognise the area as a field any longer.

Houses will have been erected, the landscape itself may have been changed, or the people may have moved on to another place.

Give them a few thousand years and they may start to affect not only their entire planet, but may move on to change other planets too.

In effect, what sentient beings are, is the smallest part of the Universe becoming aware and saying :

"What am I? What is this? What...?"

And aren't those very questions the ones which haunt us most?

I believe, that billions of years from now, humanity may well have evolved to a state of consciousness so advanced that it would be as far removed from us today, as we are from present day atoms.

Can you see then, that any race of beings that began to realise that they are the Universe becoming aware, would want to expand and awaken more?

This concept may be seen as one that challenges an individual's religious and philosophical beliefs, but in fact it does not. It is above the petty differences of race or religion. What this concept is saying is not : "The Christian is wrong", or "The Moslem is wrong", or "The Jew is wrong."

What it *is* saying is that each one of them only has a very small part of the truth, and this, from the perspective of a Universal consciousness, is a small and somewhat twisted truth, merely a necessary step at the beginning of the journey of consciousness which all sentient life must eventually embark upon.

A person's religion is merely a matter of dogma when seen from this perspective, and not so much a matter of right or wrong.

If these points are valid, then it makes perfect sense, that upon reaching a certain point, any intelligent race would begin to travel from system to system and try and do all it can to see that life evolves and grows wherever possible.

It may be that if we reach a certain critical amount of consciousness, our whole reality and Universe will change completely. Perhaps this is the long term goal of such beings, to change our Universe from the coarse physical matter that it is today to some more ethereal but paradoxically more 'real' and conscious form. Literally, to awaken the Universe from its slumber of form and function to a state of pure thought, where we can take form or leave it as we choose, where we may create a whole planet of form merely for our own amusement, in much the same way we do when we dream.

For these reasons then, the Ancients would go about 'creating' humans. Why humans? Because in all probability, the Ancients are themselves humanoids. the quickest and surest way of creating intelligent beings is by using some of the DNA of beings which are already intelligent. Variations would occur from planet to planet, and if some local DNA was also used, as surely must be the case, then these variations could amount to quite considerable differences. While our distant cousins living on a far away planet may also have two arms, two legs and two eyes, they may not require nostrils or ears, perhaps picking vibrations up through their skin, or they may be blue skinned or whatever, but I would guess that some humanoid features are in most cases quite evident.

## Back to the Neanderthal Problem and the Reasons for War

If you recall, the problem we had was that of explaining how come Neanderthal man (as well as possibly Homo Sapiens Sapiens) was running around dressed in furs at the same time that some relatives of his were flying in spaceships and destroying whole planets.

I would suggest that what happened is that a colony of humans living on a planet which orbits a nearby sun became advanced enough to travel between the stars. On arrival to this solar system, these humans may have settled Mars which may or may not have already been a home for less developed hominids.

For whatever reason, perhaps after many years of peaceful existence in the Sol system, these more advanced humans went to war with one another. The reasons are surely lost in the mists of time, but there may have been disagreements on local policy. Some of the Martians for example may have wanted to leave the Hominids of Earth alone, while others wanted to genetically improve them, resulting perhaps in mass genocides of Neanderthals, either intentionally or by ignorance.

But more likely: it could be that while this race of humans evolved quickly and learned the art of space travel between the stars, and while their creators may have made it known to them what their purpose was to be (to evolve life wherever they found it) they may have been evolved so quickly by their original creators, that while they were full of good intentions, they were still a rather primitive race at heart. And while possessing the know-how for the control of vastly powerful technology, they did not possess the wisdom necessary to use it beneficially.

This is a path we can already see in our own history; right up to the present day.

The reason so many African countries have failed economically, is largely because people that had little or no history of governing a country were suddenly placed in charge. Even where some training was given, (and this was rare in most cases) most of the people concerned simply had no history of self-government.[8]

If South Africa succeeds as a country in Africa, it will be, at least in part, because it had the benefit of seeing the mistakes of other African countries that underwent a rapid change. This may be true not only of South Africa, but of humanity as a whole.

If we can unearth the truth about the ancient past of humanity, and it happens to be that we very nearly wiped ourselves out, we may be more cautious this time around!

Whatever the reasons given at the time then, the events which led to the Martian War may not have been so much a matter of politics, as a matter of genetics. The humans in question were simply too quickly evolved. They did not go through the necessary process to reach the fantastic technology before them, but instead it was almost given to them at no cost.

Michael Crichton's character Ian Malcom, in the best-seller *Jurassic Park*, makes a similar observation when he states that basically, the park cannot succeed, because the people concerned did not *earn* the experience to control large numbers of genetically engineered creatures; but rather it was given to them on a silver platter. He also makes an interesting analogy to a karate expert.

By the time one has become proficient enough to be really dangerous in a martial art, in most cases, he no longer *is* dangerous. The things he had to go through that make him able to perhaps kill another human being bare-handed, will have changed his character to the point where he will not use such skill indiscriminately. If however you

---

[8] It needs to be pointed out that I am specifically referring to governing a country by the accepted Western ideologies. Before the Caucasians conquered and enslaved most of Africa, large settlements of native people were quite successful, but their way of government and the one of the Europeans was vastly different. Secondly, I am in no way implying here that any people or race is implicitly or genetically superior to any other people or race. If the Europeans had been conquered and enslaved by the Africans and then hundreds of years later, when they finally managed to gain independence, they had been forced to govern themselves according to rules dictated to them by the Africans, they would no doubt have failed just as miserably as some African countries have done. I would hold this to be self-evident, but since racism and intolerance are (unfortunately) by no means dead yet, I thought it wise to actually spell it out, hopefully minimising any possibility for misunderstanding.

had a magic pill that gave anyone that ate it instant Bruce Lee ability, the world would be a very dangerous place indeed!

And this is what happened with the Martians.

They had the power, and they may have been dimly aware of the responsibility that went with it, but not enough so that they really understood the root of it.

Stan Lee, creator of Marvel Comics perhaps said it best when he coined the corny phrase: "With great power comes great responsibility."

Most of the Super-heroes of the Marvel Universe live by this code; those that don't, do not last long; which is why the good guys usually win in comic books, while the bad guys can die in droves.


It is for this reason, that I do not believe that the people that destroyed Mars are the same ones that terraformed it.

The very process of terraforming and genetically engineering whole races is so utterly vast and complex, that even with incredibly advanced technology, it would still be a very major undertaking.

This in turn would make it (in my opinion) very difficult for those same creators to destroy all their newly engineered lifeforms (along with their entire habitat) in a brief and violent flash of anger.

I suppose some could take the idea that the Martians, in their unwise ways, thought they might be ready to 'take on' their creators. In this case, it might be acceptable to say that Mars could have indeed been destroyed by the Ancients rather than by infighting between Martians. It would still leave the problem of the Neanderthals being around at the same time as the Martians though, which would still suggest that separate races were created by the Ancients.

Perhaps it was the case that the Martians were more advanced than the Earthlings, being older, but not necessarily originating from another Solar System. In this case, the only extra Solar aliens would be the Ancients.

At any rate, for whatever reason, the Martians either destroyed themselves or where destroyed by the Ancients, and the few survivors escaped to Earth, where they set themselves up as 'Gods' and became what I call the Atlanteans.

This theory of destruction by the Ancients could be said to have a few advantages over the previous one, since it might explain why the Ancients are apparently no longer around.[9]

After destroying what could be classed as a 'failed experiment' — perhaps disgusted even, with this sector of space— they left our Solar System to return at a later date, when the more primitive Earthlings might have evolved a little further.

The handful of Martian survivors may or may not have escaped the notice of the Ancients. It seems there might be a case for saying that the Ancients observed them for a time, but the Atlanteans (or Ex-Martians if you prefer) attempted a second uprising, at which point, the Ancients destroyed Atlantis and then left, wisely realising that any remaining survivors would be tempted to continue rebelling in the future if the reminder of their destruction continually hovered above their heads.

If however, the Ancients left this Solar System altogether (or at least observed it in a way that would not give them away), and considering that the destruction of Atlantis was probably cataclysmic enough to give the Earthlings more to worry about for the next few hundred years than the detailed recording of history, it would not be surprising, if in a few thousand years, almost all traces of these momentous events became lost. Furthermore, the destruction of the Great Library of Alexandria by Amru, the Moslem conqueror,[10] would have gone a long way towards ensuring that Atlantis survived more as myth than as fact.

Despite this, it is interesting to point out, that many *thousands* of books have been written about Atlantis. It lives with us still. Why? Why for example are the ancient Babylonians not written about with as much fervour? Or the 'myths' of the ancient Greeks? If all that Atlantis ever was is a myth, why does it persist so in our imaginations, folklore, and ancient history? The same can be said of the 'Gods'.

---

[9] Of course, it could well be that the Ancients indeed *are* around. We may just not be advanced enough to detect them properly yet.

[10] Charles Berlitz, in his book, *The Mystery of Atlantis, The 8th Continent*, writes that the books of Alexandria were burned to heat up the city's 4000 public baths for a period of six months; a loss of literature amounting to over a million books. The reason given by Amru for this act of barbarism was that any writing already contained in the Koran was superfluous, while any material not included in the Koran was of no use to true believers. So, once again, thanks to religion as opposed to spirituality, the human race lost out.

They are truly everywhere in our mythology, no matter where on Earth you originate from. Why so many 'Gods' and not just one overall power? I maintain it is because there is some grain of truth to some of these legends.

In any case, regardless of whether the Martians originated on Mars or another Solar System, I would hold it that they are the ones responsible for the destruction of Mars either as the result of infighting, or as a result of rebelling against the Ancients.

Furthermore, I would hold it too, that the humans we find on Earth today are the distant descendants of those Martians. Undoubtedly some mixing between Martians and original Earthlings must have taken place, probably as a result of genetic experimentation at first, and later as a result of naturally mixing the races.

A considerable period of time is likely to have elapsed between the point during which genetic manipulation occurred and the time during which it become at least somewhat acceptable for the Martians to breed with the Earthlings.

The evidence for this having taken place is again deeply rooted in our past, and especially our most ancient texts.

Half-man creatures, such as minotaurs, centaurs, medusa and others can be found in the myths of all human tribes, once again, regardless of location on the planet.

Furthermore, in the Bible it states that God (or the Gods?)[11] decreed that if a man shall lie with a beast, or a woman lie before a beast, then they shall be put to death (both the human and the beast).[12] Surely, this is a pretty weird thing to say!

The practice of having sexual intercourse with animals is discouraged in the strongest possible terms. This in itself is interesting, because a couple of thousand years ago, there was no 'Green Movement'. Animals suffered the most appalling of conditions and no one thought much of it at all. The same is true of humans.

While several countries today have a death penalty for violent crimes and some Arab nations still hold to the practice of cutting off limbs to

---

[11] See Appendix E for more references concerning the disturbing appearance of the plural word 'Gods' in the supposedly monotheistic Bible.

[12] Leviticus 20:15 and 20:16.
There are also ample references here (Lev. 18, and Lev. 20) to discourage inbreeding practices, (incest) in the strongest possible terms. Once again, a useful thing to do if one is trying to produce physically and mentally sound offspring.

punish theft, and while too, torture of political prisoners still goes on in several countries in the world, the practice is generally frowned upon, and certainly, no country in the world would accept the public crucifixion of even a single person as a means of punishment, much less great numbers of them.

Nor do we normally burn people at the stake any longer. What I am trying to point out here, is that the reason behind the instructions given by 'God' with regard to sexual intercourse between humans and animals, is not rooted in some sort of morality.

In view of the numerous atrocities in our history, it seems that the reason was a far more practical one, and that would be to keep the human genetic pool as clean as possible. Today of course science tells us that human sperm or ova will not produce a zygote if the other half of the equation if of animal origin. It is interesting to note though, that occasionally, some pretty revolting reports are made to the press. These involve the birth of certain animals that have human features. This author has seen at least one photograph, of a calf that had human feet instead of the hooves one normally associates with cattle. Similarly, there are reports that certain European governments, and possibly others, have undertaken secret experimentation in this field. The story goes, that somewhere in a secret installation, a half human and half gorilla creature was successfully 'created' and it survived for a number of years. It remains to be seen of course whether such claims are elaborate hoaxes or not; but during the Nazi era of the second World War, certainly Germany performed some horrific experiments on people, particularly on Jews and Gypsies. Japan too, used war prisoners to perform the most abhorrent 'operations'.[13] The prisoners were not even considered human and were referred to as 'logs' by the 'doctors' involved in the experiments. Perhaps a less well known fact is that at the end of the war, the American government agreed not to prosecute the 'doctors' involved for war crimes, in exchange for the data concerning the experiments.

---

[13] The Japanese Army Biological Unit 731, staffed by about two thousand volunteering doctors, undertook experiments on live prisoners of war in a medical experiment camp at Harbin. The experiments were also done in order to develop biological weapons, and some of these doctors are still alive today, holding high office and being respected members of their communities. The results of these experiments were at least in part passed on to the Americans, whom, in payment for the information, refrained from pressing charges of war crimes on the Japanese doctors concerned.

The full extent of what some of these experiments were may never be known, but I would not at all be surprised to discover one day that some of them involved the creation of some half human and half animal creature.

If long ago humans could reproduce with certain animals, then it would make sense, that the 'Gods' would impose certain restrictions on them with regard to the mixing of their genes. If the humanoids could be kept from breeding with less evolved creatures for a time, they would stabilise eventually, at which point, perhaps the 'law' could be further enforced by additional genetic alteration.

## After Atlantis

With the destruction of Atlantis, either once again by war or perhaps this time as a result of a natural catastrophe, the Atlanteans were even further reduced in number. Furthermore, their mixing with the less evolved hominids would eventually result in their absorption anyway. Realising this, they may have decided to create certain artefacts that told not only of their passing, but that perhaps even predicted the return of the Ancients. Some of these artefacts last up to the present day, long after the last Atlantean has been assimilated by the more primitive ways of his cousins.

Prominent cases in point are:

1) The great structures found in South America. These include huge pyramids, enormous statues, walls and carvings so large that they can only be seen in their entirety from aboard an aircraft. Most of these amazing formations have properties which are similar to the Great Pyramid from the point of view of accuracy, and they can be used to determine the precise occurrence of particular astronomical events such as summer and winter solstice.

2) Similarly, enormous pyramidal constructions in the Far East, particularly Thailand and China, although I know little of these structures, not having researched them at all.

3) Structures found in Europe, particularly in the British Isles, such as Stonehenge. Incidentally, Lemesurier points out that the circle inscribed by Stonehenge is identical to an Egyptian Quarter Aroura circle. This is also confirmed by the fact that Stonehenge has been found to be some sort of an ancient astronomical computer, and the use of the Quarter Aroura circle (which

represents one year) would of course lend itself particularly well to this purpose. Keep in mind too, that the British inch is so close to the pyramid or primitive inch in size that it's obvious it is a descendant of it. Such a similarity in phenomena and measuring units between places as distant as Egypt is from the British isles, would suggest a common origin. And this in turn fits in rather well with the theory of Atlantean survivors being scattered around the globe.

4) Perhaps most important of all, especially in view of the code defined by Peter Lemesurier, the Giza pyramids of Egypt and in particular the Great Pyramid. The many other much smaller (and newer) pyramids, seem to be merely attempts to recapture the glory of this titanic creation, and as such are of little value for the purpose of finding out more about the Atlanteans.

Taken in this view, Peter Lemesurier's claim that the Great Pyramid predicts the *appearance of a great sign in the sky* to herald the return of the Messiah somewhere around the year 2034 AD becomes rather interesting. While I don't hold it against people who accept such predictions on faith, I find it very strange that anyone could predict the future with such accuracy as seems to be detailed in the Great Pyramid.

The code used by Lemesurier could perhaps be brought into question by some, but this is actually the least of the arguments against the Great Pyramid being some sort of oracle. Regardless of the correctness of Lemesurier's code, what is absolutely clear about the Great Pyramid, is that it unquestionably *does* have a mathematical code locked in its very bones, and *obviously its architect intended for it to be so.*

Furthermore, it does indeed seem that this code (whatever its true meaning) is almost certainly tied to the future of the human race.

For these reasons then, the question of the correctness of Lemesurier's code is not as important as the question of how anyone can predict the future!

There are basically only two ways in which one could predict the future, the first is by sheer magic, or some equally mysterious power. The other is by using scientific methods.

While certainly we are nowhere near knowing all of nature's secrets, and while also some type of 'magic' does seem in fact to exist, I hold it that *utterly unexplainable* magic does not exist.

I do believe however, that there are certain processes, which, while perhaps being somewhat able to use, we do not yet know how to fully explain. Like Einstein, I too am absolutely convinced that "God does not play dice with the Universe."[14] Where it appears that he does, all that is in fact happening is that we are not advanced enough yet to comprehend the absolute order behind events so complex that they seem to be random in nature.

In short, what I am saying is that magic as such does not exist, but what I do believe exists, is technological, or even superconscious 'magic'. That is, processes so advanced, either in terms of technology or in terms of thought, or a mixture of both, that they are literally able to produce miracles.

The subject is rather complex and I don't want to get bogged down in it, but certain advanced branches of maths maintain that this physical Universe we are familiar with is only a lower aspect of one of several higher dimensions.[15] If we had access to these higher dimensions, either directly or perhaps by use of certain mathematical means, we might then be able to discern the pattern that our futures would take, especially if in such higher dimensions we were to find the order of the Universe belittled that dimension we call time, by greatly reducing its apparent effect.

If you recall the basics of Chaos Theory, you will be able to grasp the concept that it might actually be a possibility, that the Universe is *so orderly, that every single event is completely pre-ordained!*

What we think of as our 'free will', may in fact be the result of our very narrow perspective not allowing us to see the big picture!

In such a case, looking down on ourselves from a higher dimension would give us the opportunity of seeing our future. To use the analogy of the supposedly random moving atoms, if they somehow became conscious enough to develop the idea of Chaos Theory and fractal

---

[14] Although I reserve the right to use this phrase *away* from the context in which Einstein conceived to use it!

[15] People who want more information on it are advised to look up very heavy tomes of certain branches of topographical mathematics and enlist the help of a genius or two in deciphering them.

movements to a degree higher than our current one, atoms would then be in a position to predict where they would be in the future.

In a sense, we already predict the future to some extent with respect to large groups of atoms, by using our weather computers. This is merely a mathematical process, and if we had access to infinitely more precise measurements and infinitely more complete data we would be able to predict the weather for years and years in advance.

Similarly, if we had the same sort of infinitely precise and infinitely complete data on human evolution, and if we furthermore had some way of processing it all, we too, I believe, would be able to predict the future with respect to humanity.

In this way then, the Atlanteans may have been able to predict certain key events with regard to the human race as a whole.

A kind of weather prediction for a whole bunch of humans instead of a whole bunch of atoms. The key storms would happen at around the birth of Christ and the World Wars, as well as at the return(?) of 'tornado' Jesus in the year 2034 AD

Incidentally, I believe that the computers able to process such incredible data would not be electronic devices, but rather, biological ones. Namely our brains.

The brain, to my knowledge, is the only thing capable of processing untold billions of variables every hour of our lives.

Naturally, before our brains can handle such complexities as long-term future trends, our minds have to develop quite a bit. The key, I feel, is not in ever more complex machinery, but on approaching the problem from a completely different point of view.

Some might call the approach a mystical, occultist, or spiritual one, but the name is unimportant. For me, the answer lies in simplifying things, not making them more complex, and in this respect, I have found that the intuition plays the key role. The problem with this of course, is that our intuition is not generally accepted as being a valid scientific method for discovering answers to our problems.

I maintain however, that our intuition or higher instincts, are far superior to any computer we have ever designed, or are likely to ever design in the future.

It would take billions of dollars to design a computer that could learn how to use language correctly so that it could hold up its own end of a conversation. Even then, it would probably be a very limited conversation and in any case the computer would not fit in the space found in the skull of a human.

We on the other hand, use language every day and take it for granted, but we consider ourselves 'stupid' because we cannot perform complex mathematical operations in our head. Yet this ability is available to us, as is evidenced by certain people, some of whom are idiot savants, who can perform ridiculously complex operations in their heads and give the correct answer almost immediately after it has been posed, and certainly a lot sooner than one could do it in if using a scientific calculator. Nor are all the people with this ability idiot savants, some are perfectly normal in other respects, with the exception of this extraordinary ability.

What one human being can do, so can others. Therefore it follows that such abilities are available to all of us potentially, and it may be that all that is required to achieve them is some rewiring. Literally.

Our neurones form new bonds with other nearby neurones whenever we learn a new skill or think in new ways. The longer we practice the new skill, the stronger become those new bonds, so by expanding our range of experiences, we quite literally increase our intelligence.

The average person uses only the smallest part of his or her brain. The percentage varies according to whom you speak to, but in all cases I know of, the experts agree that we use less than 10% of our grey matter.

Imagine what we might accomplish if we could use say even just 50%. If we are able to make rough predictions in terms of economics, weather and so on with less than 10% of our brain, surely if we were using 50% of it we could 'read' the future. Especially since a person that uses 10% of their brain as opposed to one that uses only 5% of their brain is not *twice* as smart, but *several times* smarter!

I have no way of proving this, but I just *know* it to be true, and I challenge any scientist to prove me wrong on this point.[16]

---

[16] They can't anyway, because you would first need to be able to measure intelligence, and while IQ tests are commonly held to do just this, the reality of the matter is very different. I suppose I have no *scientific* way of proving that a doubling of brain usage would result in far more than a doubling in ability, but perhaps a way to 'prove' this would be to point out that neurones can make several connections each, so a doubling in the number of active neurones in the brain would result not in a doubling of connections between neurones, but an increase of several times the number of connections.

Such ideas as I have presented, particularly with respect to Chaos Theory, our intuition, and its role in long-term predictions of the future, may be held to be too speculative for science. Nevertheless, they all point in the same general direction, adding fuel to an already considerable fire.

The Great Pyramid after all, is such a work of genius precisely because it says so much with such humble simplicity.

Such a 'complex simplicity' is found constantly in nature. A blade of grass knows nothing of DNA, genetics or much of anything else, it just quite simply *grows*. And yet, the plan behind this apparently simple growth is incredibly complex, so much so that man is still unable to fully grasp it.

It has been said that we could build the Great Pyramid today given our technology and a rather large amount of cash. These statements of course are totally false. What they really mean is that we *may* be capable of *duplicating* the Great Pyramid,[17] which is a very different thing from building it for the first time.

It is simply not in the human consciousness of today to built a structure like the Great Pyramid. A work of such precision and idealism, quite apart from any predictions of the future which it may or may not make, would just not be built by us in this day and age. To put it bluntly, it just would not occur to us to build a structure which embodies such fundamental measurements as to represent both Earth and the humanity on it. Especially since it is not just a plaque of metal like the one we might stick on the side of the Voyager spacecraft, but an enterprise which would require incredible commitment and cost from a vast number of people. Nor can we compare a 'structure' like say NASA, to the Great Pyramid.

Organisations such as NASA for a start are not as idealistic as they would have you believe. And while it's arguable whether the builders of the Great Pyramid were all doing it out of love and higher spiritual concepts or not, the fact remains that the pyramid is still there, their accomplishment cannot ever be denied.

The placing of a man on the Moon more than twenty years ago however, symbolises our failure more than anything else. The idealistic notion attached to NASA should be the furtherance of space

---

[17] Something I already feel is doubtful in any event, as merely its settlement rate shows, never mind all the other details of precision and engineering difficulty.

exploration, and this of course has obviously NOT been happening since the last moonwalk. If NASA and other organisations like it were truly fulfilling their idealistic requirements, by now there would be permanent bases on the Moon and probably a small one or two on Mars too.

Secondly, while the expense of organisations such as NASA is great (especially in view of their relatively lacklustre performance) it cannot compare to what the expense of the Great Pyramid must have been in terms of human labour given the supposed lack of technology of the times. This point of course is somewhat diminished if we accept that the Great Pyramid was built by means of an advanced technology, but then we have to at least admit that the builders of the Great Pyramid left behind a creation that baffles our understanding of history, which I doubt will be the case of NASA a few thousand years down the line.

The point being made here, is that regardless of all other considerations, the fact that the Great Pyramid is a work of idealism and/or spirituality cannot realistically be denied, and this in turn shows not only that the people who built it were technologically advanced, but also that they were spiritually advanced. So much so in fact, that their spiritual or idealistic notions surpassed our own today, since we would not build a structure like the Great Pyramid.

If we assume that the more intelligent a person becomes, the more likely it is that spiritual considerations (such as the welfare of others, the betterment of society, etc.) become increasingly important,[18] then of course this ties in extremely well with the fact that anyone able to predict the future, in view of their higher intelligence[19] would also have a higher level of spirituality.

It would appear then, that the builders of the Great Pyramid not only had a well developed technology and mind, but also a well developed spirituality, making them basically a superior kind of people, which of course fits in exactly with the theory of Atlanteans, or 'Gods' having built the Great Pyramid rather than the ancient Egyptians.

---

[18] There is ample independent evidence supporting this view by the way, since those persons in history whom we have tended to identify with high levels of intelligence, all believed in a higher purpose to life than mere survival.

[19] Due to their using a higher percentage of their brain.

# The Religionists and Human Origins

For certain people, to think that the entire human race may be the genetic creation of a vastly advanced race of aliens may seem disheartening, even blasphemous. Certainly, the more narrow minded religious fundamentalists will never accept such a theory even if the aliens land with the most detailed, videotaped documentary footage of our pre-history having started in their laboratories.

Such people (and I fear their numbers are vast) are too weak to consider such a hypothesis. For some reason, they would feel insignificant or almost a cosmic joke if the human race was proved beyond the shadow of a doubt to be an alien experiment in genetics.

In a way, this very fact proves the weakness of their beliefs, for if there is a higher power with good intentions, what some may choose to call a God, then he, she, or it, loves us all, regardless of whether we are cockroaches, laboratory rats bred for testing, cashew nuts, or humans created by yet another higher form of life than ourselves.

The existence of an alien race that genetically 'evolved' mankind over a period of thousands, perhaps even billions of years, in no way threatens the position of God. What it *does* threaten (for some rather limited individuals) is a person's own idea of self.

What it *does* threaten is our own ego, but it does not threaten our soul, it does not threaten the very real fact that *however we came to be, we are here now!* And most certainly, in no way does it threaten God. In fact the idea that it does is ridiculous.

If we assume that the aliens responsible for the human race not only manipulated the hominids on Earth, but that they did and do so all over the Galaxy, then these overseers would have to be so advanced compared to us that a comparison may be made between genetically altered peanuts and humans.

We genetically alter peanuts so that they grow in more plentiful quantities and reach maturity in a shorter period of time. Over several decades, the peanuts begin to become dimly self-aware, in a very limited and narrow way, and they begin to question the Universe around them.

At first they might just state the obvious: "Here we are!", but soon they would start to wonder where they came from, and when (in digging up the nearby soil) they find several different species of peanuts grew in the same fields long ago, they would no doubt be confused.

Perhaps the peanuts would even go from believing that a faceless, nameless *something* created them, to believe the 'rantings' of a particularly Darwinian peanut that proposed they evolved to their present state from a long and varied peanut evolution.

The peanuts would undergo this shift in consciousness in their efforts to discover truth, to balance, to reach a perfect state; because of that compelling Universal urge to make order of things, which every single atom or quanta of energy in the cosmos is inherently imbibed with.

But would the peanuts ever suspect that humans actually shaped them into their present configurations? They would first have to begin to grasp, at least faintly, what a human is. And how would they do that when they have only dimly begun to understand themselves?

Imagine the fear of peanuts world-wide, if they knew they were bred and genetically shaped by more powerful beings simply for the purpose of being recreational food (one can hardly call peanuts serious food). What a mortal insult. What shame. How lowly and insignificant would peanuts feel?

Such is the dilemma of the religionists, of the Popes and priests.

But is this feeling of indignity justified?

Not at all. From a logical point of view, it has no basis whatsoever. The place of peanuts in the Universe may be to be recreational food. That may simply be a universal constant. In which case they would not feel bad about it (as I think is the case in reality).

But maybe the peanut is not to be recreational food. Maybe indeed, it may have a higher and nobler purpose; in which case, it would have to grow and evolve to the point where it becomes say self-determining and fully cognisant.

Now, what would spur a peanut on to become so much?

Why, the indignity of knowing that if it *does not* evolve, it will remain *forever*, merely recreational food!!

I'd say that's a pretty good incentive, and without it, peanuts would no doubt remain recreational food; so can you see, that the fact that they presently *are* recreational food is a very necessary step in their path to enlightenment? For without it, they would have nothing to spur them on to become all that they should become.

Funny, isn't it, how a situation the peanuts may feel as an intolerable indignity, is in fact merely a wrong perception of theirs, which they nevertheless need to have, in order to grow and evolve.

That, my dear reader, is because, the Universe *is* perfect! And if that's the case, why bother being angry and frustrated about it? The clever peanuts would simply, quietly and happily, begin to go about the process of evolving into more than just a recreational food.

Meanwhile, the ones ranting and raving about the injustices of being recreational food, or the ones that plain refuse to believe in humans altogether, remain inevitably at the same level and consequently, remain merely recreational food.

Besides, it is my belief, that if we do have alien 'creators' their purposes are far more high and lofty than our own in respect of the peanut, in which case, we might not even need to feel all that affronted after all!

The point I have made here with respect to the religionists, is intended not so much for them (especially since it is highly doubtful that such a person would have persevered this far into this type of book) but more for those that are somewhat uncertain as to the origin of their feelings of unease when the concept of genetic engineering is mentioned in relation to the origin of humanity.

It is also intended to show, that there is a vast difference between what I call religionists and truly spiritual people.

Firstly, the religionists outnumber truly spiritual people by several hundred times to one, and secondly, a truly spiritual person, although they may or may not be associated with a particular religion, is more interested in the bringing of goodwill to all creatures, intelligent and otherwise, than in the details of how this is to be brought about. You will not usually see a spiritual person condemning another's behaviour or beliefs, directly or otherwise.

It has also been my experience, that spiritual people normally focus on the positive side of things, on what *can* be done instead of what should *not* be done, and while the idea of being a genetic experiment would trouble a religionist no end, a truly spiritual person would not even see it as being relevant, because regardless of our origin, to a holy person, our purpose is clear: be all that you can be, and bring as much joy and happiness as you can to as many people and animals as you can during your time here.

On this note then, I would like to point out, that the critics to whom I will feel obliged to reply or possibly make apology to, will not be the ones that base their criticism on 'fundamental' religious beliefs, but to those that go about showing me the possible 'errors of my ways' in a (truly) scientific and logical manner.

As I have stated repeatedly, my concepts are by no means complete, and certainly a refinement of this topic needs to be done by many others, but it is necessary to point out, that we climbed out of the brutal times dating from before the birth of Christ, mainly thanks to a few brave martyrs that were fed to lions and crucified for expressing their theories. Eventually, as more and more of the world came to see the benefits and 'truths' of the New Order, we then reached the point where the very same 'New Order' became the establishment, ruling by fear and greed to benefit the few at the cost of the many. The very same Christianity that was responsible for the freeing of people two thousand years ago became the tool of the Conquistadors, the Crusades, and the Inquisition.

Similarly, we climbed out of these more recent (and aptly named) Dark Ages, thanks to a few heretics, whom were often burned at the stake for their trouble. People like Galileo, Michelangelo and Copernicus had the unenviable task of being the midwives for the birth of the so called 'Modern' era.

Today, we find ourselves once again in the position, where the once new and fresh Scientific Approach, has become the same blind behemoth that Christianity was five hundred years ago.

Some people have gone so far as to actually call the present era of science the era of 'Factless Science'.

Science nowadays, is more interested with being politically correct on one side or downright greedy on the other. The great Medical profession, which has certainly improved our lot over the last hundred years, is nevertheless a greedy monster.

There exist rumours of people whom have found a cure for cancer based on dietary remedies that is more than 90% effective. Over twenty years ago, a vaccine for tooth caries was invented.

In Britain, one Paul Cook whom also developed the missile guidance system for Tornado jets has invented a laser system, that projects lights on a screen in such a manner that it trains the eyes of people with eye defects such as myopia (short-sightedness). It does this to the point where people who had been wearing glasses or contact lenses for years, no longer needed them at all after a month of being subject to the treatment a few times a week.

Yet most of us do not know about these things, because people like Paul Cook are either legislated into silence, made lucrative offers that keep their inventions under wraps or in the worst cases, even silenced permanently.[20]

Why would anyone want to suppress the cure for cancer, you may ask. The answer is simpler than you think, if slightly more abhorrent than you might have guessed. Money.

The great God of our generation. Cancer, you see, is the most lucrative business of all. Cancer drugs and chemotherapy treatments amount to many billions of dollars each year. That's why doctors in the Western world are well off. Don't misunderstand me; there's nothing wrong with making lots of money. It's not wealth we should be against, it's the making of that wealth over the bodies of other people that we should be against. And I really mean *"over the bodies,"* literally.

The doctors, 'realists' and other 'men of intellect' that doubt my claims, should remember two things.

First of all, the so called 'reasonable' people have always been against the 'wild' claims of heretics. That of course, is not to say that every wild-eyed madman that comes up with a crazy idea should immediately be given Messiah status, but it does mean, that however sceptically, we should at least investigate any claims which might, if found to be based in truth, have an impact on the entire world.

This of course brings me to the second point, which I have already stated in the prelude to chapter three. Namely, silence is equivalent to acceptance; and if I say : "You are all being kept in the dark; we need not wear glasses, have tooth caries, or die of cancer. Speak up!" are you really so apathetic that you will stay silent? Will you not investigate my claims at least a little? Will you not, in short, speak up?

---

[20] Perhaps this does not happen all that often in the medical profession, but certainly in view of technology which applies to the governments of countries this is true. See for example, the case of Doctor Jessup as described in the *Philadelphia Experiment* by his friend Dr. J. Manson Valentine (See also footnote 1 in Chapter 11). Or more recently, the killing of the inventor of the Supergun (which was originally intended to be a cheap satellite delivery system rather than a weapon of war) by the Israeli Mossad.

"But what can I do as an individual?", is the question that springs to mind for most of us. Quite a bit actually.

Firstly, simply by becoming aware of such things, your frame of mind will have changed.

This alone may be already worth more than you think. Especially if a man called Rupert Sheldrake is right. Because if he is, then your merely knowing that these technologies are available, may in some way catalyse the process of their becoming realities. More on this in the next chapter.

In the second instance, as an individual, you can join a number of organisations that are involved in the process of bringing such things to light. At the end of this book I have included an order form, which you can complete and send in with membership fees. Joining this particular organisation will keep you in touch with developments in such fields by way of a bi-yearly newsletter.

The balance of the funds will be used to sponsor private enterprises which are involved in the researching, publicising and the making available to the general public of advanced technologies such as the ones mentioned in this book.

In one of the later chapters, I also give a list of references, and through some of these it may be possible to contact other organisations which are similarly involved in the bringing about of new and much needed technologies.

Lastly, the truly dedicated can change their lifestyles to the point where they are actively involved in the bringing about of such developments, either by direct research and development work, or more indirectly (but no less importantly) by the sponsoring of research groups.

# Summary

In this chapter I have tried to place the ideas brought up in the previous ones into some sort of context. The idea that we may be nothing more than the genetic creation of an alien race may offend and enrage many of us, but the fact remains, that if the Face and pyramids on Mars are accepted as being artificial in origin, this is more than likely. It is also a fact, that throughout history, every new discovery concerning the origin of mankind, has at first been rejected by the majority as the work of madmen, frauds or heretics.

Later though, what at first seemed to be heresy becomes the driving force behind a general betterment of humanity. And although shaken at first, eventually the human race again settles into its complacent ways, until it once more stagnates; at which point the cycle repeats itself.

It would seem too, that the cycles become shorter and shorter. Christianity began two thousand years ago, while science started only at the beginning of the 19th century, and the current revolution of the 'New Age' is happening a scant two hundred years later. If the factor of ten is applicable again, the next twenty years should prove to be interesting indeed!

The ancient Greeks seemed to have a healthier attitude towards their own origins, stating that the Gods created and meddled into the affairs of men amid fits of laughter. We on the other hand, while having almost rid ourselves of the more tyrannical aspects of religions such as Christianity, have yet to free ourselves of the more insidious idea that we are somehow the pinnacle of the Universe.

The more we discover the less likely this seems to be. If one looks at the progress of humanity over the last two thousand years, we find the consistent trend that we have always held a certain predilection for placing humanity at the centre of importance. We also find the trend, that the closer to our own time we come, the more this notion has been found wanting.

It is not that I hold humanity to be unimportant, quite on the contrary. It is just that up to now, most of us have failed to place humanity in the proper context.

Sure we are important, even terribly so, if you like, *but so is everything else!*

The tiny sparrow is incredibly important, so is the earthworm or the cockroach or the grass we so carelessly trample. And not only those things we consider living, because here too we have made a mistake. There is no such thing as *dead matter*.

A rock, is not a dead lifeless thing, as most of us believe. It is a veritable dance of atoms. Every part of our Universe, including that 'lifeless' rock is dancing and vibrating at close to the speed of light, the molecules in it vibrating imperceptibly to our coarse and dull senses. An analogy may be useful in explaining what I mean.

Let us imagine the Universe as a closed sphere of some impenetrable, unbreakable material, with only water in it.

Let us also say, that for some reason, perhaps a drop of temperature outside the sphere, a small quantity of that water congealed to ice. The part that turned to ice is very small, and the sphere is extremely big. Now let us suppose, that the minute ice particles started to say that they were different and separate from the water in the sphere.

From our vantage point (on the outside of this sphere filled with water) we would be amused by their ignorance and simplicity. In fact, we would probably endeavour to explain to them that they are merely water in a different form. Stronger, for they would be able to retain their shape while water in its natural state cannot, but water all the same.

If we explained to the ice particles that essentially, there is no difference between them and their surroundings, they may feel dejected, small, meaningless.

If this happened, we would then probably try to explain to them, that although they are still essentially water, they are indeed worthy creatures.

After all, are they not solid? Do they not have the ability to freeze a small amount of water at their surface, and thus of changing their environment and of gradually growing? Becoming bigger and more aware?

We would explain to them, that simply because they had made a mistake about their own nature, it did not mean they were any less. And if the ice particles were receptive, they would get together, consider what had been told them, examine it, test its truth and once this was done they would probably say : "Oh well! it looks as if we have lots to learn", and would then carry on with their lives. Which thanks to their new awareness would now be instilled with a feeling of kinship with the rest of their Universe, the water in their sphere, because they would now know they are themselves like the very water they live in, and are from it and part of it.

Should one day the ice particles be able to manufacture a useful machine for themselves, but which polluted the water by producing strongly ionised particles, they would make very sure that the machine was perfected until it could operate in harmony with the water. After all, they themselves are only another form of water, and they would not want to pollute or destroy their environment or themselves, since, they are one and the same.

It is easy to see the similarities between the water particles that had frozen into ice in the sphere and us humans in our own 'sphere'.

One may argue that there is a lot of difference between a person and a piece of ice, or a boulder. A boulder cannot walk or drive a car, it is (as far as we know) incapable of conscious thought or communication as we presently understand it. Generally, a boulder can perform a lot less than a person.

Their basic make-up however is exactly the same.

I am not talking of DNA. Obviously rocks do not have DNA as plants and animals do, but DNA is after all merely a long and complex chain of molecules. Molecules in turn are made up of atoms, the so called building blocks of matter, and a rock has atoms. The very same type of atoms a human has, in different proportions and alignments, but atoms all the same.

And have we not seen that atoms are merely a form of energy?

Therefore at the root, there is no difference between a man or a rock of similar mass. They are both created out of what we call energy.

The only difference is the way in which that energy is sculpted.

If we accept that matter and energy are one and the same, and there should be no reason not to, since it has been proven time and again in our everyday lives, in fact, our very existence depends on mass from our Sun being converted to energy; then, some radical new concepts also become true.[21]

For a start, we can immediately see that absolutely everything in the Universe is ultimately made up of the exact same "stuff".

---

[21] Einstein's equation, $E=mc^2$, (Energy is equal to Mass times the velocity of light squared) is famous just exactly for the reason that it proved correct in saying that matter and energy are one and the same.

Everything in the Universe is made up of either energy or matter or a mixture of the two, if we like to differentiate, but that differentiation is not a true reflection of the fact that actually, there is no difference, since matter *is* energy.

This means that inherently, there is actually no difference between the basic make up of a rock and a human being, or of a planet and a grasshopper. Absolutely everything is energy, and the Universe is merely energy in all its many different forms.

The only difference between sentient beings and so called inanimate objects, is merely a level of awareness. Something we call consciousness.

If humans are being observed by aliens, we are sure to go down in history as the most arrogant race in the Milky Way.

We are so proud of being sentient and of being conscious we seem to think we are the *only* sentient life.

We like to think of ourselves as different and separate from the rest of the Universe. In fact we are so egocentric we see ourselves not only as different and separate from objects and other creatures, but even as different and separate from fellow man.

War is the sad truth that reflects this fact.

Just like the ice particles, we refute the initial premise that actually the whole Universe is like us and alive. That we are made from it and are of it.

If the ice particles (whom originated in a closed sphere filled with water only) were sentient, then we can only assume that the water in the sphere is also sentient, perhaps not to the same degree, but if it was *not* sentient to *some* degree, then the ice particles, having originated from it, could not possibly be sentient either.

Whatever component gave the ice particles what we call "life", must also be present to some degree, in some form, in the water which surrounds them, and from which, *ultimately*, they came from and are composed of.

Similarly, since ultimately everything in the Universe is merely energy, and since we come from this Universe, and therefore are of it, we can only assume, that the whole Universe is in some way sentient.

If this was not the case, then we could not be sentient ourselves.

Not too long ago, it was thought ridiculous to consider a plant as sentient. Today it is known as a fact that a plant is not only undeniably alive, but also that it is sentient in some way.

Typically, we describe a plant's form of awareness as 'primitive', simply because it takes the form of an intelligence we are not familiar with.

But then, as history shows, humans are masters at classing all that they do not understand as inferior to themselves.

The spiritual beliefs of the Red Indians for example, were thought to be similarly inferior and 'primitive' by the Catholic bigots of the sixteenth century (as well as by a large number of bigots still around today!) and yet if you take the trouble to investigate the matter a little, you find that the Red Indian attitude towards their environment was a lot healthier than the one exhibited not only by the invaders of the time, but by those exhibited by the average American even today.

We don't consider plants as our equals. Why? Because they seem to be less conscious than us. But is this truly so? Who is to say that plants do not communicate telepathically and are in fact more evolved creatures than we are? Sounds like science fiction? Surprise, surprise: When some industrious scientist decided to attach lie detectors to plants, he discovered that they do have a 'primitive' range of emotions.

Furthermore, when a plant is cut down in a forest, similar apparatus attached to it and to nearby trees registered not only the 'scream' of the dying tree, but the sympathetic screams of his nearby companions. Proof that some type of telepathy, or at least communication, is indeed at work.

As for the idea of plants being more evolved than us, consider the following concept:

Pretend you are a young alien being, living on a planet far away, that knew nothing of Earth and its life forms.

A friend of yours, that has visited Earth, tells you that there are two main types of creatures on it.

One set of creatures, he tells you, has reached its nearby moon and sent probes to other planets of its system and even outside of their solar system altogether.

The other species from that planet however, is immobile and rooted to the spot.

So far, you may be inclined to believe that the species that reached their moon is the more intelligent of the two.

Your friend, however, being a wiser member of your race as a result of his many travels, cautions you from jumping to conclusions.

"The race that can move around," he says dispassionately, "goes about destroying and killing every other creature on their planet, including themselves, and making extinct at least 70 species of creatures on their planet every day. They are so aggressive that they destroy millions of creatures every second that goes by.

They are particularly violent towards the creatures which are rooted to the spot. In one area, called the Amazon by them, they destroy about one acre of these peaceful, immovable beings every second.

The still ones on the other hand, produce oxygen which is essential for life on that distant blue planet, and many of them also produce a kind of growth they call fruit, maize or by several other names, that can be eaten by many creatures; including the ones that kill everything, (whom are known as Humans). The still ones (known as Plants) give of this food freely, asking for nothing in return."

Shocked by this new bit of information you may be inclined to ask your friend what relative use the humans have when compared to the plants.

"Well, if we were to destroy all the humans, no other creature would suffer. In fact, many would thrive, and eventually find their balance, as is the way of nature.

If however we were to destroy all the green ones, nothing on that planet would survive."

So.... Who do you think is the better or more evolved creature between the two now?

A young and rash alien, such as the one you may have imagined yourself to be a short while ago, may be prone to say :

"Let's blast all these ugly Humans away. After all they are a nasty breed of critters, and at least we'll be saving the Plants!"

A wiser view however, would be to see that while indeed the Humans are presently about as useful to their planet as a case of the bubonic plague is for an individual, they have the potential to become at least as useful as the plants, and perhaps even more so, because while plants are excellent at helping life in their immediate vicinity, they are not very good at propagating it across the stars.

Of course, before the Humans can be considered even marginally useful, they would have to learn something from the plants with respect to the treatment of their surroundings.

Furthermore, the idea of letting Humans travel to other stars and their corresponding planets (possibly inhabited at that) before such a change in attitude has taken place, would be anathema to a race of beings that wants to promote life (such as the Ancients would be if my theories concerning their motives are correct)[22]

In view of the importance of our surroundings then, it may be wiser on our part to take ourselves a little less seriously. Who knows, after all the ancient Greeks may have been right. We may after all, be a mere form of amusement for the 'Gods'.

The peanut, recreational food that it is, may be a closer relative to us than we think!

Let us then, at least have the grace to fulfil our role, as eminently as the peanuts fulfil theirs.

In conclusion then, the important message of the Face on Mars is not so much that we may have been genetically created by aliens, because in the end this is irrelevant.

But rather, by showing us what at first seems to be a piece of information that diminishes our own view of ourselves, the Face is in fact showing us a far more important thing. It is placing the human race in context with its surroundings. And the point which perhaps stands out most of all in this view, is that we need to learn how to co-operate with our environment, since we are a part of it.

In fact, if we were to accept that we are the genetic creation of a more advanced race, we would not be diminishing ourselves at all. What would be happening instead, is that we would now have been made aware of more of the amazing Universe we reside in, and in comparison to this vastness, we suddenly seem very small.

Our new ability to see a bigger piece of the Universe, while seemingly shrinking us, has paradoxically made us come closer to the truth and hence made us more capable and aware than before.

This is a process that is familiar to any serious athlete. At first one moves the body in an uncoordinated fashion, but soon learns the correct way of moving. As soon as this realisation is made, the new athlete suddenly feels very incompetent. But actually, it is now that the athlete has begun to learn the correct body movements!

---

[22] This of course may also be a clue as to why the Ancients may have destroyed Mars and/or the Atlantis outpost.

The reason he feels incompetent, is because at this point he has actually grasped one aspect of his discipline and is awed by how much he still knows nothing about. In learning one movement he suddenly becomes aware that there are thousands more which he still needs to learn of, never mind perfect. Over time, he learns more and more movements, and at the same time becomes more and more able, but paradoxically, he also becomes more and more aware of his own lacks and limitations.

It is for this reason, that the best teachers of a physical art are often the most humble, although possessing vast knowledge.

The human race as a whole is undergoing a similar process of growth. There is no need to feel small simply because your entire race may have originated in the test tubes of some other beings. Quite the contrary. Feel awed. Feel joy at the amazing vastness and complexity of the Universe, as well as its perfect and beautiful simplicity.

To borrow the words of Wayne W. Dyer: Universe is a composite word, made up of *uni* —meaning *one* from the Latin root— and *verse*, which of course is a line of a song.

In other words, Universe means *Onesong*. And truly, each tiny particle of our Universe is a note, each with its own unique frequency and tonality. Together, all these particles and rays create the most beautiful song of all, the Universe we live in.

# 7      Other Ideas

## Race Memory

The main theories concerning our true origins could be said to have been explored at this point, but there remain certain loose ends to tie up, that while not necessarily being evidential in nature, certainly merit mentioning. Perhaps foremost among these is what Rupert Sheldrake calls Morphic Resonance.[1]

In his book *A New Science of Life*,[2] Sheldrake reports on experiments with mice performed by W. McDougall at Harvard in 1920. These experiments, in my opinion, proved quite conclusively that indeed there is such a thing as race memory. But this should not be the discovery of the century. After all, if one follows Darwin's evolutionary arguments, a sort of memory of the cells can be allowed for. It is the theory of evolution after all that states that were a mutation is useful to survival it will become more and more pronounced in the later generations.

What is astounding about McDougall's experiments, is that they also proved conclusively that there is such a thing as transference of consciousness between sentient beings.

A specially bred strain of white mice were used because they reproduce rather quickly, and thus offspring are readily available for next generation testing.

Basically, the experiment involved the placing of a number of mice in a containment area which incorporated two levels. The bottom layer being water filled, so that mice placed in it would have to swim, and in order to reach the upper level, climb up one of two ramps. One of the ramps however was illuminated and connected to a weak electrical circuit which would deliver a small shock to a mouse that tried climbing it.

The idea was to try and teach the mice that they should not climb up the illuminated ramp.

---

[1] For a definition of this term see footnote number 10 later in this chapter.

[2] A New Science of life, pg. 185 to 191. See bibliography for details of this book.

The first batch of mice, took as many as 300 or more tries before they learned to associate the lighted ramp with the electrical shocks, but once they learned to do this, they almost never made the mistake of climbing up the lit ramp again. The interesting thing, is that the offspring of these mice learned to climb up the unlit ramp much quicker than their parents, and the next generation, in turn learned quicker than both their grandparents, *and* their parents, and so on. Eventually, a point was reached where offspring would learn the skill after a little more than 30 tries. An improvement by a factor of almost ten, or put in percentages, an improvement of nearly 1000%.

Statistically, this kind of relationship is considered to show a definite relationship. In other words, the chances of this sort of thing happening by chance alone are practically zero.

But the story does not stop here. W.E. Agar and colleagues, decided to repeat the experiment with an important addition, some of the mice from the control group (which received no training) were to be tested at regular intervals, in order to see how quickly or slowly they learned the skill when compared to the mice being trained.

If a group of untrained mice was tested, they would not be reintroduced in the experiment, since otherwise, their offspring would be similar to the offspring of the mice who underwent training.

The results were astounding.

As already shown by McDougall, the mice that were being trained produced offspring which would learn the skill quicker than their parents, but amazingly, *the exact same tendency was shown by the untrained line of rodents!*

In fact, this trend had already been observed by McDougall to some extent. McDougall had also used a control group of mice, but he tested these only infrequently, and not with the same regularity as did Agar.

Despite this, he remarked that it was disturbing for him to have found a tendency in his control groups for the years 1926, 1927, 1930 and 1932, that showed: 'a diminution in the average number of errors from 1927 to 1932'.

In related experiments, performed by one of McDougall's critics, (F. A. E. Crew) some of the untrained mice learned the task immediately when tested, never receiving a single shock! And right from the beginning, his mice (which were of the same inbred strain used by McDougall) all learned the skill much quicker than McDougall's mice originally did.[3]

If evolution or individual experience was all there was to learning new modes of behaviour, then these experiments would not have yielded these results. In fact, the mice were behaving as if they could somehow communicate to one another the danger of climbing up the brightly lit ladder.

In advancing the theory of morphic resonance, Sheldrake came to the conclusion that the mice were communicating on some level which, while not being fully conscious, nevertheless was useful in passing new skills to a race of creatures.

In short, he put forward the theory that whenever a creature learned a new skill, or whenever a new event occurred, this would set up a field of some sort[4] that would make it easier for the next creature to learn the same skill or for the new event to occur again under similar circumstances. The larger the number of creatures that learn the new skill, the more powerful the field becomes, thus facilitating the next generation's learning of the skill.

A similar phenomenon has been written about in a book called the *Hundredth Monkey*[5] where supposedly, monkeys residing on islands of the pacific learned to eat sweet potatoes that were covered in oil by first scrubbing them in the sea.

At first, none of the monkeys would eat the potatoes, until one intrepid ape decided to scrub the tuber in the sea before biting into it. After this, the other monkeys that saw the event followed suit, and

---

[3] Crew's experiment had some shortfalls though. His practice of letting mice breed only between brother and sister encouraged inbreeding of his mice to the extent that several generations later, many were deformed and unusable for testing. In view of this, the fact that the mice were still able to learn the task much quicker than the original batch of McDougall's experiments, vindicates the theory of morphic resonance to the nth degree. Crew's mice on average, right from the beginning, had scores comparable to McDougall's after more than 30 generations of training.

[4] Sheldrake calls them morphogenetic fields.

[5] I have not personally read the book, but the person who described to me the contents of the *Hundredth Monkey* is one I consider scrupulously honest, and at any rate, Rupert Sheldrake's experiments are well enough documented that the book *The Hundredth Monkey* becomes almost irrelevant.

pretty soon, quite a few had learned to eat the oil stained sweet potatoes.

Interestingly enough, after about one hundred monkeys had learned how to do this, other monkeys from entirely different troops began to do the same. But even more importantly, *monkeys on other islands altogether that had never seen this method of cleaning food, began to do the same, very shortly after the first troop of monkeys learned the skill.*

The interesting thing, is that for a relatively long period of time prior to this, no monkey would eat the potatoes, but once a critical number of monkeys learned the skill, baboons all over the place began using it.

This is not science fiction, it's real. And inanimate objects are as prone to it as monkeys. It has been found that once certain crystals which were very hard to produce have been successfully synthesised once in one laboratory, in a short period of time, scientists all over the world suddenly start synthesising the same crystals, *without any information having passed between the scientists.*

Once a crystal forms in one way, it becomes easier for crystals the world over to form in that manner. The more crystals you have, the easier it becomes, the same goes for people learning a new skill.

It would be interesting to know for example, how quickly illiterate people learn how to read and write today as opposed to illiterate people a thousand years ago, when reading and writing was far less common.

Morphic resonance is also evident in sports.

For many years, there used to be a ten second barrier for the hundred meter dash. No one seemed to be able to run a hundred meters in less than 10 seconds.

Eventually, on June 20, 1968, James Hines broke that barrier by running one hundred meters in 9.9 seconds. Shortly after that, the same record was achieved by Ronnie Ray Smith and Charles Greene, who also ran one hundred meters in 9.9 seconds.

Today, we know that several people have run one hundred meters in under 10 seconds,[6] but for years, this was not the case. It really does

---

[6] No doubt some are bound to point out that increases in living standards allow longer and more intensive training periods, and that drugs such as steroids play a part, but even so, there is enough evidence to suggest that the trend is a real one, quite apart from these factors.

seem that consciousness is one. After one person manages to break a barrier, it suddenly becomes easier to do so for many others.

It also follows, that if long ago our ancestors had certain skills which we have since lost, they should still be buried somewhere in our subconscious or into our cells, or perhaps in the make up of our electromagnetic aura. In other words, skills which we have lost are not truly lost, just buried deeply.

And this too, is a fact that any one of the people that was involved in the border wars with Angola and Namibia in the South African Defence Force will attest to. The men that undertook these forays into enemy territory could not wash, they could not brush their teeth and of course they could not smoke, for the very simple reason that they would be smelled by the enemy if they did so.

I have met some of these people, and each of them has an uncanny sense of awareness. To some lesser degree, the same sort of thing happens to people who train in martial arts, although, since in this case, one's life is not actually in mortal danger every day, to develop the same sort of skill via martial arts takes considerably longer, as well as a particularly sensitive personality to begin with.

At any rate, apart from the psychological scarring which these ex-soldiers have undergone, which in certain cases is quite severe, even to the point of being pathological, they have some uncanny abilities. In pitch darkness or wearing a blindfold, they are able to know whether a person is in the room they have just entered or not.

The infamous Tunnel Rats of the Vietnam war are another example. These people's senses, developed to such an intensity that they would be able to sense another human being at a distance of several meters in pitch darkness and total silence.

But does it take the constant threat of death to regain these lost skills? I do not think so. It may take an unusual amount of concentration, or certain conditions, perhaps the pursuing of a particular interest for many years, but I believe that we have access to race memory under conditions that do not necessarily need to be life-threatening.

One such condition perhaps, could be a deep kind of relaxed concentration, during which time, although relaxed and not tense or stressed, a person's mind is nevertheless very focused.

The inspiration of genius of course, has long been said to arrive at just such moments.

In taking morphic resonance, race memory and such things into consideration, it becomes very interesting to look at certain 'amazing coincidences', as well as certain social peculiarities, which may have some bearing to the Face on Mars with all its relevant implications.

It is a well known fact for example, that for whatever reason, since the dawn of science, Mars has been held to be the home of Martians.

In the early seventies there were plenty of jokes going about concerning little green men from Mars, and even today, when we should know better according to orthodox science, the term Martian is immediately associated with a little green man.

The reason that Mars has been held to be the home of such unlikely creatures, could in large part be attributed to Percival Lowell, whom, late in the nineteenth century, claimed the channels on Mars first described by Giovanni Schiaparelli in 1877, were artificial in origin and built by Martians.

For whatever reason, the idea that Mars was populated by intelligent beings, remained very prominent until relatively recently. Less than fifty years ago in fact!

The Guzman Prize is a case in point. This prize was established on the 17 of December of the year 1900 and was to be awarded to the first person that established some sort of contact with beings from another world. It consisted of 100,000 francs, which was quite a sum back then. There was however one condition: the planet Mars was excluded, as it was thought too easy to establish contact with Martians.

When Mars was first photographed by a probe in 1962, the then director of the Jet Propulsion Laboratory, made a speech where he stated that he was convinced, as he'd always been, that we would eventually find some form of life on Mars.

Is there some race memory however, quite apart from anything Percival Lowell may have said, that makes us particularly inclined to believe that there is (or at least was) life on Mars?

Even now, when we 'should know better', and quite apart from the Face and Pyramids, isn't it easy to imagine that thousands of years ago the Martians were real?

Doesn't the idea of lithe Martians constructing Titanic monuments and living in idyllic palaces surrounded by beautiful gardens stir something in us?

Why is it, that these sort of images, some of which have been immortalised in the works of fiction of several writers, appeal to us so?

One may ask too, how is it that, presumably acting on intuition, since he lacked the necessary equipment at the time, Jonathan Swift, creator of *Gulliver's Travels*, wrote in his *Journey to Laputa* of 1726 that Mars had two tiny moons more than one hundred years before their discovery. But even more astonishing, is that he described their motion around the planet with uncanny precision.

For me however, one of the most impressive kind of 'coincidences' or perhaps, as Jung called it, I should say synchronicity, is a short story I read quite a few years ago.

It was written by a man that I consider a master of the science fiction novel, Arthur C. Clarke.

Arthur, along with his now deceased friend Isaac Asimov, could be considered to be the grandfather of modern science fiction. Both he and Asimov, wrote a type of novel that is seldom seen today. Some connoisseurs call it Hard Science Fiction.

Whenever I enter a bookshop I shudder at what you find nowadays in the section labelled science fiction. All too often there is too much fantasy[7] mixed in with it, and what there is that is not of a fantasy genre has little to do with science, and quite a lot more to do with fiction. More often than not, seriously warped fiction at that, with little basis in reality, never mind science.

It is my sincere hope, that in the near future, Hard Science Fiction will not only make a comeback, but finally be awarded its own shelf, separate from the drivel that it's now associated with.

At any rate (forgive my rantings, science fiction is a subject close to my heart) the point is that, if I recall correctly, the short story in question was written way back in 1950-something. I can't recall if it was 1952 or 1959, but it concerned a Moon landing. I cannot even recall if it was supposed to be the first manned Moon landing or not, but I suspect it was.

---

[7] For those unfamiliar with the term, fantasy books usually deal with magic, dragons, lost princesses etc.

The story started and continued in that fascinating manner that is Arthur's. He described the technology used with such precision and simplicity of style that you were transported into the novel.

Arthur being Arthur, you more than half expected some twist at the end, especially since it was a short story, but the description of the events taking place was so interesting, that they lulled me into a sense of relaxation, which almost made me forget to be on the lookout for the final twist. The fact that Arthur's story could do this despite the fact that I was born only days after Armstrong set foot on the Moon, is a testament to his ability to predict the future.

The equipment he described was almost exactly identical to the one actually used several years later in the *real* Moon landings.

The story came to an abrupt end when the astronaut in the story hops over a small rise and comes face to face with a small pyramid.

That's right, Arthur C. Clarke placed a small pyramid on the Moon more than twenty-five years before pyramids were found on Mars! Of all things, why a pyramid? What made him choose that particular shape?

Of course some people will put it down to coincidence, but I for one am not so sure, and certainly, if ever there was a candidate for receiving inspiration from some kind of race memory with respect to real Martians in our ancient past, then Arthur C. Clarke would be it, considering how many books and short stories he's written which focus on space travel and alien customs.

Clarke also wrote a story in which the human race, through the aid of benevolent beings that look like devils, reaches a critical point where eventually all the individual minds of people coalesce into one overall mass-mind. There remain no humans as such, because they all die from a physical point of view, but they evolve to a higher plane of consciousness, where they continue to exist eternally as a whole race.

Isaac Asimov presented a similar idea in a short story of his own, where through the use of ever more powerful computers, the Universe is evolved to a higher plane of existence. The one question the supercomputers cannot answer however is: "What will happen when entropy has destroyed even the last star?"

Over the aeons, the computers and people all evolve and are brought to this higher, hyperspace dimension, until nothing is left in the physical realm. At this point, the computer finally answers the question by creating a physical universe once again; making this perhaps the first theory of an exploding/imploding Universe, long before astrophysicists came up with the refined version.

Both of these concepts, fit rather well with the one of an advanced race of beings going about trying to evolve intelligent life wherever they can in the galaxy, for the purpose of reaching a similar critical mass of consciousness.

Another example of some sort of race memory, is Star Wars.

The movie that is, not the military plan of the Reagan era. Why was a series of science fiction movies so popular? Especially since there was such a large gap between the films.

After all, although the special effects were amazing for the time, and are still pretty good even for our own day and age, the plot was relatively simple. Bad guys against good guys. But Star Wars struck some deep chord in many of us.

"May the Force be with you" became a classic phrase. Why?

I maintain that it's because the concept of the Force is pretty close to the way things work in reality, and furthermore because the idea of a huge conflict in space in which a planet perishes (Alderaan) once again struck a chord somewhere in our primitive race memory.

Of course, if one accepts my comments in the previous chapter with regard to chaos theory, and the idea that every single action may be pre-ordained by the infinite accuracy of the Universe, as well as what this may say about the purpose of our brain as a supercomputer to which time is not a factor, then race memory becomes no longer a mere possibility, but a virtual certainty. And if there is even only a part of truth in such ideas, then it would make sense if occasionally, our mere 5% or so of 'connected' brain matter, gave us an intuitional flash of what things may have been like on Mars untold thousands of years ago, or perhaps of what things will be like in our future, which brings me to a final example of this sort of synchronicity.

One night in August of 1877, Asaph Hall, the director of the United States Naval Observatory, pointed his telescope in the direction of Mars. His intention it seems was to indeed look for any satellites Mars may have, since the idea had been around for a long while, but of course, no satellites had been found to date, despite repeated observations by astronomers. On that night though, Hall discovered the two tiny moons of Mars, and as he saw them for the first time, shortly after his observations had begun, he was suddenly grasped by an intense feeling of 'fear and terror'. So strong was this feeling, that it was he that choose their names: Phobos and Deimos.

Is it not strange though, that two small asteroid-moons could create such irrational emotions in the mind of a man that was in no way threatened by them in any way, being millions of miles away from them, and in fact stood to gain a great deal from the discovery in terms of prestige and acknowledgement?

Not so strange though, if one considers the emotions that may have been felt by some prehistoric distant ancestors of ours at the sight of the incoming asteroid-missiles that spelt a sudden, horrible and complete destruction for their entire planet.

# The Principle of Morphic Resonance in Nature

Regardless of tests and experiments performed to prove or disprove the reality of morphic resonance, it is also very useful, as always, to use our own powers of logic and observation in this regard.

I call the process whereby I reach a conclusion logic, but perhaps, a better name for it would be 'patterns of relationship'.

One should keep in mind that our entire life, is not composed of rigid inflexible rules, but by patterns of relationship.

These patterns of relationship, do follow certain strict rules, but we must be careful not to think the rules are the reason for being.

The reason for being, is the pattern, not the rules.

Mathematics is considered to be the most inflexible of disciplines, as well as the foundation of logic, and quite correctly so. Yet, maths is not a dead, inflexible discipline, in fact far from it, maths is the one subject that perhaps shows the dynamism of the Universe best of all.

Those of us that disliked or were bad at maths in school, found it difficult not because we were stupid, but because of the way it is taught.

In the classroom, we are mostly presented with artificial, static, and dead situations. Although these are necessary building blocks to learn, many of us find them boring or lifeless. So much so, that we end up rejecting the whole discipline.

An analogy may help to underscore what I mean.

1) $1 + 2 = 3$

   One plus two equals three.

   This is the lowest order of mathematics. An inflexible, dead statement. It has no mobility, no dynamism to it.

2) $x + 1 = 3$    therefore $x = 2$

   If x plus one equals 3, then x must be equal to two.

   This statement, in contrast to the previous one, has added an element of dynamism.

   At first, x is an unknown, it is only by its *relationship* to the number one that we are able to discover what x actually is. But in the bigger patterns of life, it is more likely that we do not have any such clear-cut arguments.

Although an improvement on the first one, this second statement is still pretty dead and lifeless when taken in a *real life* context.

3) $x + y = 3$

x plus y equals 3.

This statement begins to approach more closely the way that life presents itself to us in an everyday context. Taking this statement all by itself, it seems impossible that we would ever guess the correct identity of either x or y.

For example, x could be equal to any negative number and y to a positive number that is bigger than x by three units, or perhaps it is vice-versa, or maybe, both x and y are positive.

In the present form of the statement, there is no way to know, but life, always gives us more information than this, because life is not something that occurs in isolation, but rather in relation to all that surrounds it.                                                   If for example we knew that x and y referred to people being picked from two different teams, then:

if $x = 0$ then $y = 3$
if $x = 1$ then $y = 2$
if $x = 2$ then $y = 1$
if $x = 3$ then $y = 0$

So, from having an infinite number of possibilities, we have already reduced the outcome to one of just four cases.

We might also be inclined to say that since both team x and team y are mentioned, it might be reasonable to assume that at least one person would be picked from each team, and this would immediately give us a better idea of what the results might be, since the equations were x or y equals zero would now carry less weight than the other two.

If we furthermore knew that team x had (in our opinion) a better class of player, then we may even go so far as to hazard a guess that the third equation, were x is two and y is one, is the most likely outcome.

The process used in the above example to get at a value for x and y is indeed rooted in mathematics, and hence in firm logic, but nevertheless it is a dynamic process, with several possible results.

But in fact, the process of life is so dynamic, that one can never be completely sure of an outcome.

If, for example, a hurricane comes along and wipes out both teams, not only is our guess for the values of x and y meaningless, but the whole equation becomes invalid.

So, in fact, it may be more accurate to say that in real life, we are most often presented not with one of the three equations mentioned so far, but with a fourth one, which incorporates an even higher level of dynamism.

Namely:

$$x + y = z.$$

Every decision we make in life is just an extension of this process, and hence just a 'best guess'. If however we are careful to stick to the rules of logic, we can still come up with some pretty amazingly accurate outcomes, despite all the dynamism involved.

There have been some people, eminent scientists among them, that have rejected Sheldrake's theories concerning morphic resonance outright. Some going so far as to say his book may well be a candidate for burning!

Surely such an attitude, so reminiscent of a witch-hunt, is far from scientific. But at any rate, I would like to point out two things.

Firstly, Sheldrake's theories are receiving more and more support from the rest of the scientific community, and it may be proper to point out here, that Sheldrake is a competent scientist himself, being a biochemist and having studied in Cambridge and Harvard, and having received a string of letters to append to his name for his troubles.

Furthermore, he's been a member of more than one emeritus organisation, and while some people have taken a dislike to his ideas, none can seriously dispute the man's competence in his field.

Secondly, and more importantly, when keeping in mind the points mentioned a little earlier with respect to the dynamism of life and how despite it one can still make some pretty accurate deductions, Sheldrake's theory of morphic resonance, falls squarely in the very centre of the probability circle that he is correct when we examine the idea from our own point of view.

Allow me to expound a little.

The concept of resonance is one that is found constantly in nature. At this point then, it may be wise to define resonance in easily grasped terms.

A good example of resonance, is the shattering of a glass by sound vibrations of a particular frequency. Generally speaking, unless the sound waves are so powerful that they actually break the glass out of sheer brute force, one can play any number of notes in the vicinity of a glass and it will not shatter.

If however, a particular note is picked to which the glass responds and it is kept steady, the glass will begin to vibrate at the same frequency as the note. But resonance does not end here.

By some means that are not fully understood yet, the action of setting up sympathetic vibrations in the glass, actually amplifies the vibrations to the point where the glass breaks.

In other words, it would seem as though we are getting more work out of the glass than what we put into it.

This principle of resonance does not just apply to glasses that are shattered by the powerful lungs of some opera singers, it applies to virtually everything.[8]

The word laser is an acronym for Light Amplification by Stimulated Emission of Radiation.

The way in which a laser works, is by stimulating a number of particles to vibrate at a particular frequency. This is done by passing light through a prism of a particular colour, which makes some of the particles vibrate at the frequency of that colour.

The light is bounced back and forth through the prism by mirrors several times, each time, increasing the number of particles that vibrate at the required frequency.

It is not necessary to bounce the light back and forth enough times to 'convert' each particle to the new frequency however. It is only necessary to bounce the light back and forth until about one thousandth of the particles are vibrating at the required frequency, because after this, all the other particles 'convert' to the new frequency almost at once.

---

[8] Almost any substance can be set up to vibrate sympathetically, just like the glass, by similar means, but I do not refer only to sound waves here, I maintain that the process of resonance is applicable to anything, and for those things that show no evidence of this, I would theorise that we simply haven't yet determined to what *process* the object responds in sympathy to (i.e. magnetism, sound, light etc.).

In other words, by changing the frequency of only one thousandth of the particles involved, we find that the other particles automatically begin to vibrate at the new frequency. And in doing so, amplify their overall potency to the point that they can cut through steel as if it was butter.

Do you recall once more how if we reduce a human being to his component atoms the overall magnetic influence of these atoms is far less than that exhibited by the same atoms structured in such a way as to produce a living person?

Well, this too, is a form of resonance.

Whenever you have something that gives you more than the sum of the component parts, what you have is resonance. The same principle applies to electricity, and in particular to the inventions of a man which we will be discussing again later, Nikola Tesla.

A transformer for example, works by resonance.

What happens in a transformer, is that a relatively small current is passed in a circuit. This in turn induces a current in a second circuit which via a process of resonance amplifies the initial current to a much higher one. Depending on the transformer, this process can be repeated several times, giving a larger and larger current each time. Also, the same method can be used in reverse to lower the initial current so that it can be used for supplying power to an item requiring a small current only.

In this way, a normal wall socket outlet, can be used to provide power to machines that require a high current as well as to more delicate ones that require only a few volts.

◆◆◆◆◆

Scientists are prone to saying that nothing can give you more than its component parts.

Certainly, this concept was fundamental to 19th century science, and in large part, it is still so today, although physicists in particular, are quickly changing their minds with regard to this topic.

Receiving more out of life than what you put in is supposed to be impossible according to the black and white world of science, but in fact, this happens every time we encounter the principle of resonance. From a small initial input, we can receive a larger output.

It would appear too, that resonance is, like gravity, one of the subtler phenomena of life, and yet a fundamentally important one.

Gravity cannot easily be observed in say a piece of rock, even if the rock is several tons in mass, and yet, gravity most certainly exists. Without gravity, none of us would be here.

Resonance, it seems, is another effect that is little understood and usually difficult to observe, and yet, without it, magnetism would not exist, and of course, without magnetism, once again, life would be impossible.[9]

On a subatomic level, every particle exhibits some kind of magnetic properties, electrons float around the positively charged nuclei of atoms, supposedly as a result of their negative charge.

If we were to eliminate the force of attraction between positive and negative, (which is all that magnetism is) the Universe would literally collapse at its most fundamental level.

In short, resonance, or the principle of getting more out of a whole than its component parts add up to, is fundamental to the structure of our Universe.

If this is the case, then it would seem that the theory of morphic resonance is not only correct, it's very important for our understanding of how the Universe works.

---

[9] Magnetism in a bar magnet is due to the orientation of its particles. Once again, by aligning in one particular way or frequency, the component atoms of a bar magnet produce an effect superior to that of a similar item whose particles are randomly aligned. Similarly, the particular alignment of molecules in a person produce a magnetic field of a particular frequency, which is different for each individual.

# The Place of Science in the Scheme of Things

The problem with morphic resonance,[10] is not that it is an invalid proposition, it's just that orthodox science is only now approaching a point where it can begin to measure it.

Science, you see, likes to give the impression that it can and will eventually measure, and explain everything in scientific ways, that is, in ways that can be tested, proved, and nicely compartmentalised in a little box somewhere on the vast shelf of knowledge it represents.

This unfortunately is not the case. Science has a long way to go before it can explain everything in the Universe in scientific terms, and personally, I do not believe it ever will achieve this unreachable goal. Not even if the human race were to exist for another fifty trillion years. The reason is simple.

Science, you see, is only a tool that we use to probe at the unknown, and that is *all* that science is, which means that sooner or later, we will find a new and better tool.

Orthodox science is not the holy grail of ultimate truth, it's just a better method of discovering the truth than the oppressive religious system of the dark ages was; that's all. It's a system for expanding our knowledge, but as we progress, we are bound to find new and better systems.

Let me make it clear, that I do not intent to vilify the scientists. The scientific method, has served us well, and undoubtedly will continue to do so for quite a while yet, but it's imperative, that we realise that science is only a process that we have learned to trust more than blind faith in a preacher of dubious character.

If we compare the travels of humanity to that of a single person, we may come upon a useful analogy.

It may be fair to say that three thousand years ago, this man, who represents the entire human race, was not only blind, but also trussed up like a salami. He could only wiggle from side to side a little and tell even less of his surroundings. The era of Jesus is when this man

---

[10] It may be time to point out that morphology is the study of the formation of plants and animals and their parts, hence, the term Morphic Resonance means the process of resonance when applied to plants and animals (living things).

finally found freedom from his bonds, but he was of course still blind with no idea of where he was or even *what* he was. Gradually he learned to use his hands and body to feel his way around. One day, relatively recently as time for this man goes, he found a walking stick, which helped him a great deal because he no longer needed to crawl around on the floor anymore in order to go places, also, he didn't need to bump into things with his hands anymore, risking to loose a finger if he touched something unfriendly. He liked the stick so much he gave it a name.

He called it Science, and soon, the man became obsessed with this new toy and he improved it and refined it until it was straight and narrow, weighing much less than the original crude stick he'd picked up. But this man is now on the point of discovering a curious protrusion on the side of a wall. That protrusion is a light switch, and if and when he presses it, the man will find out that he's not blind at all, but that he's just been living his life in an underground room that is pitch black.

Just like the stick was a major discovery for him, so will the light switch be an even bigger surprise for him. Once he's found the use of his eyes though, he'll no longer need a walking stick, except perhaps to poke at things that may be too dangerous to touch with his hands right away.[11]

I cannot stress enough, that all that science is, is a way we have chosen to interpret our reality that affords us a view that is more in keeping with what we are familiar with in our everyday lives, but it by no means describes that reality completely, or sometimes even accurately. In fact, the only *real* way of discovering the truth is by using *our own minds.*

It is our *minds* that make us think science is a better way than the authoritarian religion of the dark ages. It was our minds two thousand years ago that made many of us believe that Jesus's theories were a better way than the previous system, and it will be our minds again, that will shortly begin to convince us that our intuition is a far better guide than we have previously suspected.

For this reason, it may not be too far from the mark, if we were to call philosophy, the only real 'science', the only really meaningful way of

---

[11] More importantly perhaps, is the fact that only once he throws the light switch will this man finally be able to truly see himself for what he is.

Wayne W. Dyer's conclusion that we are a spirit inside a human shell rather than a human with a spirit within, may be a hint as to our true nature.

discovering the elusive, and always evolving truth, because philosophy exercises the mind.

The Longman Dictionary of Contemporary English, describes philosophy as:

"**1.** The study of the nature and meaning of existence, reality, knowledge, goodness, etc."[12]

Under this definition, all branches of science, are merely small branches of philosophy, but even more importantly, *the whole of science is merely a branch of philosophy.*

Philosophy concerns itself with questions that science cannot answer by their very nature, as well as with all those questions that science *does* concern itself with. But does this make science more or less important than philosophy? The answer, which may come as a shock to some, is that science, in the final analysis, is less important than philosophy, because philosophy *encompasses* science.

Science is only a piece of the pie which composes philosophy, and it is a piece —which important as it is— we have been focusing on almost to the exclusion of the other means for probing our environment that philosophy makes available to us.

It is science that may have made us aware of the Face on Mars, but it is philosophy that must be used to discover the relevance of it.

Resonance has been proven conclusively by science in several ways, usually when applied to inanimate objects, which are the forte of science in general, as are all static systems, but when we try to discover the truths of resonance in living things, *Morphic Resonance*, it seems science may fall just a little short.

Morphic resonance is a topic at the edge of science. Science may tell us that there is something happening, but it cannot explain how or why in terms that are adequate by the black and white meter it favours so much.

Yet, if we examine morphic resonance using the whole of philosophy, not just that branch of it called science, then we find that the pattern repeats itself throughout nature, and consequently we can arrive at the startling conclusion, already reached separately by science itself with

---

[12] Longman Dictionary of Contemporary English, Longman group, 1978, (reprinted 1981). Point **4** also describes philosophy as "Calmness and quiet courage, esp. in spite of difficulty or unhappiness" a quote that those of us challenging the hallowed halls of academia should do well to remember.

respect to chaos theory,[13] that the universe has a purpose, and that purpose is to organise things in a *particular* way.

Perhaps an example may be useful in order to show just *how* particular the Universe can be.[14]

For certain complex substances, several ways of formation are possible. In other words, the molecules and atoms that construct a particular substance can conceivably come together in a large number of ways, each of which can be just as stable as the next.

It is interesting to note then, that when certain of these substances are formed, they always form in one particular way.

Conceivably, they could form in several billion different ways, and yet, they always form in just *one* way.

In Sheldrake's book, *A New Science of Life*, a description of this process is quoted from another book,[15] where it is shown, that if there are only two possible states in which individual 'branches' of a polypeptide chain could exist (in fact there are many states in which these branches can exist) then, the number of different outcomes for a chain of 150 'branches' is $10^{45}$.

If each of these possibilities could be explored in turn, even at the speed of a molecular rotation ($10^{12}sec^{-1}$) then the process of going through all possible permutations would take about $10^{26}$ years.

In fact, the process of formation of such a molecule actually takes place in about 2 minutes, so it's clear that all the possible permutations are *not* examined.

What happens, is that the molecule chain 'settles' in one particular way out of billions and billions of possible alternatives. The mechanists (that would be so fond of burning Sheldrake's book at the stake) scream blue murder at the suggestion that fields of morphic resonance play a part in this, or even at the suggestion that the Universe has a purpose. For them, the answer is simple: the molecule chain forms in just that particular way simply because it's the most stable alternative.

---

[13] As well as by science in the field of quantum mechanics.

[14] This example is a somewhat simplified version of one found in Rupert Sheldrake's *A New Science of Life*, on pages 64 to 71.

[15] Anfinsen C.B. and Scheraga H.A. Experimental and Theoretical aspects of Protein Folding, from *Advances in Protein Chemistry* (1975).

But no one can deny that the atoms that form such chains can indeed come together in an untold variety of ways, each of them stable in itself, so why should they always form in just one particular way given certain conditions. Surely at least in some of the chains, even if all external conditions are the same (temperature, pressure, etc.) there should be some variation? Those who say "No" simply on faith, have not understood the complexity of protein molecules.

To say that there would be no variation in the shape of one polypeptide chain from the next, would perhaps be equivalent to saying that if you had an alternate Earth, with exactly the same geography and just as many people on it as this one has, things on it would be exactly the same as on this one.

All our common sense tells us that it's impossible to predict how some six billion people would all behave on a carbon copy of our planet. In fact, if it was shown that such a planet did exist, and that indeed everyone on it was just a carbon copy of each one of us here, few of us would remain atheists. And yet, although the analogy is an approximate one, this is what happens in the minute world of large molecules.

There are an almost infinite number of ways in which the atoms that compose a polypeptide chain or a blade of grass can join with one another, and yet, they almost always build just a polypeptide chain or a blade of grass.[16]

Why?

The mechanist's answer (it's the most stable state) is failing in all fields of science, in particular in quantum mechanics and the higher aspects of physics, but now, also in biology (where the mechanist theory that living things are just like machines, never really had much success in explaining any of the more complex functions of living beings anyway) and of course, its cousin chemistry.

Similarly, the religionist's answer (because God ordained it so) leaves one with a sense of frustration.

A time is arriving, when regardless of whether they like it or not, religion and science will not only be pushed together, but will have to permeate each other's beings, shedding the lies and ignorance of both sides so that something new and powerful can be born.

---

[16] Of course there is a small percentage of mutant blades of grass, but this is also an orderly process. Occasionally, those mutant blades of grass, go on to produce a new type of grass altogether, which may be better suited for the changing environment, so the process of mutation, is a necessary one.

The scientists will not hail it as a good thing while it's happening. The priests will not hail it as a good thing while it's happening.

A few individuals though, will.

And their task will be somewhat easier than it has been for heretics in the past (the burning of people at the stake or feeding them to hungry lions is somewhat frowned upon in our day and age).

But even more importantly, the human race as a whole, perhaps a generation or two from now, will most certainly hail it as a good thing; but then, for them it will not be happening, it will have already happened.

It is probably for this reason, that some wise person once said:

"It is not necessary for human beings to suffer in order to grow, but it is seldom indeed, that a person grows without some measure of suffering."

# The Origin of the Ancients

Although in this chapter I have strayed somewhat onto some slightly more philosophical musings than before, I would like to come back to one more practical concern, the origin of the Ancients.

To make a guess at where the Ancients come from of course will have to involve a high degree of speculation, and therefore, it is very doubtful that any conclusive or even useful decision can be arrived at.

Nevertheless, it may be possible to perhaps confine the star systems where we may expect to find some remnant of intelligent life to a few possibilities.

It must also be stated, that there is bound to be some confusion in our history between what are Ancients, what are Martians and so on, so for the purposes of this section, it may be wise to define the Ancients as the advanced race that had something to do with the terraforming of Earth and/or Mars.

There are three basic possibilities with respect to the Ancients. These are outlined below.

## Case 1

If we assume that spaceships using fusion/antimatter engines were used to reach speeds close to that of light, then it would be reasonable to assume that the stars closest to us are somewhat better candidates for having intelligent life orbiting them than ones which are, say, on the other side of the galaxy.

In keeping with this idea, it would be sensible to assume that the Ancients may have had or still have, a base of operations within twenty light years of Earth.

Giving the Ancients ships that are capable of travelling at 95% the speed of light, would mean that a one way trip to Earth from a place 20 light years away, would take a little over three years for the ship's crew.

A return trip, which included a ten year period spent in the Sol system, would take the travellers a total time of about seventeen years. But for their friends back home of course, more than fifty-two years would pass between the time they saw the travellers leave and the time they saw them return.

If we assume that the aliens possessed technology so advanced as to radically change the appearance of a planet over a span of a few decades, then these travel-time figures fit rather well.

In making the assumptions necessary for case one, we are in a sense making the Ancients very similar to us. For a start we are limiting ourselves to using technology that we have at least conceptualised, and are also assuming that they have lifespans that are somewhat similar to ours, although we could reasonably expect the average Ancient to live say two or three hundred years, due to their advanced technology.

In a sense, this very idea of making the Ancients so 'understand-able' to us, tends to weaken the probability that we are correct.

In truth, we know very little about the Ancients. This is clearly evidenced by the fact that the very vast majority of the population of this planet, has no knowledge of them. The few who have come to the conclusion that the Ancients (whatever their nature) did exist, are quick to admit that their technology, their motives, and most of what they left behind in the form of monuments and/or legend, is unfathomable to a large extent. It is doubtful then, that with such scant knowledge, the Ancients would fit our own prejudiced ideas of technology, lifespan or culture.

The most likely usefulness of case one then, would not be its accuracy in predicting the star of origin of the Ancients, their technology, or the extent to which they are present in our galaxy. Instead, the usefulness of it, would be to give us a frame of reference. A conceptual model which we can grasp because it is closer to what we know, rather than what the facts actually are in all likelihood. In a sense, for us to talk of ships powered by antimatter engines, is only a slight improvement on what our ancestors did by calling the same ships 'chariots of fire'. It helps us to talk about the subject in a way that is understandable to us, but of course, this does not mean that the facts are exactly as we say, in fact, the history of science itself tells us this is quite improbable.

Any person who has studied chemical engineering (or indeed any of the science subjects to a level equivalent to a university degree) knows that atoms and molecules, have in actuality, little in common with the way they are described to pupils in highschool.

The models used by highschool teachers, are merely tools to allow the students to understand the basics of the subject.[17]

For this reason then, although case one may be the one we feel most comfortable with, the likelihood is that it's quite a ways from the bull's-eye; even if it's aimed in the right general direction.

## Case 2

If we assume that the Ancients have technology of either so advanced a nature, or whose roots are based in a mode of reality with which we have little experience, then the question of distance and travel times between stars becomes rather flexible in the best of cases, and totally irrelevant in the worst of them.

In this instance, the 'best' of cases may be a way of travelling between stars akin to some form of hyperspace, where, while time and distance may still be factors, they are greatly reduced. Or perhaps, the Ancients have some way of placing themselves in some sort of hibernation, making accessible to them stars much further than they would otherwise travel to.

A 'worst' case scenario instead, would place their technology so far removed from ours as to be inaccessible to us even on a conceptual level, in effect making the Ancients truly more like Gods than like us humans.

If we assume that time and distance *do* play some part, then we can assume that the Ancients at least originate from this Galaxy, and from a point in it, that is probably relatively close to us when taking the whole Galaxy into consideration.

Of course, since the Galaxy is some 100,000 light years across, that leaves a lot of room. Even if we limit our search to a sphere just 100 light years in radius, we are already encompassing a large number of

---

[17] Unfortunately, few of us realise that by teaching conceptual models, we affect the way in which children (and hence society as a whole) perceives reality.

If all we teach are conceptual models, in the end, what we perceive as being our Universe and reality, is in fact only a more complex conceptual model, founded on those conceptual basic 'blocks' we got taught in school. Just like the weather predicting computers, if our basic data incorporates a small error, our first, second, third and subsequent guesses, which are based on that data, will in turn incorporate an error which becomes bigger with each guess. The end result, is that we are bound to have made some serious mistakes in the way we interpreted the very nature of those things we hold most dear and which are based on multiple extensions of the basic information, namely, the nature of our existence, reality, God, and even the physical Universe around us.

stars, but for the technology of the Ancients to be truly superior to antimatter driven ships, the difference it would have to make to travel times is considerable, and so it might be more realistic to think of a sphere that is say one thousand light years in radius, and this of course would have to take into account a number of stars large enough that it would ridicule our attempts at choosing one over another as a possible point of origin for the Ancients.

## Case 3

In this instance, the Ancients are more akin to Gods than humanoids, at least with respect to their technology, making this point an extension of case two, but it is a possibility interesting enough to make it worthwhile of separate analysis.

In this scenario, the difference between humans and Aliens is of a magnitude so large as to make the analogy of the genetically engineered peanuts a close approximation of the facts.

If this is the case, the purpose of the Ancients is truly unfathomable, in fact, they may be all around us and examining us as we go about our everyday lives, but we are so dim-witted by their standards that we are blissfully unaware of their existence. Just like a group of ants in a laboratory is ignorant of their condition with respect to the human observers.

Keeping these three possibilities in mind then, our way of deciding where the Ancients originate from is bound to be shaky at best. Since, as is often said, the truth usually lies somewhere in the middle ground, it may be useful to take elements from all three cases and reach a somewhat hazy conclusion that would probably fall somewhere in the middle of case two.

This theory at least has some supporting evidence. A race of beings capable of terraforming an entire planet (and possibly two in our own system) certainly must possess technology more advanced than ours, and if their ships indeed did destroy Mars by meteoric bombardment, then they must have had some ingenious machines, for radically changing the course of a boulder a few kilometres in diameter is no easy task.

Such feats, and to some lesser degree also the building of the Great Pyramid, seem to hint that some type of gravity control may have been known to them.

Gravity in turn is intimately tied in with the whole of the physical Universe, and mastery of it could conceivably have important implications for interstellar travel.[18]

If we are to hazard a guess as to where the Ancients may have a base (or at least had in the past) it might be useful then, to discard at least to some degree any preconceived ideas we may have had, and base that guess on a mixture which includes not only the distance of the stars in question, but also the frequency with which such a star may have been mentioned in legends as being somehow connected to the 'Gods'.

This of course is a process that can be gone into at considerable length, and this author certainly does not profess to be an authority on mythology and/or legends relating to the 'Gods', nevertheless, a reasonably informed person should be able to hazard at least a guess, and with a bit of luck, such a guess may come quite close to the truth.

In particular, we might want to study the stars mentioned most often by those tribes in our ancient history that we may have tentatively associated with the Atlanteans. On doing this, we find that certain stars are particularly prominent in our ancient legends, making them good candidates.

On the next page is a table that gives the stars which in my opinion are most likely to have hosted the Ancients at some time or other.

The table was arrived at by a combination of factors, among which, (but not exclusively) :

1) The prominence of its appearance in ancient Egyptian, Babylonian or Mayan legends.
2) Its distance from us in light years.
3) The type of star with respect to radiation output, variability, spectral class etc.

---

[18] Black holes, worm holes and other such phenomena, have been cited as possible means of interstellar travel not only by science fiction writers, but also by scientists. In each case, gravity plays the key part, and in any event, the fact that gravity curves space itself, and the fact that time is inextricably connected to space, suggests that time may be affected by a machine capable of controlling gravity. In turn, if the passage of time can be reduced or controlled in some way from the point of view of the traveller, interstellar travel becomes a very real possibility.

| Table 6: Stars that may have (had) inhabited planets | | |
|---|---|---|
| **STAR NAME** | **Distance (Light Years)** | **Notes** |
| Sirius | 8.7368 | Binary System. Main component of type $A_1V$. Secondary component is a small white dwarf star. |
| Zeta Orionis | $1532 \pm 65$ | Binary System. Main component of type $O_{9.5} Ib.$ Secondary component $B_0III$. |
| Epsilon Orionis | $1679 \pm 147$ | Type $B_0I_a$. |
| Delta Orionis | $1549 \pm 82$ | Type $O_{9.5}II$. |
| Alpha Draconis | $218 \pm 13$ | Type $A_0III$. |

Ultimately, the process I used in composing this table is one based far more solidly on intuition than on any factual basis, but then again, perhaps that is not such a bad thing.

In any event, although Sirius appears at the top of the chart, it must be pointed out that this is a binary star system and hence it seems unlikely that conditions on any orbiting planets would be conducive to life, despite this, the Appearance of Sirius as central to most ancient cultures places it high on the list of 'possibles'. Also, the fact that the Dogons knew of its binary companion, which is invisible to the naked eye, and attributed this knowledge as having been given to them long ago by the 'Gods', must carry considerable weight for the purposes of this exercise.[19] The Orionis stars are the three found in Orion's belt, and my introducing them here is mainly as a result of the positive aspects of Robert Bauval's book, *The Orion Mystery*. Despite this, Zeta Orionis must be an unlikely spot for the location of a habitable planet.

---

[19] The Dogons are an African tribe which possessed this knowledge prior to its confirmation by modern telescopes.

Alpha Draconis was also an important star, being one of those that periodically becomes the Pole Star, and had an almost universal association with death and diabolical significance, but it was also associated with Lucifer and a 'fall from grace' so perhaps, in view of the destruction of Mars and also of the arguments expounded in appendixes B and E, it may be this star that should head the table.

It is also interesting to note, that both Sirius and Alpha Draconis, are much closer to us than the Orionis stars.

# Summary

The aim here has been to place science in perspective.

The wonderful thing about science, is that once it can measure something, it's pretty accurate, and in fact is hard to beat as a tool for examining our reality.

The problem arises when science *cannot* measure a phenomena, or when it does not have enough data, in a familiar format, to come to a solid scientific answer.

As mankind progresses, this is going to become more and more true of every aspect of science, because the Universe is not a static, cold, dead thing, but it is alive and dynamic, and while it seems to follow specific rules, they are far more complex than man has previously suspected. For this reason, we should not make the mistake of saying that simply because something cannot be measured, seen, or felt in scientific ways, it does not exist or merit investigation.

Consider radio waves. Science a couple of hundred years ago or so would have dismissed the possibility of researching into radio waves because surely, the expense of searching for or manufacturing something which no one can see, taste or hear for the purpose of supposedly transmitting the voice over thousands of kilometres would have sounded ridiculous.

And yet radio waves exist, and today, anyone who tried to deny it would quickly be labelled a fool.

In the context of this book, I have suggested some ideas which, if correct even only in part, would have serious implications for the entire race. Not only would we need to revise our history, the theory of evolution, and our own religious beliefs, we would also have the adventurous task of looking for and trying to re-invent, the technology that the Ancients and/or Martians once had.

This is technology which may still be (at least in part) locked into some hidden chamber not only of one of the Egyptian pyramids but even more likely, in one of the structures on Mars, which have lain undisturbed on the red planet for untold aeons.

It would be a shame to ignore these potential boons for the human race at a time when a new kind of technology (alongside with a correspondingly eco-friendly attitude) is sorely needed, simply because I have not made my case exclusively through the use of a strict code of scientific procedure.[20]

My method has been to supplement scientific fact with a logical process of analysis, which although somewhat unorthodox, has the advantage of being firmly based on a branch of dynamic mathematics which I choose to call *patterns of relationship* for want of a better definition. I have also tried to show why in some ways, this process is superior to relying merely on "pure" science, and I hope to have had some measure of success doing that, because I truly believe, that this is the way the world is moving.

Our minds are the only "computers" powerful enough to crunch through the enormous range of variables that this process requires, because remember that in using patterns of relationship (which could also be called an advanced form of common sense) we are not acting on fixed values of x and y to arrive at a probability z, but we are in fact acting on a range of values for both x and y to arrive at the forming a landscape for probability z.

A computer would have to go through each step in turn to create this landscape point by point, but our minds work in an analogous fashion, that is, they need not go through this procedure, the landscape is formed almost instantaneously once the variables have been considered.

It is true of course, that if all the variables have been correctly entered along each step, and given enough time, a digital process (step by step, like a computer) will eventually give a precise answer, while an analogous process will always only give an approximate answer.[21] But when the variables are known only to a certain degree of accuracy, the analogous process is not only faster, it

---

[20] It is also worth pointing out, that if the argument that by using more of our brains we would have access to new knowledge (and this must be hard to refute!) then a way to train ourselves to make more of our minds accessible is by trusting our intuition more and more. Like any other faculty, the mind improves with training, and if we practice 'guessing' with our intuition every day, in a few years we will soon have developed some new ways of thinking that no doubt will improve our lives. Edward De Bono, the famous author and philosopher, as well as coiner of the phrase 'lateral thinking' has much to say on the importance of teaching people how to think in new ways.

[21] Analogous processes are capable of giving answers as precise as ones derived from a digital process, but when talking in terms of the human brain, this is only true of those events we think of as common-place. The act of catching a squash ball on the head of a racket is an analogous process. It's imperfect (sometimes we miss) but try having a computer do it for you and you'll soon be glad your brain works analogously.

is infinitely more accurate, although, like I've already said, it's probably not going to be exactly spot on.

In this respect then, science could be compared to a computer; it likes to do things in stages, step by step and without skipping any steps. This is a very useful way of doing things, especially for the accumulation of precise data, but it can be a somewhat blind process, and in any case, it is rare indeed that a new discovery is made in such a fashion. New discoveries, as is evidenced by the very vast majority of all inventors, are almost always made by an analogous process. That is, a "skipping of steps" sort of way.

Einstein for example, did not sit around crunching numbers together until he just happened to write the theory of relativity in mathematical symbols. His insight came intuitionally.

He *felt* something must work in a certain way, and then he went about finding out if that was indeed so, and if so, why.[22]

The very word *insight* (made up of *in* and *sight*, meaning sight inside oneself) indicates that the process is an analogous and intuitional one; and which scientist would deny that it's the spark of insight that leads one to reach a new conclusion?

In any event, no great truth was ever discovered by doing things as they have always been done before. In fact if we did things always as we'd done them before, then we would not be able to grow, change and evolve.

Luckily though, life cannot help but evolve, so we have little choice in the matter.

For all these reasons then, I hope I have adequately shown that although perhaps a little different, my logic is still valid, and while some of my conclusions may be strange, new or even upsetting for some, if the methods used to arrive at them are valid, then the conclusions merit investigation.

Besides, simply because something does not necessarily fit with the way we'd like things to be, this does not make it any less true.

---

[22] It might be worth pointing out that it was Einstein that said that : "Imagination is more important than knowledge".

# 8      Conclusions

Over the course of the last seven chapters, I have presented some facts, theories and ideas, that tend to support the general concept of extraterrestrials having been involved to some degree or other with the human race of Earth.

In addition I have broadly presented what may amount to a general reconstruction of the origin of mankind when seen from this new perspective.

What I have *not* done yet, is give a detailed reasoning of what this general theory would mean for the human race if it became accepted. Of course, before the theory is broadly accepted, we may want to find some definite evidence that it is indeed valid.

Happily, this is something we can do relatively easily by sending a probe to Mars to take more accurate pictures of the Face on Mars along with its nearby pyramids. If these proved to be as obviously artificial as I (and others) have suggested and believe, then the rest of the ideas presented in this book must be valid to some extent or other.

         *Unhappily* though, the sending of a probe to Mars for this purpose, is also something that NASA seems uninterested in doing, as I will discuss in more detail in section two of this book.

Furthermore, if NASA (and others) have an interest in fooling the world into believing that the Face does not exist, then that is precisely what they will do.

While photographs and video up to a few decades ago could be said to be reasonably hard to fake, this is sadly no longer the case.

Anyone with a relatively sophisticated computer and the right software, can alter a photograph to show absolutely anything.

While this may not be true of the original video cassette, or photograph negative, it is important to remember, that data from NASA probes is received by computers.

There is no 'original' negative shipped back to us by the probe; meaning that even if NASA did take pictures of the Face and nearby City of Pyramids in all their glory, they can very easily doctor them to make it look like a bunch of hills, or even quite simply just make them disappear altogether.

This is particularly true when you consider the very suspicious and shocking statement by NASA that even if pictures of the Cydonia region of Mars *are* received, they would have to be analysed independently first and it would be at least six to twelve months before they could be made available to the public.

This set of circumstances, as Richard Hoagland has said before me, is totally unacceptable.

The reason NASA gives for this minimum moratorium of six months, is that they do not want 'nobodies' (i.e. the public at large, or even other scientists) to scoop their discoveries.

Once again, this basic premise is flawed, because it's implying that NASA has a right to knowledge before the rest of the human race, which is not only wrong from an ethical point of view, but it goes against the very principles which NASA is supposed to stand for, namely, the advancement of science for all of humanity.

But even if we wanted to give the NASA scientists their dues, there may well be ways around this problem.

For example, an independent non-governmental body could be set up to also receive the data at the same time as NASA, except that instead of being scrutinised, copies of it would be made, and each would then be locked away in a safe place for the necessary six months, after which, it could then be released to the public, along with the NASA data.

Of course, this sort of set up would be very expensive and certainly far from completely safe from tampering.

After, all, a government that can kill its own president and then cover up the fact that Kennedy was shot by at least two different people and probably three, despite the massive evidence for it, would find it child's play to alter or replace computer data, no matter where it was kept.

Another alternative may be to sponsor a privately owned space exploration firm to send a probe to Mars. Unfortunately though, such a firm does not yet exist, although some people are indeed trying to create a viable one.[1]

For these reasons, it may be that we have to continue using our own powers of deduction for quite a while before the matter of the Face on Mars is finally resolved, and in any event, I for one, will not be

---

[1] So interesting is this possibility, that I shall devote myself in the near future, to investigating such claims of independent space exploration firms being set up, discovering whether they are indeed feasible concerns, and if so whether they would be interested in such a project as the further exploration of the Martian surface.

satisfied with a simple explanation of the non-existence of the Face unless I was personally present to the events that supposedly conclude this.

But these concerns aside, we can still ponder what effect the Face on Mars may have on our society if it is indeed found to be an ancient alien artefact.

# The Practical Implications of the Face on Mars for Human Society

The main impact, would seem to be:

1) The general restructuring and reconstruction of ancient history along with its resulting effects on human society in the present.

The "*resulting effects on human society in the present*" would in turn be mainly composed of:

a) The realigning of each individual's personal beliefs with respect to religion and its attached dogma. This would include such highly emotional issues as the reinterpretation of 'holy' books like the Bible, Koran, Torah etc.

b) The rediscovery and implementation of old technology such as the one which was used by the Atlanteans/Martians at a proper pace and with the proper respect for it by all, so that we do not wipe ourselves out as a race because some fanatical and idiotic 'leader', along with his faithful followers, decided to use it to show some group of heathens/infidels/dissenters, the light/error of their ways/their subhuman nature, or whatever.

The first part of point number one would be perhaps the easiest to do in terms of simply practical discoveries which can be made once historians, archaeologist, palaeontologists, geologists and others decide to approach their field of endeavour with this new perspective in mind.

Part a) instead, is perhaps the hardest to accomplish, and paradoxically it influences part b), because as long as we have people on this planet that are willing to destroy others and themselves, in the name of some benevolent God (no less) then it would be relatively unsafe to make certain technology accessible to them.

Nevertheless, if the Face on Mars proves to be an artificial construct, religious dogma would most certainly take a knock, and a serious one at that.

Note that I said religious *dogma*,[2] and not the basic concepts of all religions, which are really all the same, namely, the principles of Universal love, harmony, respect for all life and so on.

For example: The Pope at present is held to be God's representative on Earth, and as such is supposed to be infallible. This is a clear case of religious dogma, and has no basis on reality whatever.

It can be argued, that as human beings and exponents of life, every single one of us (along with the cockroaches, ants and all the others) is God's representative, and as such, we should all be infallible.

From the point of view of those that believe the Pope is the only representative of God alive at present however, such an argument would meet with an incredible amount of criticism.

And yet, in my view (or indeed in the view of any truly *thinking* individual) either we are all God's representatives, (as I believe is the case since ultimately he's our creator) and we are just too unevolved to appreciate this; (why for example, are there such things as mass murderers among God's representatives?) or none of us is, in which case, the Pope is just as fallible as the next person.

In this respect, I'd like to point out a curious matter of fact.

If the human race continues to reproduce at its current rate, by the year 4700 AD, which lies only some 2700 years ahead of us, the weight of the human population on the planet, would outweigh the weight of planet Earth itself.[3]

Clearly an untenable position.

But even without having to look so far ahead, the world's population has more than tripled over the last 80 years. Which means that if this trend continues, by the year 2080 there will be some 18 billion-plus people on this planet.

Once again, a position that would make it very difficult indeed for humanity to ever reach a level of technology in the near future that would permit us to colonise the nearby Moon and planets, since our resources would be tied up in feeding ourselves, while the planet's

---

[2] It has come to my attention that quite a large number of people confuse religious dogma, with spiritual concepts, to this end, it may be useful to point out, that by religious dogma, I mean that part of a religion which people are supposed to accept without question or reasoning and which has little or no basis in logic or fact.

[3] Assuming an average weight per person of only 65 kilograms.

ecosystem rapidly degenerated to a point that would make it difficult for *any* life to exist on Earth.

Recently, the Pope (speaking for God you understand) has reiterated that the church's position with respect to family planning is unchanged. That is to say, birth control pills, condoms etc. are all unacceptable according to him and hence, according to God too, we are presumed to believe.

To me, this is not only a clear indication that the Pope is by no means infallible; it is also a clear indication that the man is somewhat of a fool.

Surely God did not intend it for us to destroy Earth?

Surely God did not intend it for us to reach such levels of population explosion that we end up destroying the entire planet and ourselves in the bargain while starving and dying by the millions due to disease, malnutrition, war and poverty levels that would make even the bare necessities for life an unreachable goal?

Surely, when we are told: "Help yourself and heaven will help you" we are to understand by it, that no sudden miracle will happen that will somehow prevent such things in a benevolent manner, but that rather, we are to actively halt and prevent the current population explosion by means of some very urgent and prominent campaigns of birth control and family planning amongst all the world's people.

As for point b), some will undoubtedly feel, and possibly quite correctly so, that mankind is still far too savage and primitive to be presented with some of the technology of the Ancients (with which I shall deal a little more fully in the next section).

For example, if we just hypothetically assume that something like a teleportation machine could be built, we have to point out that, while such an item would certainly have beneficial implications for the entire human race, as well as all other life on the planet, it would also have an unparalleled potential for the perpetration of nefarious deeds.

On one hand, a teleportation device would remove the need for roads, cars, trucks, ships and planes, reducing the need for fossil fuels to an almost negligible level, which in itself would not only result in a complete restructuring of human economics, but would also result in a most beneficial outcome for the ecosystems of planet Earth, which we are presently taxing well beyond their capacity for regeneration.

On the other hand, if one of the oil-rich nations, like the Sudan, or Texas, decided that they would not be willing to restructure their economy in line with a world which no longer required fossil fuels, they may use the same teleportation device to deliver groups of terrorists at any point on the globe, which could perform their evil deeds and then just teleport out again.

In fact, there wouldn't even be the need for the teleporting of terrorists. The teleporting of an explosive device with a very short fuse would be sufficient.

As we shall see a little later, these problems concerning advanced technologies, are far more real and threatening than is at first realised.

It might be for this reason then, along with a certain amount of greed for money and power, that the governments of most countries seem to have come to a tacit agreement regarding these technologies, and that is, to keep very, very quiet about them for as long as possible.

Which brings us back to why NASA may be far from willing to admit that the Face on Mars is indeed artificial in origin.


As I shall point out more clearly a little later then, the main problem of the Face on Mars is not the personal struggle that each of us would have to endure in considering its implications as far as one's religious, philosophical or ideological beliefs are concerned.

The main problem would be that the Face on Mars would be a catalyst for the development of certain technologies that may well place the entire human race in danger of extinction by virtue of the fact that, while their potential for good is enormous, so is their potential for evil; as is, of course, always the case with *any* knowledge.


In a sense, once the Face is pondered on for a little while, we find it's telling us that we really don't have the time to indulge ourselves as far as coming to terms with our own misinterpreted, little truths and bigoted beliefs go.

We have a much more real problem at hand, and that is:

If we don't learn to curb those very same childish and selfish ideas and emotions, we will not survive as a race.

Nor will our planet.

The human race is shown to be a little like a couple of children that have been fiddling and fighting over a favourite toy on a railway track.

The toy is now stuck and entangled under one of the railway sleepers, and the train is fast approaching. If the two children waste further time arguing about whom the toy belongs to, they will perish alongside it.

There are only two alternatives, to the rather depressing one already indicated.

Either the children abandon the toy altogether, which is unlikely to happen in the human race context because some of the religious beliefs are too deeply rooted; or, they can work together to try and save both themselves *and* the toy.

This last option of course, would seem to be the most ideal one.

There is however a little problem associated with it.

Even if the children co-operate, they will find that the toy is firmly trapped, and at best, they can come away with only a piece of it.

If we pretend that the toy is a catapult complete with a handle for the storing of ball bearings, then, at best, by acting quickly, the children may be able to smash the handle and retrieve the last ball bearing inside it, before having to jump off the tracks to run away to a safe distance.

It is no coincidence that in this analogy I took the favourite toy to be a weapon. Nor is it a coincidence that, while a ball bearing, if shot from a catapult, can be a deadly instrument, by itself, in isolation, it is a perfect and nearly indestructible sphere.

And the sphere, of course, is nature's way of expressing perfection.


The dogmatic and nonsensical ideas of all religions could not survive an artificial construct such as the Face on Mars.

The *basic truths* and concepts of *all* religions though, most certainly could; and in fact would probably welcome it, since at last, after thousands of years of people and animals being butchered in their name, these basic concepts of truth and love could finally come out for themselves and be self-evident. As indeed they have been throughout history. But perhaps for the first time in *recorded* history, these base concepts will be more obvious than before, since the crippling mask of religion (the catapult) as opposed to spirituality (the perfect sphere of a ball bearing) will have been removed in large part.

No doubt, the process of losing the catapult, which could be said to be the vast majority of the toy in question, would be traumatic even for those of us that have no intention of using it to hurt or injure others; but growth, as always, involves a measure of pain as well as the wonder and thrill of adventure.

One cannot exist without the other.

Some will face the loss with a courageous and accepting attitude, while trying to move on towards a better and cleaner future; perhaps even encouraging the change.

Others will put on a brave face and at least not complain out loud, while not necessarily enjoying the process.

Some of course will not like the direction this whole voyage is taking and will have to gently be helped across either by the braver ones or by virtue of the fact that, like sheep, if enough of their number head in one direction, the others will follow.

And some, will have to be dragged kicking and screaming from the railway track, and although they will be glad to have survived, they will never truly accept the loss of the catapult.

While others still, will choose to perish rather than leave the catapult behind.

This last group is actually a small number of individuals, but in effect, they hold the rest of the human race to ransom; because it is they, that would rather blow the whole planet up rather than accept the possibility that their sacred religions (be they scriptures such as the Bible, the greed of power and money, or whatever warped belief system) are actually flawed in some way.

It is this last group that would perhaps use any new technology to further their own ends rather than to benefit mankind as a whole.

A perfect example of the type of individual I am talking about is Hitler.

Adolf Hitler is generally accepted by the majority of people in our day and age as having been the scourge of the twentieth century. Even so, for all the evil that he can be said to have brought upon the Earth, Adolf Hitler's intention was not to be an evil man.

I do not believe that Hitler, in his quiet moments sat around thinking: "Oh what joy I take in being such a complete bastard to so many people." The fact is, he didn't actually appreciate just what kind of a monster he was.

In his warped mind, Hitler thought he was saving the human race. In his psychotic brain, by killing off all the 'subhumans' he was ensuring that the human race became an 'ideal' utopian dream.

Let us ignore for the moment the simple fact that Hitler's utopian dream would have been a fucking nightmare for the rest of us, the point is that although his methods were obviously wrong to the nth degree, he believed he was doing the right thing; not just for himself you understand, but for humanity as a whole.

The question which arises in the context at hand of course is:

How many potential Hitlers are there out there, and if we did begin researching and developing Ancient technology, could we keep it from being used by such beings so that planet Earth is not reduced to a lifeless slag heap?

It is important to note here, that with any new development there are bound to be some negative side effects. Either by mistake or by design, someone is most definitely going to die as a result of new developments in technology.

People have died as a result of failed operations; my maternal grandmother being a case in point; sometimes people have even died as a result of *badly performed* operations, but the fact remains that without modern medicines and surgery techniques, *a whole lot more people would have died instead.*

In short, even if another Hitler type did come to power somewhere in the world and did bring about the deaths of millions of people as a result of using new technology to further his own plans, ultimately, if he was defeated and the planet not too badly damaged, humanity would recover and continue along its path of progress.

We now reach a final conclusion then.

Even if we accept the basic premises I have outlined, we are left with two points: a real dilemma and also an apparent dilemma.

The second is apparent, because as I hope will become obvious, it is not a dilemma at all. It is a simple matter of choosing a path that offers a slim chance over one that ensures certain destruction.

The first dilemma is that it would appear (at first anyway) that there is no real evidence for such Ancient/Atlantean/Martian technology, and even if we assume that such technology is possible, it would seem, by a general and superficial look at the world as we find it, that we are currently nowhere near developing such amazing technologies as teleportation machines or even 'just' whatever it was that allowed the Atlanteans to build a pyramid using huge blocks of stone aligned in such a way as to present negligible cracks between them.[4]

As it turns out though, if one begins a somewhat deeper investigation into the matter, we once again find the amazing possibility that such technologies may not only be available, but that in certain cases they have already been developed.

This is once more, a long and involved part of this ever expanding topic, and it is dealt with in the second part of this book.

The second point instead, which I have dubbed an apparent dilemma, is that some people, after having weighed all the evidence and arrived at the conclusion that indeed the basic theories set out here are valid, may reason that it would be better to keep our mouths shut anyway.

After all, if there is such potentially dangerous technology around, the safest thing is to keep it hidden.

The world certainly is far from being in a perfect state of harmony, and someone, somewhere, is bound to use such technology in a devastating manner.

The problem with this mode of thinking, is that it ignores a very important and very real point.

Namely, the simple fact, that if we do not begin trying to use this new technology to improve humanity's lot, very shortly, we will no longer have the ability, even with this new technology, to save ourselves from ourselves.

Any new technology, particularly of the type which will be introduced in part two, takes time to implement in a society, and this in turn brings about socio-economic changes which also take further time to settle down to a somewhat stable and non-inflammatory level.

This whole period of time is automatically one of great change, and as such, by its very nature, an unstable situation.

The more rapid the change, the more likely that the result is more akin to a violent explosion than to a gradual and peaceful growth.

---

[4] It is worth pointing out again, that in order to achieve a crack width of 0.5 mm, this must be twice the tolerance level of the whole surface of both adjoining blocks of stone. Furthermore, even if the blocks where so precisely cut, they would still have to be placed, and it goes without saying, that once a fifteen ton block of rock is set down, it's pretty much down. It's unlikely you're going to be able to 'tap it' into position without risking to alter that very same tolerance of 0.25 mm.

At the current rate of reproduction, the human race will reach the 20 billion mark in the next 100 years.

Even if we allow some greater shift towards 'green' principles and the introduction of new, but not 'Earth shattering' technology, the fact remains, that this planet cannot support 20 billion people and also hope to have the human race develop into something meaningful relatively soon in our future.

In fact, the more of us there is, and paradoxically, the less our chances of 'making it'.

There was a time, not too long ago, when humanity did not have enough members to allow it to become a meaningful contribution to the intelligent races of the Universe.

Today, we are very rapidly expanding past the already sometimes uncomfortable middle ground of about 5 billion people.

The optimal level, is probably something like 4 billion people, but of course, I am not suggesting that we wipe out a billion or two persons.

Realistically speaking though, we cannot hope to amount to much if we expand past the point of say about 10 billion people, and presently, we will reach such a point in about 40 years.

Most people reading this book will still be alive then (bar tragic accidents either of a personal or global nature).

I for one would not want to bring children into a world where by the time they are twenty they will have to coexist or survive (one cannot really call it living) with another 10 or 12 billion people.

Especially since it would mean that before they reach the age of their theoretical death of natural causes, they will have had (if they somehow miraculously survived that far) to witness either the destruction of the human race on a large scale as the result of war, which may well escalate to the point of being nuclear, or through the more relentless, and perhaps more likely, cause of natural means. These would not in fact be *truly* natural means, because we would have brought them upon ourselves, but they are the Universe's way of telling us we have overstayed our welcome.[5]

---

[5] Considering that humans are supposed to be self-aware, and thus supposedly self-motivated and self-actualising, the process of 'natural' culling by disease, starvation etc. is in fact artificial, because as "intelligent" beings we theoretically have the opportunity to control our own destiny.

If the Earth's ozone layer continues being depleted, the very same process of water disassociation which took place on Mars will begin here too, along with the killing of the seas' plankton, which in turn would deal a crippling and possibly final blow to the whole food chain of this planet.

It is worth pointing out, that certain types of lichen which grow in the northern latitudes, have already been found to be perishing as a result of the ozone depletion allowing greater doses of ultraviolet rays to penetrate to the surface. This in turn has resulted in a steady decline of the reindeer which survive on this lichen, and that is but one simple example. The reality is that what is happening to the reindeer, will pretty soon start happening to all sorts of creatures, including us.

In short, we've run out of options, and every year we delay from taking action, is a year closer to certain catastrophe.

In the end, even if the development of new technologies results in Armageddon, it may be a preferable end to the far more drawn out collapse of society along with the Earth's ecosystems.

If properly introduced, even at a relatively rapid pace, technology such as anti-gravity machines, free energy systems and the like, could turn our situation from one that is rapidly approaching desperation to one where the mythical garden of Eden could be made into a reality right here under our feet.

♦♦♦♦♦

In view of this overall conclusion then, the priorities would seem to be:

1) Send a probe to Mars to further investigate the Face and nearby pyramids, while ensuring that whatever it photographs or does is made instantly available to the public at large as well as any interested scientific groups.

2) Research into the possibilities of developing some of the technologies which seem to be 'under wraps' at the moment.

3) If it is found that at least some of the new/secret technologies could be developed with a degree of risk that is minimal, or at least acceptable, then begin such development alongside mass

education with respect both to the new technologies, as well as family planning techniques and harmonious ways of conduct.

4) The mass education should not be done in a brainwashing fashion, because as history has proven time and again, one cannot force the human spirit into directions it is unwilling to go. Education should take the position of showing the correct direction to be taken through the use of pure logic.                In fact, if a training for thinking in logical ways is promoted, people will then eventually arrive at the correct conclusions by themselves.
While this is a long term process, it would eventually enable the introduction of the more advanced (and also more dangerous) technologies in a more peaceful way than would otherwise have been possible.

5) A further long term goal might also be the establishment of self-sufficient colonies on the nearby planets or moons suitable for it, so that even in the eventual destruction of Earth, some chance for the continuation of the race is ensured.

At present, the last suggestion may sound like science fiction, but if you recall the ideas mentioned in chapter six concerning the ultimate purpose of life in general, and of intelligent life in particular, the eventual expansion of humanity in space is an inevitable fact if it manages to survive long enough.

A final point worth noting, is that the average person, even assuming he or she agrees with the principles set out in the previous page, will usually feel quite powerless to do anything positive as an individual.

Certain green movements that rant and rave about the destruction of the planet, have in fact helped to perpetuate the problem.

So successful have some of these organisation been in fostering an idea of all-powerful governments and corporations going about the business of destroying our home planet, that they have left many of us with a feeling of helplessness.

In a sense, such movements are their own worst enemy, because all they have done is highlight the problem to such an extent that the average person has become despondent.

The general feeling, seems to be that, since things are so bad that they cannot actually improve, why bother. We're all going straight to hell anyway, right?

As the years pass and life pretty much seems to carry on as before, the previously despondent person may rise out of the general depression concerning planet Earth's ecosystems and conclude that the green fanatics were wrong after all.

This 'new' attitude in turn does not help the green movement at all of course.

The reality of the matter though, is that the Earth *is* rapidly degenerating. We are currently making several species of animals extinct every day. The tropical rainforests are being destroyed at an unprecedented rate, and pollution has reached the point where the world's oceans are starting to become seriously affected, even though they had been thought too vast to be influenced in any measurable way.

The problems we have created up to now are not irreversible however. Given a general goodwill and a change in attitude, as well as some hard earned money and the sweat of some willing brows, the damage *can* be repaired.

Perhaps even in a relatively short time.

The problem is that we *cannot repair this damage if we continue reproducing at the current, or even a slightly reduced rate*.

As our numbers grow, any advance in technology that reduces the damage to the environment becomes inadequate, because even if the pollution per head is reduced, the increase in numbers will still result in a higher order of pollution than before.

It does not take a genius to figure this out.

But if you have any doubts, once again, don't take my word for it, research the subject yourself. Even just a few hours at the local library, should give you a decent idea of the basic concepts I have outlined here with respect to population growth and world pollution.

But as I have pointed out, the answer does not lie in pointing out all the problems. The answer, as always, lies in pointing out the solutions.

And happily, there are several of these.

The truly dedicated can choose to enter fields of endeavour which have beneficial outcomes for the various ecosystems, be they the education of ignorant people, the joining of conservation groups, or even the forming of new organisations dedicated to such ideals.

It is worth pointing out, that Greenpeace, which was started in the late 1960s and early 70s, has grown from what was often perceived as a hippie, flower-power movement, to a force to be reckoned with; tangling with the governments of several countries, and more often than not coming out on top.

The incident where the French government destroyed one of Greenpeace's ships, killing some of its members in the process, was one that resulted in an embarrassing situation for France, to say the least.

More recently, Greenpeace could be said to have possibly played a pivotal role in the ruffling of feathers and almost a change of government which took place in Britain as a result of (at least partly) the Shell incident.

Obviously though, the majority of people are unwilling to give up their job and family life in order to place themselves between a whale and a whaler's harpoon cannon, like the members of Greenpeace do on regular occasions.

All the same, the individual can make a very real and meaningful contribution to a cause by several means. Voluntary work and monetary donations being the most obvious. Nor is the donation of a few dollars to an institution like Greenpeace an action too small to make a difference. Let me highlight why.

If we consider that of the 5 billion plus persons on this planet, only some 1% can realistically afford to contribute a donation of ten dollars; that still amounts to 50 million dollars.

While perhaps this is not a sum of money that will allow anyone to save the world, it will certainly go quite a ways towards it; and if for example, an antigravity machine might be developed for, say, 40 million dollars, then indeed, such a sum of money may well have saved the entire human race.

The donations of money and unpaid voluntary work done by the mainly anonymous individuals, are the very lifeblood of organisations such as Greenpeace, and without them, they could not survive or indeed have reached their current level of prominence.

The interested person though, is often wary of giving his hard earned money to an organisation that is often a relatively large unknown. Some unscrupulous people after all, have built small empires on the pretence of helping the poor, the sick, the mentally disturbed or any other number of emotional issues.

How then, does one know when his donations will be put to good use and when they'll end up lining the pockets of the organisers? The answer, sadly, is that often one will never be entirely sure of just where the money will have gone, but becoming a little more familiar with the group concerned will usually give an indication of the general tendencies. The more involved with it, the more likely that you will have a better idea of just where your money is going.

Organisations which publish public financial statements and have a non-profit charter for example, are generally speaking, safer bets than ones which do not.

A concept which might have some merit, is to keep regularly in touch with an organisation one has selected as a beneficiary.

In this way, the donor can not only select perhaps only one or two organisations as opposed to the giving indiscriminately to any number of charity organisations, some of which may not be as honest as others, but also, it allows the donor to become more familiar with the organisation in question, and in so doing, can gauge for him or herself whether he or she wants to continue contributing to it or not.

The choice of whether to give or not to give though, sometimes also boils down to a gut feeling. A sense of intuition or empathy.

Nor should this be lightly dismissed, especially in view of the role which the intuition may have to play in the advancement of humanity.

Listening to our intuition on a daily basis, while being very still and quiet will help develop us into more powerful beings, while ensuring that it is indeed the intuition that is speaking, and not that fearful impostor, paranoia.

# Appendix A

## Catastrophe in Human and Earth History

There is no doubt that humanity has suffered greatly as a result of various catastrophic events which took place over the course of our planet's history, the only questions which are doubtful are the extent of that damage, the true origin of the catastrophes, and what level of technology any previous race or group of humans had achieved before such events reduced them to an existence which was based on mere survival for an unknown period of time.

Generally speaking, the common view is that the damage caused, while occasionally being serious indeed (such as the extinction of Neanderthal man), did not destroy any fantastic technology or seriously advanced group of peoples. It's a nice and comfortable world after all, our forefathers simply could not deal with nature as well as we can today, and a major earthquake (for example) was a problem of much vaster magnitude for them than it would be for us today, who populate every corner of the globe. The idea that there may have previously existed a time when advanced technology was in use; that well structured and far-reaching societies may have been wiped out without leaving anything but the faintest of traces, is generally anathema to our common view of the way things are and work.

Before we examine anything else, it's worthwhile to question this very basic premise that the Universe is (relatively speaking) quite a safe place after all; and that although bad things might happen, it's unthinkable —or at least extremely unlikely indeed— that we might be washed from the face of the Earth without but the faintest trace of our existence remaining to mark our passing.

◆ ◆ ◆ ◆ ◆

In this century alone, there have been at least three asteroid hits on our planet that could have resulted in widespread death and destruction if they had landed on populated areas, the Tunguska explosion of 1908 levelled trees in Siberia in a radial area of about 2,150 square kilometres, and exploded with about 2,000 times the destructive impact of the atom bomb that was dropped on Hiroshima. The size of the object is still unknown, but it sent pressure waves around the planet, and the shock was felt more than 800 kilometres away. Had it landed just three hours earlier and a few thousand kilometres away, it would have hit Moscow.

In 1947 Vladivostok escaped a similar situation by merely 400 kilometres or so, when a meteorite with the explosive power of an uranium bomb came crashing through.

There have been others to be sure, but perhaps not as big, or maybe, they just landed in water and we only 'felt' their shock waves as sea-quakes.

The Arizona crater, which has sometimes been cited as an example of what may have resulted in the extinction of the Dinosaurs, is 1.2 kilometres across, but this pales in comparison to at least ten recognised meteorite craters which range in size from 11 to 64 kilometres across, three of which, including the 64 kilometre monster of Manicougan are found in Quebec alone.

And recently, a buried crater with a diameter of 180 kilometres is being investigated at Chicxulub on the Yucatan peninsula of Central America.

But one of the largest suspected Earth craters I am aware of is more commonly known as the Gulf of Mexico, which is over 700 kilometres across in its smallest dimension and over twice that if measured from an East-West perspective.

On Thursday July 21 1994, most newspapers around the world carried pictures of the most massive celestial collision that has taken place in our solar system within living memory.

The encounter of Shoemaker-Levy 9 with Jupiter.

Lucky for us that it didn't land here on Earth; because if it had, that safe, illusionary idea that we surely could not be wiped out in a flash, would have been severely shattered.

But a quick end may not necessarily have to be induced by a meteorite. We are perfectly capable of destroying almost all trace of human existence (and life) on this planet ourselves, as was made perfectly clear by the nuclear tensions of the 1970's, 1980's, and —if the stupidity of the French government goes unchallenged— in the 1990's too.

Then again, if we all just keep driving around in cars and spewing billions of tons of various fumes into the atmosphere, we may not need to have a nuclear war in order to be literally washed from the Earth.

The greenhouse effect, may well result in the melting of the ice caps in an unforeseen and incredibly short time, perhaps even less than one hundred years, and if so, water levels could rise by up to 300 feet (about 100 metres) which may not be enough to drown us out quite as completely as the dramatists of the Hollywood movie *Waterworld* may like, but it certainly would be bad times indeed, and a very large portion of our technological achievements would indeed be lost at least for a considerable length of time, as the race adjusted to the new situation.

And in any event, the levels of the world's oceans have already risen by just such an amount in our distant past. Not only is this in evidence geologically, and archaeologically, but also anthropologically, as is clearly shown by certain island populations that as close to our own time as the fourteenth century, were amazed to discover that there were any other people left on Earth, as they were convinced that they were the descendants of the few and only survivors, of an ancient and catastrophic flood.[1]

For a more complete (if somewhat depressing) list of catastrophes which could destroy the human race quite efficiently, one can peruse Isaac Asimov's *A Choice of Catastrophes*,[2] or, perhaps more simply, just ponder on the following paragraph, which is to be found in a report written by the Jet Propulsion Laboratory, as a result of a US Congress directive to NASA, to look for ways to reduce the threat of an Earth collision with a large asteroid:

> Concern over the cosmic impact hazard motivated the US Congress to request that NASA conduct a workshop to study ways to achieve a substantial acceleration in the discovery rate for near-Earth asteroids. This report outlines an international survey network of ground-based telescopes that could increase the monthly discovery rate of such asteroids from a few to as many as a thousand. Such a program would reduce the time-scale required for a nearly complete census of large

---

[1] Namely the Guanches of the Canary Islands, whom by the way, used to have a higher technological base further back in their history than closer to the present, as has been noticed by several historians and investigators.
The people of Easter Island would also qualify as survivors of some catastrophe, and if the stone giants they built were in fact a form of 'worship of the Gods' as seems likely, then perhaps we even have an echo of those very same Gods abandoning the people of Easter Island to their fate, as they did with the rest of Humanity long ago.

[2] Hutchinson & Co. 1979. ISBN No. 0-09-141240-4.

Earth-crossing asteroids from several centuries (at the current discovery rate) to about 25 years. We call this proposed survey program the Spaceguard Survey (borrowing the name from the similar project suggested by science-fiction author Arthur C. Clarke nearly 20 years ago in his novel *Rendezvous with Rama*).

On a last note, on 23 July of this year (1995), a giant comet was spotted for the first time by two amateur astronomers, Alan Hale, and Thomas Bopp, designated 1995-01, but more commonly known as Hale-Bopp, it is expected to arrive closer to us in the next year and a half or so. This object's size has not yet been precisely determined since it's still beyond Jupiter's orbit, but estimates range from about 100 to 1000 times the size of Halley's comet, and reports that it's about 1600 kilometres in diameter have been made.
Scientists have all assured us that it will miss the Earth by a good margin, so we should have nothing to fear, but it's interesting to note that they also estimate that it will be close enough, and bright enough, to be seen very clearly by the naked eye...in daylight!
Whatever the case, this object is definitely unusual, and has literally come out of nowhere.
Let's hope this time around the reminder that our existence is a fragile one is a gentle one!

In view of these points, it should be obvious that despite how much we'd like it to be so, the Universe is far from a 'safe' and orderly place.[3]
Perhaps, the best way to put it is still as the car bumper sticker (and the producers of *Forrest Gump*) says: Shit Happens.

---

[3] I do believe in an implicit order to the Universe, but I feel too, that the magnitude of this order is so vast, that we are often, perhaps even always, unable to grasp even the tiniest part of it. In other words, if the Earth does get hit by a giant asteroid, it may well be that it is quite simply, *meant to happen*, but of course, this is of little comfort to those that will be affected by the event.

# Appendix B

## The Gods and Technology in Mythology

### Another Interpretation of the Old Testament

The interpretation of the original Hebrew text of the Old Testament is even today hotly disputed by biblical scholars, and indeed was already disputed long ago, by the scholars of the time, as is evidenced in several, already ancient, Hebrew commentaries.

Zecharia Sitchin (among others) is one of those that feels that the Old Testament has been even more badly interpreted than most of us already assume in most cases and this time (perhaps unlike with respect to the Vyse markings) it is difficult to find any fault with his position on the matter.

Although I do not profess to be in any way proficient in the reading of Hebrew or the study of biblical documents, or indeed even the Bible itself, I have undertaken a little research into the matter, that convinces me that Sitchin may indeed have more than just a point; indeed, he may well be on his way to winning the match.

He may be killed for it by his critics, for his side is not the popular one in the game at hand, but win the match he may just do.

What, in my opinion, Sitchin's long years of study have uncovered which is of paramount importance however, is not so much his theories on the extra-terrestrial origins of the human race maybe, but the way in which he came to conclude this startling fact.

Unlike me, Sitchin did not have the Face on Mars with nearby Pyramids to start him off on a long search for the truth.

Instead, he versed himself in the far more lengthy and studious task of examining the most ancient written records of human civilisation. Namely, the translating of Sumerian, Assyrian, Babylonian, Hittite and Canaanite writings and tablets, which lie at what may be called the 'dawn' of modern history.

Such a study of course would require a lengthy investigation indeed, and unless one is interested in such things as Sumerian tablets for their own sake, it would not be something that just anyone might undertake. But in studying such work, Sitchin became aware, that the old Sumerian texts, appeared to vindicate the Old Testament to the $n^{th}$ degree, if in a somewhat unorthodox manner. The names of cities and towns found in the Old Testament, were revealed to have been factual names of places which in fact existed in ancient times.

And indeed, some of the very same stories told in the Bible, have even older Sumerian counterparts. Sitchin then, came across what has since been generally accepted even by the most sceptical of archaeologists, and that is, that at least in part, the Bible is a factual document which relates to factual events and places.

Knowing this, it becomes more acceptable to think that perhaps even some of the more outlandish stories found in the Bible have a factual basis.

Sitchin, however, has the quirk of having been fascinated at an early age by the writings of the Old Testament, and in particular by the interpretation of certain words in it, such as *Nefilim* and *shem* among others and hence proposes, that even creatures as unlikely as the Titans (or Nefilims) may have had a real counterpart.

At first, the suggestion that at one time in the distant past of our planet, giants roamed the Earth, seems an outlandish and foolish one. But on closer and deeper investigation, this may not be as incredible as one might suppose.

Let us however, confine ourselves here, to showing just two alternative interpretations of parts of the Old Testament. The word *shem* as investigated by Zecharia Sitchin and the destruction of Sodom and Gomorrah.

Sitchin points out, that already over a century ago, G. M. Redslob, published a study in which he correctly pointed out, that the word *shem* stems from the same root as *shamaim* (heaven); namely, the word *shamah* (that which is highward).

Hebrew, as most languages, (even if it is only dimly recognised in newer ones such as English) has the characteristic of having meanings for names and words. Meanings which are sometimes much deeper than those already implied by the word itself. A good example in English of this sort of 'hidden meaning' is perhaps the word Wednesday. We all know that this, generally speaking, refers to the name of one of the week days, but not all of us realise that each name of the week, as well as of each month, has primitive roots indeed, and that those roots are well embedded in the worship of multiple 'Gods'.

In the romantic languages this is more evident, since the week days have names which are very close to the original ancient Roman or Greek names of the relevant Gods; but one cannot at first easily see how the Italian *Mercoledì*, which is in honour of the God associated with the planet Mercury (*Mercurio* in Italian), has anything to do with the English Wednesday. If one is made aware however, that the Germanic people called this same God Odin, or Woden, and hence the Dutch *Woensdag*, we can easily see, that even though not at first apparent perhaps, the name Wednesday, has indeed connections to the very same ancient 'Gods' as does a lot of our present day language, even if this is seldom recognised.

In any event, because Hebrew is an old language indeed, these 'hidden meanings' are perhaps a little more evident than in a relatively new and ever-changing tongue, such as English.

In this respect then, Zecharia Sitchin points out that if *shem* is taken to mean 'sky-ship', as he believes was originally intended, instead of the normal meaning attributed to it by biblical scholars ('name'), then certain passages of the Old Testament suddenly begin to make a lot more sense.

The passage in Genesis 11:4 to 11:8 for example, concerning the tower of Babel, which reads:

4.  And they said: 'Come, let us build us a city, and a tower, with its top in heaven, and let us make us a *shem*; lest we be scattered abroad upon the face of the whole earth.'
5.  And the Lord came down to see the city and the tower, which the children of men builded.
6.  And the Lord said: 'Behold, they are one people, and they have all one language; and this is what they begin to do; and now nothing will be witholden from them, which they purpose to do.
7.  Come, let us go down, and there confound their language, that they may not understand one another's speech.'
8.  So the Lord scattered them abroad from thence upon the face of all the earth; and they left off to build the city.

It would seem unusual for mankind to want to build a 'name' for themselves in order to prevent their being scattered all across the globe.

If however we read 'sky-ship' into *shem*, it suddenly makes sense. A sky-ship would be useful indeed to travel from place to place and ensure that contact is kept amongst all the people. Such a set-up would in fact be extremely closely reminiscent of the Vedic hymns in which the 'Gods' travel from place to place, along with their armies and humans, in just such sky-ships.

It would explain too, why the Lord feels he has to come down to confound humanity.

And once again, that disturbing plural tense is evident. The Lord says: 'Come, *let us go down*, and there confound their language,...'

The commentators of course try to brush this aside as being the "plural of Majesty" or perhaps, like in the earlier passages of Genesis,[1] as God talking to his angels. But the fact is that if the Lord is not talking to someone other than himself, then the case for there being 'Gods' in the past, suddenly gains a lot more weight, and if he *is* talking to himself, then it would seem that as well as paranoid, he's schizophrenic too. How else can we explain why God would be affronted by mankind's rapid advancement? And particularly why would he be upset at them for wanting to make a 'name' for themselves? Are these in fact the actions of a good, just, and divine God?

It is their trait to repeatedly act in less than divine ways that marks the Chaldean Gods of old as being a volatile bunch overall. And of course, if the Bible is a collection of older Sumerian myths, then it makes sense that the traits of those Gods are also present in the Old Testament, and particularly in the older parts of it, as Genesis, being part of the 'beginning' stories, surely is.

If Sitchin is right however, the people of old also built effigies of the original sky-ships, and in later stories, the word *shem* is also applied to such structures. If the ancient Egyptian culture is anything to go by of course, with their massive needle-like monoliths, and depictions of unlikely structures that could similarly be taken as effigies of rocket-ships, then this is indeed a valid statement.

As for the destruction of Sodom and Gomorrah, recounted in Genesis 18 and 19, it is easy to read this as if it were the tale of three futuristic spacemen bent on the destruction of these two wicked cities.

---

[1] See appendix E

They appear to be men, yet they are instantly recognised by all that see them as more than mere men, and in fact are described later as angels. Yet they eat, wash their feet, and behave as normal humans.

So startling is this fact that it is noted by the scholars as the only passage in the Bible where celestial beings eat food.

One of them is in fact the Lord himself, which tends to give weigh to Sitchin's idea that there was a 'God' which seemed to lead the other 'Gods' and whom has been misunderstood as also being that Higher Power which surely exists and orders the Universe.

Off to see if indeed the two cities deserve destruction, they are held back by Abraham, whom asks the Lord to spare the city if but fifty honest people are found in it.

The Lord listens and as Abraham continually adjusts the figure, continues to agree that he would spare Sodom even if only the number of honest people in it is as stated by Abraham, whom finally stops altering it when he reaches the number ten.

The three then set off again towards Sodom but in the next chapter we find only two are left, the Lord no longer with them.

These two arrive at Sodom and are promptly invited into Lot's home, where once again they partake of a feast, but before they can bed down for the night, the Old Testament informs us that Lot's house is surrounded by the populace, whose intentions towards the two strangers is far from friendly. Nor is it surprising if we assume that news of their city's fate had somehow trickled down the grapevine to them.[2] Lot is in such despair that he offers the crowd his two virgin daughters rather than the two 'Gods'. Commentators put this down to the hospitality and protection one must show a guest in the East, but in view of the general tendencies prevalent in Sodom, it is highly doubtful that such an offer would have been made without good reason.

And if Lot was aware of the power of his 'visitors', which he surely was because he prostrated himself before them when he first saw them, it makes sense that he would try anything to avoid the

---

[2] Which is indeed the case by the Old Testament's own admission in Genesis 19:9 : "And they said: 'Stand back.' (To Lot) And they said: "This one fellow came in to sojourn, and he will needs play judge; now will we deal worse with thee, than with them.' And they pressed sore upon the man, even Lot, and drew near to break the door. In other words, 'That man has come here and presumes to judge us, now stand aside while we bash his head for him' it appears that one of the 'angels' was near Lot, while the other remained still in the house when the crowd began to press upon them.

conflagration which would be sure to follow, and which may well result in the destruction of his entire city with him in it.

Indeed, the two beings, whom are described as 'angels' at the beginning of Genesis 19, but are talked of as men later in the same chapter, have soon had enough, and they *'smote the men that were at the door of the house with blindness, both small and great; so that they wearied themselves to find the door.'*

This passage is extremely interesting if it's compared with the effect of certain weapons the 'Gods' used according to the Vedic hymns mentioned in chapter two. It would appear that among other instruments of destruction, the 'Gods' had some kind of weapon at their disposal that is somewhat reminiscent of a kind of laser, but with slightly different properties, that seem to induce a sort of radiation sickness, tiring troops out to death in a short span of time and simultaneously blinding them if they do not avoid it's 'beam'.

After this little encounter, which can be easily imagined as the plot of a science fiction movie, where the righteous heroes, outnumbered but possessing technologically advanced weapons, fend off a marauding horde, the 'angels' lead Lot, his wife, and two daughters to safety, by directing him out of the city and telling him to flee without so much as a glance backward and to do so speedily, because the Lord (the third man that was not with these two at the beginning of Genesis 19) will soon destroy Sodom.

Lot's wife seems to have faltered behind him and was turned into a pillar of salt for her trouble. It is interesting to note, that similar pillars of 'salt' were left behind at Pompeii when the nearby volcano erupted and mummified the entire town along with its inhabitants, and that some corpses resulting from the Hiroshima and Nagasaki atomic explosions looked more like bundles of salt than human beings. Some were so close to the blast that all they left were 'shadows' on the concrete walls.

And if the 'Lord' used a smaller version of the meteoric bombardment on Sodom and Gomorrah as I believe was employed to destroy Mars, then the resulting covering of acidic ash, would probably envelop bodies that were not immediately destroyed, with a soot that would prevent decay much like in the case of Pompeii.

Genesis 19:24 and 19:25 says: *Then the Lord caused to rain upon Sodom and upon Gomorrah brimstone and fire from the Lord out of heaven; 25. and He overthrew those cities, and all the Plain, and all the inhabitants of the cities, and that which grew upon the ground.*

Which of course seems to fit rather well with a meteoric bombardment of the cities concerned. The conflagration was no small affair, and could be seen from quite a ways indeed, because Abraham, whose tent was presumably at least a half-day's march away, could see the huge columns of smoke rising from the destroyed cities.

Reading the Old Testament, and perhaps Genesis in particular, from the point of view that extra-terrestrial beings of human appearance are involved, makes for interesting reading indeed, and perhaps, it is high time that such an 'unorthodox' reading of it should receive a wider audience, which is why it is mentioned here.[3]

## Other Legends

The appearance of Gods with technologically advanced equipment is not however, by any means, confined to the Old Testament.
In particular, a recurring theme is the use of certain 'staves' by the various 'Gods' of several cultures.
In each case, these staves are more than just sticks, being capable of various destructive functions, as well as beneficial ones occasionally, such as the healing of wounds.
The staff is a recurring motif amongst all the legends of the 'Gods', whether they are of European, Middle Eastern, or South American origin.
        The God Hermes, which seems to have got around quite a bit, seems to have been particularly fond of his 'staff' which to this day remains the symbol of the medical profession.
The Greek Hermes, was the Odin or Wotan of the Germanic people, the Thoth or Tehuti of the Egyptians and the Votan of the Maya.
It is of course, not possible in a book of this size to give a complete cross-reference of even just a tiny part of the overlapping myths concerning this God, but we can say with some certainty, that he was venerated as one of the true founders of civilisation by all the people who knew of him, and in each instance, his staff is with him and is responsible for various miraculous feats.

---

[3] It is interesting to note that even today, God is generally thought of as existing in 'heaven' and this in turn is taken to be somewhat synonymous with 'sky' as is clearly shown by the attitude of looking 'heavenward' when indicating the location of this 'heaven'. God in other words, even today, although in a much watered down form, still 'comes down to us from the sky'.

The Aztecs have stories of similar staves being used by Gagavitz and Zactecauh to shoot down flying birds at a distance as well as part the waters of the sea, much like Moses is reputed to have done at the crossing of the Red Sea, using his own mighty 'rod'.[4]

In the Isis-Osiris Egyptian myth, where Osiris is chopped up by his brother Seth, Isis (Osiris' widow) impregnates herself with Osiris' seed (or at least his genetic matrix) after the fact, by using an artificial phallus in order to produce her son, Horus. This is perhaps one of the most graphic examples of artificial insemination in mythology, and of course it involves celestial personages. It is also interesting to note, that Isis is named as the mistress of the Great Pyramid by the Inventory Stela discovered by Mariette, and that the King Lists, which for some reason are taken as factual by Egyptologist only from the beginning of the First (Human) Dynasty, contain Dynasties preceding this one, in which the rulers are demigods, and earlier ones still, in which the kings are Gods; the very same ancient Gods the echoes of which are still with us today.

But perhaps most damning of all such legends are the Vedic hymns found in the *Samarangana Sutradhara*, where detailed descriptions of what can only be called 'flying machines' are given.
Such 'sky-ships' were often called 'flying-horses' or 'flying-chariots' in other parts of the Vedic stories, and considering that the Red Indians referred to locomotives as 'iron-horse' one can see that this was just a way in which a people that did not have a word for spaceship or aeroplane, would name such an object.
After all, even today, the word 'space-ship' is only a mild improvement on 'flying-chariot'.

So detailed and interesting are the Sanskrit texts, that there are rumours that modern scientists have been studying them in detail in the hope to unlock the secret of anti-gravity machines, and a translation made by Maharshi Bharadwaja, called *Aeronautics* and described as *A Manuscript from the Prehistoric Past*, comes to mind in this regard. Considering that according to the Sanskrit texts, some kind of mercury was supposedly a component required for the

---

[4] Exodus 14:16

operation of these crafts, we should then perhaps take stories of that quasi-mythical substance, red-mercury, more seriously.[5]

The reading of almost any myth with an eye set on the lookout for the uses of advanced technology by the various 'Gods' makes for very interesting reading indeed, more so especially, when one realises, that each 'God' was given a different name by each group of people that became aware of him or her. The multitude of Gods found in the mythology of our planet, stem not so much from their vast number as such, but rather from the multitude of names which each of them acquired. And when this is fully grasped, the picture of a handful of superior beings leading, controlling and also abusing the inferior masses of humanity emerges, which of course, fits in exactly with the theory of a limited number of space-faring individuals in some way colonising a technologically backward planet.

---

[5] Red mercury is a substance which would supposedly allow (among other things) the manufacturing of nuclear devices the size of hand-grenades. Publicised the most perhaps by Peter Hounam, a British investigative reporter, in a series of newspaper reports, the existence of this substance has been consistently denied by western countries, but in Russia, there is much evidence to the contrary, including the murdering of people involved in transactions of it by use of rifles equipped with night-scopes; a decisively government-flavoured execution. Red mercury, has also been mentioned at the very highest levels of Russian officialdom; namely by Alexander Rutskoi, when condemning Yeltsin for allowing a national strategic asset to fall into western hands, and apparently by Yeltsin himself in his approval of an export licence for the substance, where his signature appears.

# Appendix C

## Technologically Advanced Items from Ancient Times

If buildings with the dimensions of the Great Pyramid, or the huge figures of animals found near Nasca in Peru[1] have been ignored as evidence of a technologically advanced race in our past, how much easier has it been for us to dismiss the occasional archaeological oddity?
Apparently we manage to ignore those artefacts that fly in the face of conventional thought quite comfortably, for in truth there is not a shortage of them around the world.

At Ancon in Peru there are clear geological examples of cross bedding in beaches that have been raised hundreds of feet by some kind of tectonic movement, but among the shells, molluscs and other fossilised sea-life, one also finds the presence of man in the form of shards of pottery, stone tools, fish nets and shreds of woven clothing. An obvious clue that some large tectonic movements have happened very rapidly.
Still in Peru, the numerous enormous works of fitted stone, some examples of which are to be found at Sacsahuaman by way of a wall, cannot be easily explained due to engineering difficulties which seem to defy explanation in term of size, weight, fitting of joints and so on. Perhaps because Peru, along with most of South America falls under the category of 'developing country' these treasures have not received the attention they deserve, but more importantly, there is a very real possibility that the more subtle forms of evidence of ancient technology that may still exist in such countries today is lost in the name of progress, and this would be tragic indeed, for in some of these items may lie the clues to technology and civilisations yet to be imagined never mind suspected.
       In the jungles of Costa Rica, Guatemala and Mexico, there have been found hundreds of perfectly spherical stone balls. Some of

---

[1] These shapes, built out of what from the ground appears to be an almost natural 'ridge' are so large that the animal or figure they represent can only be seen from the vantage point of an aircraft. The Nasca 'lines' (geometrical road-like surfaces which have been compared to landing strips) can go on for several miles.

these have a diameter of some 2.5 metres (8 feet) meaning they weigh well over 10 tonnes, but the interesting fact is that they are sometimes found in places that are far from being easily accessible, that no tools which can possibly explain their creation have been found, and that the volcanic rock from which they are usually made is generally found only many miles away.

The Kutb Minar pillar in Delhi, India, cast of iron and standing eight metres tall has stood in that spot for the last 1500 years without rusting.

While we are briefly on the topic of metals, it is worth remembering that in view of the redating that is almost certainly required concerning the Great Pyramid, the recent finds by Gantenbrink concerning possible brass objects and in any event the rediscovery by Bauval of brass objects which had previously been found in the Great Pyramid is pertinent when considering advanced technology in our distant past.

It is an unfortunate trait of this type of objects however, to suffer a fate similar to the brass objects Bauval had to go and dig out from the depths of the British Museum, as is shown by the next two items I will mention. Whether this is by accident or by design remains to be seen, but it's probably safe to assume that while in most cases such objects are misplaced due to simple lack of adequate organisation skills, in some instances, certain interested parties may have more than a little to do with their eventual disappearance.

In 1885, a steel cube was found in a block of coal at the foundry of Isidore Braun in Vöcklabruck Austria. This was donated to the Linz Museum, but today only a cast of the object survives, as the original seems to have been somehow lost or misplaced. Andrew Tomas, in his book *Atlantis From Legend to Discovery*,[2] reports that this object was described in contemporary journals such as the November 1886 edition of *Nature*, which remains a scientific publication even today. Because of its geometrical shape and a deep incision which ran the equator of its surface, it would seem unlikely that this was a fossil meteorite, but the coal it was found in obviously predated man's existence on Earth according to orthodox belief. Where then, could such an item have come from?

---

[2] See bibliography.

A similar story surrounds a geode of particularly interesting composition. I first came across this story on the Internet and reproduce here the text which can be found at the web site with the following address: **http://archon.lib.umn.edu/geode.htm**

A picture of the Coso geode next to a ruler for scale purposes is also present at this location. The caption under it reads as follows:

> "Cut in half, the Coso geode resembled a spark plug, with a shaft of metal about .08 inch thick encased in white ceramic. It was thought, however, to be half a million years old."

The dimension of the geode is approximately 1 and $^3/_4$" (4.45 centimetres) in width, while the ceramic component is about $^5/_8$ ths of an inch across (1.59 centimetres).

The body of the text is as below.

In 1961 Wally Lane, Mike Mikesell, and Virginia Maxey, co-owners of the LM&V Rockhounds Gem and Gift Shop in Olancha, California, went into the Coso Mountains, six miles north-east of Olancha, to look for unusual rocks. Near the top of a 4,300-foot peak overlooking the dry bed of Owens Lake they found a fossil-encrusted geode that proved to contain something strange.

What the geode proved to contain, after Mike Mikesell had ruined a diamond saw blade in cutting it open, was something that would later be shown to resemble a spark plug.

In the middle of the geode was a metal core, about .08 inch (2 millimeters) in diameter. Enclosing this was what appeared to be a ceramic collar that was itself encased in a hexagonal sleeve carved out of wood that had, presumably at a later date, become petrified. Around this was the outer layer of the geode, consisting of hardened clay, pebbles, bits of fossil shell, and "two nonmagnetic metallic objects" resembling a nail and a washer." A fragment of copper still remaining between the ceramic and the petrified wood suggests that the two may once have been separated by a now decomposed copper sleeve.

X-ray photographs of the object were taken, and it was after examining these that the editor of INFO Journal, Paul Willis, noticed a startling similarity between the Coso artefact and a modern spark plug.

In 1963, the Coso geode was display for three months at the Eastern California Museum in Independence. Wally Lane then seems to have taken possession of the object and in 1969 was reportedly offering it for sale for $25,000. According to the estimate of a geologist unnamed in the original report of the find, the age of the geode based on the fossils it contains is some 500,000 years.

Excerpted from *Mysteries of the Unexplained*, p. 47.
BUY THIS BOOK, usually available at used bookstores for $7.50.  A great resource for your personal collection.

If this story is a hoax, as some of the 'debunkers' of the Internet would have us believe, then it's surprising to have so much detailed information along with it. We are after all provided with the names of four people, their respective occupations, the name of one Museum where it was displayed, as well as the title of the book this information came from, which in turn, if purchased would give us also the name of the author who may in turn provide even more information.

If the details presented here are true and accurate, which until someone tracks down Wally Lane may be hard to prove either way, then this would be the first unequivocal example of advanced technology having existed in our pre-history.

It may of course be argued that even if the geode is real, then it must have been some piece of ancient trash left by some extra-terrestrials, because so far as we know Homo Sapiens didn't exist on Earth 500,000 years ago, but this of course would precisely support the basic premise we have already concluded in part I of this book.

It is also interesting to note, that a dating of some half a million years places the geode in the same time-frame as that which Hoagland ascribes to the Face on Mars.

There are numerous occurrences throughout the world of suspicious artefacts, among them the neatly round hole in the skull of a prehistoric man that lived some 40,000 years ago[3] and the similarly curious bullet-like wound found in the skull of an ancient bison.[4]

Why do we not read more often about them? The actuality is that these objects do receive some news coverage, but since no one seems able to explain them, their peculiarity soon fades, and they are quickly forgotten in the tran-tran of everyday life.

Each of these items alone is already surprising in itself, but taken as a whole, they present an unmistakable pattern.

---

[3] Still preserved in the Natural History Museum of London as far as this author knows.

[4] Again, to the best of my knowledge, this item should still reside in the Paleontological Museum of the (Ex) USSR Academy of Sciences. The skull is supposedly hundreds of thousands of years old.

One that tells us there most definitely existed advanced people in the past, and sometimes, in a past so distant that our ancestors (at least some of them anyway) were little more than primitive ape-men.

It is hoped that competent archaeologists will soon begin to investigate such 'time-displaced' objects in a new and fresh light. The time for making facts follow theory is fast disappearing and we should begin to embrace the more useful (and sane) idea of making theory follow the facts.

It has often been said (and I for one hold it to be true) that fact is stranger than fiction; if this is so, then we should not be too surprised if the facts paint for us a picture that we would have struggled to imagine, never mind suspect!

# Appendix D

## The Legend and Destruction of the Phaethon Object

In 1950, professor S.V. Orlov of the USSR Academy of Sciences, named the lost planet from which the asteroid belt supposedly originates *Phaethon*.

In Greek mythology, Phaethon was the son of Helios, the god who each day drove the chariot of the sun across the sky.

Phaethon had a dispute about his true origin with Epaphos, one of the many sons of Zeus, and as a result of it asked Helios, to grant him one wish in order to prove to all concerned that indeed he was a son of his. The god Helios called the river Styx as witness to show that he would not refuse to grant this one request, and Phaethon made his demand.

He asked of Helios, that he, Phaethon, be allowed to drive the chariot of the sun for one day.

Helios was shocked at the boldness of the request and at first tried to dissuade his son from undertaking such a dangerous task, but having given his word that he would grant the request, and in view of Phaethon's steadfast determination, he eventually had no choice but to allow the youth to take charge of carting the sun across the sky for one day.

Enthusiastic, Phaethon soon took to the task, but being young, unaccustomed to the work and not familiar with the correct route he soon lost his way and his strength, at which point, the powerful and spirited horses of the sun-chariot went wild and dragged the sun out of its correct path, bringing it so close to the earth that it set fires, dried fountains, boiled rivers and turned part of the human race black with the heat.

Zeus, suddenly alarmed at the danger both heaven and earth had been placed in, struck Phaethon dead with one of his lightning bolts, causing the corpse to fall in the river Eridanos.

So distraught was Helios at his son's death that it was only through the rousing and cajoling of the other gods that he was eventually made to resume his office and the reins of the chariot of the sun.

Far from being a prehistoric version of junior crashing the family car, this ancient myth is one that has multiple echoes in almost all of the oldest fables of all the peoples of Earth.

Immanuel Velikovsky, in his *Worlds in Collision*, perhaps presented this dramatic picture best (if somewhat inaccurately!) by having several of the planets come into contact as recently as one and a half thousand years ago or so.

He'd based his theory of collisions and near-collisions between Venus, Mars and Earth supposedly on the writings of ancient tablets of middle-East origin.

It is highly doubtful indeed that Earth and Mars could ever have possibly collided or even come close to having done so as Velikovsky's books suggested, for I think it not only almost impossible to fathom that Mars and Earth could ever have been brought so close, but also that in the event of a collision, neither planet would really survive. But Velikovsky had several even more far-fetched ideas, among them that Venus originated from Jupiter and that flies came to Earth from Venus after a near-Earth passing.

Admittedly Velikovsky also appears to have had a few good ideas, and to be sure, the treatment he received at the hands of supposedly gentlemanly scientists was far from fair,[1] nevertheless, it would not be unfair to say that although his books are very entertaining, it would be wise to refrain from basing one's concepts of astronomy on them!

In the context of this book however, the story of Phaethon seems to have clear indications of some major catastrophic event, and if we assume that Mars was destroyed by asteroid-projectiles and that survivors of that conflict —or perhaps their allies— may have also been subjected to a similar orbital bombardment here on Earth, which in addition to destroying any Atlantis-Outpost settlement would also result in widespread damage through tidal waves, earthquakes and accompanied volcanic activity, then it is not difficult to see that the legend of Phaethon may indeed be connected to such events.

Interestingly too, the reference to a darkening of the skin of that part of humanity which was closer to the events, may be suggestive of the idea that Earth was pushed closer to the Sun as a result of some

---

[1] Velikovsky's publishers were effectively boycotted so that they had to hand over printing of *Worlds in Collision* to another company or risk losing the support of a considerable portion of the scientific community.

asteroid collision. If this were the case, the rather sudden end of an ice age may then also have resulted in subsequent widespread flooding.

Admittedly the name Phaethon would seem to describe more the events which may have taken place on Earth (as an extended result of the war which by now had in all probability already destroyed Phaethon and Mars) rather than concerning itself with the planet which may have existed where we now have only asteroids; but this is not very surprising, since if there ever were people living on Phaethon, like the Martians, they perished; and in any event, the destruction of Phaethon would have to pale in comparison to the one of Mars, which of course would not be remembered either, except for its connection to the God of War.

# Appendix E

## Semitic Etymology in the Old Testament

### Genesis 6:1-4

1. And it came to pass, when men began to multiply on the face of the earth, and daughters were born unto them,
2. that the sons of God (beney Elohim) saw the daughters of men that they were fair; and they took them wives, whomsoever they chose.
3. And the Lord said: "My spirit shall not abide in man forever, for that he also is flesh; therefore shall his days be a hundred and twenty years."
4. The Nefilim were in the earth in those days, and also after that, when the sons of God (beney Elohim) came in unto the daughters of men, and they bore children to them; the same were the mighty men that were of old, the men of renown (reputation).

Zecharia Sitchin (and many others) have had much to say about several verses in the Old Testament, but perhaps this is the most indicative of the inconsistencies to be found in the Bible.

The problem is of course, that since the Old Testament was originally written in Hebrew, non-Hebrew speakers are at the mercy of translators, and of course, this is a less than ideal situation for non-Hebrew Bible scholars.

It might not be unfair to say, that a Bible scholar that is not fluent in Hebrew is in fact no such thing as a true Bible scholar. The problem with getting at the true meaning of the words found in the Old Testament, is that those who can honestly be referred to as Bible scholars, are in the main Jewish Rabbis, or at least Hebrew students of the Torah.

This of course is not a bad thing in itself, but the fact is that if one is born into a religion and a particular way of life, it is at best unlikely, that he will seriously question it. I'm not talking about questioning certain aspects of belief – the Jews are perhaps the most honest religionists around today, arguing and questioning each other's interpretations of the Old Testament and of philosophy in general, to an extend that Christianity or Islam would surely be unable to tolerate— but I am referring here to a ground shaking, nay, *foundation crumbling* type of questioning. Questions that would invalidate the whole 'spirit' behind the Old Testament.

Such as would be the case if the God of the Bible was sometimes an extra-terrestrial and at other times the more conventional all-permeating power which I feel surely does exist and which brings love and order.

If indeed this were the case, then the Bible would no longer be a truly 'holy' book, which descended direct from heaven, but 'merely' what could be described as the most amazing collection of stories about our most distant past.

Before I try to elucidate on what may be a more accurate translation for the passages of Genesis quoted at the start of this appendix, I would like to point out, that in view of the discovery that not one or two, but *several* of the stories recounted in the Old Testament are to be found in an earlier Sumerian format, is already very indicative that this is in fact the case. Adding to this the even more amazing realisation that the ancient Sumerian city-names correspond exactly to the Biblical city-names, it would seem to be not just a possibility, but a very strong likelihood indeed, that the Bible is in fact just that: a collection of even older Sumerian 'myths'.

And I use that word, *myths* in the loosest possible sense, because I do feel that there may be more truth to these 'myths' than has perhaps ever been supposed before.

In any event, the term 'sons of God' in Genesis 2 and 4 (and in other places as well) can be accurately translated as "sons of (the) Elohim" the bracketed 'the' is a fair and valid alternative, because there is some honest doubt, even amongst Hebrew Biblical scholars, as to the real meaning of this word. It is commonly translated as one of the names of God, but this is by no means an accurate rendition, because you see, elohim is a *plural* term, which might be better translated as meaning 'mighty', 'noble', or perhaps even 'majes-tic'. This is borne out also by the extensive commentaries made by already ancient Hebrew scholars in ages past as well as in the present.

There are at least two 'official' alternative renditions of the meaning of these phrases then; one is that by sons of Elohim what is actually meant is '*sons of the angels*', the other is that it stands for '*sons of the nobles*' (*nobles* is usually conventionally translated as '*rulers and judges*'). But this presents a bit of a problem does it not?

By itself perhaps, just *one* use of this word *elohim* in a plural fashion would go unnoticed, but it crops up several times in similarly embarrassing passages. But even so, we might still be taken to glossing over this troublesome spot if it wasn't for the fourth verse.
What is really meant by that term: *Nefilim*?[1]
Because whatever they are, they are the product of the '*sons of the elohim*' and of '*the daughters of men*'.
But looking in the various commentaries, the term is usually taken to mean *giants* which in English are sometimes referred to as Titans.
Interestingly enough though, the word elohim also has similar connotations, since '*sons of the elohim*' could also be translated as '*sons of the great (or mighty)*'. And of course, there are indeed fables of 'giants' roaming the Earth in our pre-history.

Most modern Hebrew scholars reject the translation of elohim as meaning angels though, because the notion of angels frolicking with human females is obviously a distasteful one from the classical viewpoint, so we could say that we are really left with only the other alternative if we want to give the religionists the benefit of the doubt.
And here of course we bump our nose into another problem, because if we take elohim to mean '*rulers and judges*', or '*nobles*', we then have to explain why such a clear distinction is made between their sons and the '*daughters of men*'. Are the nobles not men too?
Are they too, not, 'children of Adam?' (The phrase '*daughters of men*' is literally written in Hebrew as '*daughters of the Adam*' and this phrase, sons of Adam, children of Adam, or daughters of Adam, is commonly used to represent the descendants of Adam and hence of humanity.)

And if they are NOT children of Adam, then it makes it very clear they are meant to be of some other origin, and considering that the Bible deals with only celestials and humans, if they are not human mortals, then they are intended to be considered as being celestial in origin. Which would nicely explain why older commentators translated elohim as 'angels'.

---

[1] Sometimes written as Nephilim.

It would be the 'sons of the angels' then, whom impregnated the 'daughters of the Adam' (daughters of common humans as opposed to 'angelic' humans).

And of course, in the passages describing the destruction of Sodom, we have already seen, how the term 'angels' was used to describe 'celestial' beings which although very human-like in appearance, custom, mode of travel, and bodily functions, were nevertheless capable of causing great destruction by unknown means, but were not themselves privy from sustaining grievous bodily harm, as the lynching crowd that gathered around Lot's house tried to prove.[2]

It would seem then, that the translation :

2.   that the *sons of the gods* saw the daughters of men that they were fair; and they took them wives, whomsoever they chose.

which in any event has appeared from time to time in certain versions of the Bible, is indeed a more faithful one.

But of course, such a verse in the Old Testament would seriously shake the roots of the entire book. Not it's true roots of course, for the real truth can never be anything other than what it is, (whatever that may be in actuality) but the superimposed 'truth' we have placed upon the Bible ourselves suddenly becomes shaky indeed.

What is this talk of 'gods' in what should be a book about the one and only true deity? Especially since the book supposedly comes from that one true deity himself?

But the careful editing and religious censorship present in the Bible, even its Hebrew version —which although faithfully reproduced for thousands of years has had its true meaning twisted by the changing meaning of words and how these are interpreted by later generations— does not end here.

For the word *Nefilim*, which used to be translated and interpreted as *giants*, has in fact, quite a different meaning altogether,[3] as is clearly shown by the Hebrew word *nafal* which means '*to cast down*', stemming from the Semitic root NFL which stands for: '*to be cast down*'.

---

[2] See appendix B.

[3] To their credit, modern translators, having become aware of the error, are nowadays more prone to leaving the word in its Hebrew version, as Nefilim or Nephilim, although (of course) few are those willing to openly admit the etymological origin of the word and it's true meaning.

In other words, the Nefilim, where those who had been '*cast down to Earth*' or if you prefer, '*cast down from Heaven*', hence perhaps, the origin of the tale of Lucifer; the rebellious 'angel' who was 'cast down' by 'God'.[4]

And here, it must finally become quite obvious, that something quite different from the orthodox holy and allegorical meanings attributed to these sentences is being talked of.

In fact, the concept of extra-terrestrial (yet human-like) 'Gods' being involved, is the only one which fills all the inconsistencies and gaps found in the more orthodox versions of the story, in a particularly neat and simple fashion. Considering that we are dealing with ancient texts —which no doubt have undergone some considerable editing— but keeping in mind too that surely they were meant to be read in a far more literal manner than that attributed to them today, this seems to me, to indeed be a very valid proposition.

Sitchin adds a further dimension to the fourth verse by stating that if the last word was originally written as:

4.   The Nefilim were in the earth in those days, and also after that, when the sons of God (beney Elohim) came in unto the daughters of men, and they bore children to them; the same were the mighty men that were of old, the men with a name (*shem*).

then, since *shem* (as detailed in appendix B) has been incorrectly translated as '*name*', it would be normal to convert the phrase to: '*the men of renown*'. The men with a *reputation*, what else after all could *men with a name* mean?

But if the correct term, *sky-ship* is used, instead of *name* then the phrase once again makes sense, and would now read, '*the men of the sky-ships*'.

Even without this last piece of circumstantial evidence though, the situation is damning enough as it is to make a re-interpretation of the

---

[4] Interestingly enough, the name Lucifer is of Latin origin, and of course stands for '*Light Bringer*', hardly a name one would associate with the Prince of Darkness and ultimate Evil. In fact, the story of Lucifer has echoes in most ancient traditions from around the world. The Greek fable of Prometheus, the demi-god who introduces fire to mankind and is painfully reprimanded for it by the 'gods' and Zeus in particular, by being chained to a rock where a bird daily comes to peck out his liver, is certainly another, and perhaps a 'truer' version of it at that.

Old Testament a project worthy of undertaking by some Hebrew speaking mind, that is unfettered by the curse of religious dogma.

But credit must be given wherever and whenever possible, and it must be stated, that Sitchin is not the first or the only one to arrive at such conclusions either, for (as Sitchin himself kindly points out) Malbin, a noted Jewish biblical commentator of the nineteenth century, had recognised these errors of translation already some time ago, as indeed, periodically, have others before him, and he correctly stated that in very ancient times, the rulers of Earth's countries, were the sons of the deities which had descended from the Heavens and copulated with human females to produce, the much mentioned demi-gods of old.

The fact that they called themselves Nefilim,[5] would also seem to suggest, that perhaps these deities were not here by choice, but due to some other reason —on which, Sitchin has much to say— but given the amount of time that has passed, I feel it is quite irrelevant to speculate on such reasons. In part, this may well be due to my ignorance of ancient Sumerian texts, with which Sitchin is instead well acquainted, having studied them for a number of decades, but more importantly; the reason for their being here in the first place, is largely immaterial to me. The fact that they were here at all is what I find interesting, and even more so, what the implications of such a discovery are, as we shall explore in the next section.

### Genesis 3:22

22. And the Lord God said: "Behold, the man is become as one of us, to know good and evil; and now, lest he put forth his hand, and take also the tree of life, and eat, and live for ever."

Another inconsistency in the Bible is to be found in the above passage, where God, having caught Adam and Eve with apple seeds in their gullets, decides to ban them from Eden.

Quite apart from the fact that in this very sentence God admits that the serpent was indeed correct (he told Eve if she ate from the tree of

---

[5] Other than: *those that were cast down*, an alternate rendition of this name might be: *those who fell to Earth*, which is more reminiscent of ship-wrecks than of forced exile.

knowledge she would not die[6] but rather know good from evil and thus become as God himself) it also points out two important things.

Firstly, God would seem to be a rather petty guy. Definitely not worthy of being worshipped as an all-loving, merciful, just and good God at all. And secondly, he's talking to someone else!

Again, we cannot reconcile with ourselves that God is talking to himself (and in the plural tense too) unless we assume he's a bit schizoid, and to assume he is indeed talking to other entities (let us say 'angels' for argument's sake) would seem to indicate a parallel which once again comes uncomfortably close to: Gods on one side and Humans on the other. And anyway, we have already seen that 'angels' is in any event synonymous with human beings possessing great power, or indeed, with the term 'Gods'.

The fact concerning God's pettiness —as well as his falsity concerning what would happen if the apple was eaten or even just the tree touched (certain death)— is important, because the Bible scholars would like to distance themselves from the Sumerian accounts of the flood, which being closer to the actual events of this catastrophic time, (due to their preceding the Bible) are more true to the facts, and involve a number of Gods, and a lot of falsity and pettiness on their part towards each other as well as the Humans.

Some Bible scholars, using the "If you kill the chicken, then the chicken never existed in the first place" logic, go on to say that the Sumerian version of the flood is obviously false in view of the numerous deities, lack of exalted or uniform purpose and lack of reverence and restraint.

While of course, the biblical account, involves a just purpose (don't ask me...I'm just a writer...) a just and merciful God and an exalted and reverent meaning for it all.

Instead of a lying, snivelling, cruel and selfish bunch of pagan Gods, that are intent on the destruction of all mankind because they are starting to become a bit too smart for their liking you see, the Old Testament version of the flood involves a just, merciful and *holy* God. So of course this proves that the Old Testament version is the correct

---

[6] Bible scholars have hid behind the fact that Adam and Eve did indeed die eventually, and hence the serpent actually lied, but this is of course a fallacy and a belittling of that function known as logical thought, because, as God himself says in this verse, he casts them out lest they eat too from *the tree of life and live forever!* Which of course implies that they did NOT have eternal life even prior to their eating the apple, so indeed, it would seem that the serpent (Prometheus?) was indeed being truthful.

one and the ancient Babylonian one is merely the fevered imaginings of godless heathens.

Not that any of this stands up to the scrutiny of any kind of logic, but in view of the rather extreme pettiness exhibited by the God of the Old Testament in several places (apart from the one found in Genesis 3 with respect to the ousting of Adam and Eve from the garden of Eden) it would seem to me, that if anything, the Old Testament version resembles the Sumerian 'fable' quite well despite its being a much later interpretation of the facts.

One, it might be added, that was compiled only at a time when in all probability, the Jewish people were starting to dimly grasp the concept of the One Real Creator and trying to reconcile it with the various tales of 'real' Gods that had ruled in times past, and whose dim influence may have still been felt through some few and scattered descendants of theirs.

# BOOK TWO

# 9    UFO Phenomena

Once the initial shock of realisation at the extra-terrestrial involvement in or evolution and origin begins to wear off, we may find ourselves becoming intrigued by this fascinating concept and start looking for further evidence to satisfy our curiosity.

In the course of doing this, I began to wonder about a completely different facet of this topic. The idea that we had extra-terrestrial visitors long ago, fired my imagination into wondering about the countless UFO reports which are recorded regularly around the world each week.

Prior to this, I had viewed the UFO phenomena as I suppose most people do, with a healthy (or so I thought) dose of scepticism.

The concept of aliens visiting us on a regular basis was very tempting of course, I would have wanted it very much to be so, but I was inclined to believe that it was wishful (if not deranged) thinking on the part of the 'ignorant masses' that probably confused meteorites, satellites and so on as being extra-terrestrial craft.[1]

And although I have never, even as a child, believed in my heart that Einstein was right about the speed of light being an absolute limit as far as travelling through space-time goes, I certainly didn't have a better theory than his, nor did I have any knowledge of anyone who did; so the UFO sightings, to my mind, remained not only largely and extremely improbable, but also unworthy of investigation in view of their being so obviously the wrong conclusions, or even intentional fabrications, of what may be called the UFO 'nuts'.[2]

---

[1] It is not only the scientists and 'experts' that have snot-nosed, arrogant attitudes concerning 'the average man', this author is as guilty as anyone in that regard, and perhaps more so than most in the past. I can only hope that with age will come a comparable growth in wisdom, as I strive to improve my love and compassion for that which ultimately is truly the most important thing that any of us can have, namely the reciprocal love and understanding of our fellow humans.

[2] It is worth noting here that my attitude in this regard then, was extremely similar, (if perhaps ultimately not identical) to that of those 'experts' who refuse to even look at the available information on a topic before rejecting it out of hand. Richard Milton has much of significance to say on this 'unconscious' mental block which assails scientists and 'level-headed' people in general, in his book *Forbidden Science*.

After nearly two years of intensive research on the subject of extra-terrestrials in our ancient past though, and in view of the discoveries I made for myself in this process, suddenly the UFO phenomenon gained enough importance to at least merit a decent look into.

On deeper investigation, two things became very clear very quickly.
Firstly, UFO enthusiasts, like any group of people have their own subculture, and in general, this could be said to be one in which, paranoia, distrust of governmental authority, and mistrust of authority in general are fairly common and widespread.
Despite this, the classical image of a crazed schizophrenic that wears a tin-foil hat to shield himself from telepathic suggestions from Venusians, is most definitely in the minority as far as UFO enthusiasts go.
In addition, the subculture seems to be further divided into several distinct groups. Central among them is a core of well educated, open-minded individuals, with a firm grasp of logic.
Many of these individuals have a scientific background, being qualified scientists in their own right, or having trained as engineers of one sort or another. Then again, some have been involved in the military or some aspect of the defence forces of their respective countries, while others still are mainly self-educated individuals as far as the basic laws of physics and astronomy go, but having an above average grasp of logical analysis still succeed in making important and useful contributions.

Needless to say, it is this core group of individuals, that make up the real UFO enthusiasts, for while the other subgroups may be interesting from a human point of view, their contributions usually amount to no more than regurgitating information of doubtful origin with little or no detailed data that can be positively confirmed or rejected.
The largest subgroup of the UFO culture however may be termed the 'hanger-ons'. People that while not being rocket-scientists themselves, have a genuine interest in UFOs, either because they have seen one, or think they have, or for whatever reason. They generally absorb information from the core group and only infrequently make a positive contribution to the investigation of UFOs.
Lastly, there is the third and smallest subgroup that is definitely composed of individuals that are either mentally unbalanced, extremely stupid and credulous, or perhaps wilfully malignant towards

honest UFO enthusiasts and demonstrate this by behaving in an obnoxious and ludicrous manner.

There is in fact a fourth group of individuals that becomes associated with UFO enthusiasts, but I will introduce them later.

The second thing one becomes aware of quite rapidly with regard to UFOs, is that there are now some *several hundred thousand* officially reported cases of Unidentified Flying Objects on record.

Of these, the UFO community is the first to point out that yes indeed, something like 90% of these sightings can be explained away as natural occurring phenomena or even hoaxes, but that the 10% or so cases that do not fit into this category can only be one of two things:[3]

1) Either something *really* weird is happening that would (by the way) make the idea of alien controlled vehicles a much easier one to swallow than whatever actually *is* happening, or...

2) These unexplained UFOs are indeed the result of visitation from intelligent life forms that posses a highly developed technology.

Considering that these 'unexplained and unexplainable' cases still amount to many thousands of well documented reports,[4] it is not hard to see that whatever the nature of Unidentified Flying Objects, they most certainly merit further investigation.

Investigation that is government sanctioned and yet also incorporates a large civilian element; so as to assuage the natural tendency of UFO enthusiasts to mistrust governments.

---

[3] As a result of the dogged determination with which UFO enthusiasts are persecuted by the ridiculing media and scientific community in general, the figure of only 10% of reported cases being classed as 'unexplained' is perhaps conservatively low. A more accurate percentage may be twice as much, but in view of the general scepticism surrounding the phenomena, UFO enthusiasts themselves tend to be conservative in nature. The US Air Force however, which may be considerably better informed than the average UFO enthusiast, if a tad more secretive, feels that 20% of the sightings investigated under Project Blue Book are without Earthly explanation, as stipulated in Air Force Project Blue Book Special Report No. 14.

[4] As well as photographs, videotapes (complete with original negatives) and eyewitness accounts, these reports include corroborating radar tracking, sometimes from several radar stations at once, that all place the object exactly in the same place not only as each other, but as reported by the eyewitness accounts.

Governmental sanction is important though, because until it takes place, the UFOs are going to remain the easy news item to ridicule on an otherwise slow day.

But before I launch into a full-hearted support of the wrongly maligned, valiant and honourable UFO enthusiast, let's first find out if there is any reason whatever for doing so.

It is of course impossible for me to make a lengthy, detailed and decisive documentation of UFO sightings and phenomena, but luckily, there is no need for it, because at least one eminent job of this nature has already been done, and I urge the sceptics to perhaps put reading the rest of this book on hold until they have got a hold of Timothy Good's book, *Above Top Secret, The world-wide UFO Cover-up*.[5] Although I have read several similar books, I hold *Above Top Secret* to be the best of its kind for the simple reason that in addition to being extremely voluminous, it is extensively annotated with information that can be verified from the source. A large section at the back of the book for example, is dedicated solely to photographs of original declassified documents issued by the US armed forces and connected with UFO phenomena.

Copies of these documents can be obtained from the US government by application under the freedom of information act, and to my knowledge, the documents shown in *Above Top Secret* are indeed available in this manner.

If there is any drawback to this admirable effort into the investigation of UFOs, it's that it is a little monotonous. It describes report, after report, after report, after report, *ad nauseam*.

But although this may lack a certain flair, if anything, it adds weight to the argument that thousands of unexplained reports are continuously and stubbornly being publicly ignored or dismissed by the US government (as well as the governments of most other countries too) while in fact they are being investigated with great zeal behind closed doors.

As I have said, it is not my intention to make the case for the UFOs in this book, others have done so before me (and continue to do so) in a far more complete manner than I can here or even than I may be inclined to do in future.

---

[5] See the first entry in the bibliography for further details.

No one disputes that UFOs exist, not even the most critical and subversive government departments do that. The only argument that arises, is as to their origin.

There would seem to be a number of people that are interested in making us believe that *all* UFOs can be easily explained away as natural occurring incidents that have nothing whatever to do with any extra-terrestrial beings. Before we get involved in a vicious circle of paranoid conspiracy theories however, let's examine just a couple of incidents that have recently come to light.

1) With the recent collapse of the greater USSR came a startling revelation by high ranking military officers and government employees, including former officers in the KGB, that Russia was involved in an extensive UFO investigation program. These people admitted on international television news broadcasts, that indeed they were involved in such research. Video footage of several UFO objects from the extensive Russian military archives was shown on prime time news around the world.[6]

2) The Belgian government has publicly admitted it will undertake research into the ever more common UFO phenomenon.

3) The Japanese government has recently allocated 56 million US dollars for the construction of a UFO research centre.

4) There are hundreds and hundreds of organizations around the world that concern themselves with the investigation of UFO phenomena. Some of these have spent large amounts of money in their pursuit of truth concerning this topic, and many professional individuals belong to such organizations. Are all these people fooling themselves? *All* of them?

5) Since 1991, apparently after the solar eclipse, UFOs have been appearing in droves over Mexico City and have been filmed by hundreds of Mexican citizens since they first appeared.
Some of the people involved in reporting the UFOs are police personnel, government officials and members of the media, all of

---

[6] This author has an eight minute video recording of an SABC newscast that shows just such a report. Similar reports were shown in the USA, UK, Europe and the East.

whom have captured the UFOs on film.

The Mexican government has held press conferences in this regard to confirm that indeed the UFOs are actual objects but they pose no threat to the public.

The video footage includes filming in broad daylight when the UFOs can be clearly seen as manoeuvring objects, and these films have undergone scientific analysis in order to try and determine what the phenomena is. Various astronomers have been baffled by these objects and have publicly admitted that they do not know how to explain them.[7]

6) In his video, Hoagland's Mars Vol II, (extended version)[8] Richard Hoagland presents the Face on Mars in a somewhat interesting manner but one which, in my opinion, still leaves much to be desired. On the other hand, he makes up for this in the last part of his video, where with the help of qualified experts and a sound scientific approach, he analyses a very interesting piece of video footage which was filmed by the crew of the Discovery Space Shuttle while in orbit around the Earth on September 15, 1991. The interesting thing is that NASA vehemently denies that the footage shows anything other than ice-particles or some other form of debris. No UFOs whatsoever are present on the video clip as far as NASA is concerned.

Even just a casual observer, however, can tell that the footage shows some sort of flying object that moves in a very strange manner indeed. When scientifically analysed however, no doubt is left that what the NASA cameras captured is indeed a UFO. Perhaps even more interesting however, is the fact that after this incident, which was beamed live to the world from the Space Shuttle's camera, and captured by several persons and agencies,

---

[7] An excellent commercial video detailing some aspects of the Mexico City sightings is available from: Genesis III Publishing, PO Box 1362, Carbondale, IL 62903 USA, for about $ 35 plus shipping charges. Interested persons are advised to ensure the videos are in the proper format for viewing outside of the USA, which does not use PAL format videos.

[8] I only purchased and viewed this video some time after having written chapter six. It's available, along with other Hoagland videos, from: BC Mars Video, 19 Gregory Drive, South Burlington, VT 05403, USA Tel: (toll free) 800-424-0031. As for the video concerning the Mexico City sightings, ensure the correct video format for your country is specified when purchasing.

including one of the people involved with Hoagland, every video recording made by NASA Space Shuttles since this one, has been encrypted so that only NASA itself may view the video material. This video clip of what may be termed a NASA-filmed UFO, has also appeared on several news broadcasts around the world.

With respect to the last point, what may be even more disturbing is that according to Hoagland and his team's interpretation of the video, there is evidence that the US forces have in operation in Earth orbit, an electromagnetic gauss gun,[9] and that it was used to fire at the UFO object. Hoagland also appears to be of the opinion that the UFO object is not manned by extra-terrestrials, but in fact belongs to the US armed forces, and what was captured by the video clip was a weapons test against this newly devised technology.

Regardless of whether the USA is covertly undertaking weapons tests using advanced technology or just as covertly is in fact engaging extra-terrestrial craft,[10] what remains of paramount importance, is that the UFO object manoeuvred in a way that is completely impossible for any sort of conventional craft to do.

At first the object was moving along at a speed of some 15 miles per second (24 Km/s), but after a bright flash that produced at least two projectiles that seemed to be headed towards the UFO, it suddenly accelerated to a speed of over 55 miles per second (88Km/s) in a time

---

[9] A gauss gun (also known as a rail gun) is a weapon that accelerates a metal projectile along a 'barrel' composed of electromagnetic fields which by being switched rapidly on and off in sequence, accelerate the projectile to fantastic speeds (accelerations of 100,000 Gs with speeds of 24 miles per second were already being achieved by rail gun projectiles some ten years ago). Those who doubt that such weapons are even used in space, would do well to remember that project CHECMATE (Compact High Energy Capacitor Module Advanced Technology Experiment) which was well under way in the early 1980s (and which to my knowledge never came to an end yet) was undertaken to produce not only gauss guns but also to investigate the feasibility of using them in space as part of the Strategic Defence grid proposed by Reagan. For more details see *Space Warfare and Strategic Defense*, details of which are provided in the bibliography. Hoagland informs us that the project has more recently adopted the name of BRILLIANT PEBBLES.

[10] There have been several reports by former air force personnel of several countries including Russia, the USA and UK, that fighter pilots have been given the directive of crashing into UFO objects with their planes if they were unable to shoot them down with their onboard weapons, because the gains in military technology that would be made by recovery of the wreckage outweigh the importance of the pilots' lives.

period of about 2 seconds. This kind of acceleration would exert a force in the region of 14,000 G on the occupants of the UFO!

That is 14,000 times the force of gravity that Earth exerts on us. Apart from the fact that a person would be flattened into a jelly-like substance by such forces, sensitive electronic equipment would suffer a similar fate...*unless*...the craft had some way of countering such forces by controlling gravity in some way.

If an anti-gravity field could be created by the UFO, then the occupants need not even feel so much as a slight push, and could even be drinking coffee from a cup, in a normal 1 G environment inside the ship, without so much as spilling a drop.

The most important aspect of the NASA-shot footage presented in Hoagland's video, is not the UFO per se —although of course, the fact that it was NASA's own cameras that captured it adds weight to the whole issue— it was not even the idea that the USA may be experimenting with advanced weapons technology, and maybe, not even the fact that if this was not a test, then it may have been an attack on an extra-terrestrial form of life. These things of course are all extremely important in their own right, but for me, the most important aspect of this video, is that regardless of who (if anyone) was inside the UFO, and regardless of who owned it, *it showed undeniable evidence of a craft that manoeuvres by use of an anti-gravity engine.*

By the time I saw this video, of course I had already formulated my own ideas concerning alien means of propulsion, and indeed I had already concluded that an anti-gravity engine, whatever that is, and however it might be constructed, was the only way in which extra-terrestrials could conceivably travel between stars given their distances one from the other, but it was reassuring to see footage that vindicated my ideas to the $n^{th}$ degree.

We shall examine why anti-gravity is the key to interstellar travel in more detail in the next chapter, but for now, I would like to add only one more factual report concerning UFO objects.

Despite all the research, and associated problems which come with writing and compiling a book of this nature (and I assure you they are not few!) this next section is one which has perhaps caused me the most concern of all. I have literally stayed up nights worrying about whether I should or should not include it, how much I should say if I did decide to put it in, what kind of reaction I would receive from family and friends for it, what kind of reaction it would have on the viability of this book, and a thousand other questions that plagued my mind concerning this topic.

The dreaded thing you see, is that I have seen UFOs myself.

The simple fact that although I am now convinced that UFOs not only exist but that they are also controlled by intelligent beings,[11] and that despite this I have some concern about how my admission that I have seen and filmed them on my portable video camera will be received, is perhaps testimony to the efficiency with which the UFO phenomenon has been thoroughly discredited by the powers that be.

My concern is not ultimately tied so much to what people will think or say about me, for although it is true that no man is an island, I have always felt that I am at least a peninsula.

What strangers think of me is in the main irrelevant, as for my family and friends, by now they have all pretty much learned to take me or leave me as I am. My biggest concern has not been public opinion of me as such then, but rather that if I was thought of as a 'nut' by the reader, then the rest of the arguments that I have presented in this book may similarly be taken far less seriously than they deserve, despite the fact that I have strived to back my data up with rather extensive annotations.

On the other hand, the subject matter of this book, and particularly that which is found in the final chapters, is of such import to me, that I know I will remain involved in some aspect of it for the rest of my life, and if, as I hope, at least part of that life is spent talking, writing, and generally making people more aware of these things, then I cannot, in all honesty, be anything less than candid. If I were to keep quiet now about my experience for 'fear of ridicule', then it would ultimately mean that I would have to keep quiet about it in the future, otherwise the ugly question:

"Why are you mentioning this only now?", would be sure to raise its head, and in doing so place my integrity as perceived by anyone that has been listening to me up to that point, into some jeopardy.

---

[11] As to where these beings originate from, whether here on Earth in the back-rooms of the military and *intelligencia* of the world or on some distant planet in orbit around some unknown star, the question is ultimately not that important for my purposes, and I leave the argument to others, although for myself, I believe that both possibilities are likely.

After weighing all the pros and cons however, I have finally concluded that I do not want to harbour a 'UFO secret' for the rest of my life. It is not in my nature to keep quiet about my experiences in general, and the only real secrets I may have are those which other people have entrusted me with, but much more importantly, I believe in truth.

An ultimate, perfect, truth. And if I kept this to myself I would not feel completely comfortable with that part of myself that believes in this ideal.

Before I recount my own UFO experience, I want to make it very clear however, that despite the video evidence and several eye witnesses, this was a subjective experience, and I do *not* pretend that anyone believe me. I would limit myself to asking, that at least you believe that *I believe* what I say in this regard. So that, if I am guilty of any falsity with respect to this topic, it is due to my gullible stupidity, credulous imagination or some such aspect of my personality, and *not* because I have any intention to deceive in any way.

Lastly, I want to also point out, that if it was someone else that had seen and filmed what I saw, and gone through this experience, and I was one of the people he talked with about it, including the showing me of the video footage, as little as four years ago in my past, I would no doubt still have had serious reservations, and without the video evidence I would probably have thought of the person as a complete flake. Perhaps there is indeed such a thing as karma after all then, because while I have laughed at the credulous for most of my life, today I find myself forced to choose between being truthful and sounding like a credulous flake myself, or being secretive and possibly get away with the label 'eccentric'.

Whatever the consequences, I have opted for the first case, so without further ado, here is my story.

I suppose that in part, the fact that my sighting occurred in the early morning hours of April 1st, 1995 has been one of those things that helps to eat away at one's feelings concerning credibility,[12] but whatever the case, whether due to coincidence or some clever sense of

---

[12] As well as the fact that when I saw these objects, I was already in the final stages of completing this book. It all seemed just a bit too 'convenient'.

humour on the part of the UFO owners, April's Fool Day is indeed when I saw them for the first time.

On the night of Friday 31st March, I was sitting in the living room of my girlfriends' flat in Green Point, Cape Town, talking to her brother Rishi, and if I recall correctly, we were about to begin watching a video we had hired earlier. It was quite late, either close to or already a little past midnight, and Preshilla had decided she would go to sleep. Having excused herself, she'd wandered off in the direction of her bedroom.

Scant seconds later I heard her calling me while asking something along the lines of :

"What are these things?"

I remember being a little irritated while I stood up to go and see what she wanted, because I was talking to Rishi about something that held my interest at the time, although what it was I cannot say now, because the memory of it has been effectively wiped clean by the events which soon followed.

From her bedroom, Preshilla has access to her balcony via large, glass sliding doors, and these offer a perfect view of the outside world beyond her balcony, which faces a westerly direction towards Sea Point. Being situated on the third floor, one has quite a view in that direction, with Lion's Head on the left and the Green Point and Sea Point lights straight ahead, sunsets are particularly beautiful from her west-facing windows and balcony.

But it was not a sunset she pointed out to me now.

It was impossible to judge our distance to them, but quite low on the horizon[13] were several very bright lights, which outshone any of the visible stars as well as most of the street and building lights that were not in the immediate vicinity of our position.

The bright objects looked for all the world like very bright and flickering stars, except for a few 'details'.

First of all, their colour kept changing as they flickered, from red to blue, and at times it appeared that from being one solid disk of light each of them split into three tinier dots which floated a small distance apart before joining together again. Whether this was an actual effect

---

[13] One of the lights was below the top levels of the Ritz Hotel in Sea Point as it is seen from Preshilla's apartment.

of the objects or just a way in which our eyes interpreted what they saw, I cannot tell, but we all noticed it.

Most important of all though, the objects did not stay fixed in one place in the sky, but rather they seemed to hover about, making a small but quick up and down movement one second, and a side to side movement the next, followed too by circular motions. Along with these movements which were visible but not large, the objects also made larger movements by trailing across the sky for a while in one direction, before turning back in the other again. Although quite large, this movement was executed much slower than the tiny 'jumping' motions, taking several seconds to complete.

At first we were all just completely mesmerised by this incredible sight. The mind reeled as it tried to assimilate all the implications at once. I am not a qualified astronomer, nor have I ever taken so much as a single 'official' astronomy class, but the subject has interested me since childhood, and although I am not particularly expert at finding constellations or planets, I do have a basic understanding of celestial phenomena.

I have observed meteor showers in the Kalahari desert which is,   —I am told— possibly the best spot on Earth to view them from.

I have looked at the Space Shuttle flying overhead in the night sky and when Halley's comet came around I was one of the many thousands of people that were out there with a telescope in hand. I *do* know what an artificial satellite, high-flying plane, unusually bright star (as a result of peculiar atmospheric conditions) or meteoroid falling through the atmosphere looks like. And perhaps because of this, my brain may have been hit harder than the one of the other people that were now in the room, because it was instantly obvious, that no natural phenomena I could think of could in any way account for these impossible lights that I and three other people were now looking at.

I tried to force myself to think of something, anything, other than UFOs that could possibly explain these objects, including completely unlikely things like 'swamp gas' and 'ball-lightning', which are often quoted by sceptics when UFO enthusiast mention seeing strange lights.

Well, I have seen ball-lightning too from the window of an intercontinental aircraft flying over the African continent at night, and as I was 16 or so at the time, and unfamiliar with the phenomena, I was frightened out of my skin by them. For a few horrific seconds, I thought they were nuclear explosions going off on land far below me. It only took me a short while to understand what it was that I was actually seeing, and that it was unlikely that Congo or thereabouts would be the recipient of a full nuclear strike even in the event of all-out nuclear war, but I don't think I'll ever forget the feeling I had for those few seconds.

I read up on ball-lightning for quite a bit after that, and when I next saw them on a subsequent flight, I laughed at myself for having been so completely jolted by them that first time.

The lights I was staring at sometime on the threshold of April's Fool however, were not 'swamp-gas' or any kind of Earthly ball-lightning. Nor where they some advanced form of fireworks.

As I ran through all the possible explanations I could think of, I soon realised that to even suggest any of them would be ridiculous, because each of them was so far from being a possible alternative explanation, that it would have rocked my senses even further if they had been ball-lightning or fireworks than UFOs.[14]

It would be like trying to explain a runaway train that derails and cuts a half kilometre path through the surroundings, scattering carriages, goods and bodies all around, as a newspaper boy falling off his bicycle, or the result of a loose hubcap spinning off from a passing car.

As the realisation sank in, I found myself utterly speechless for a while.

Eventually, almost in unison, we all remembered my small telescope, and having dug it out and set it up, we took turns to observe the UFOs. There were four of these mysterious 'lights' that we could see from our position, but two of them seemed either closer, or at least more prominent in some way, being brighter and larger than the other two lights, so we focused on observing these two almost exclusively. Even through my modest telescope, we could see that the objects had a definite discernible disk. That is, they were not stars, who even at high magnifications continue to appear as mere points of light, the shape seemed to be very slightly ovoidal, and the 'jumping' movements were much clearer in the telescope.

The UFO being observed looked as though it was making some kind of ritual dance. The up-down, left-to-right and semi-circular

---

[14] Besides, I saw these objects for a period of about 4 hours. That would be the most impressive display of 'fireworks' or 'ball-lightning' that I ever heard of!

movements, executed in rapid succession even seemed to have some kind of rhythm to them, almost as if they were quick manoeuvres that had to be executed in a particular sequence.

We took some pictures with an instamatic camera to fix their positions, and then, as our excitement mounted and we forgot all about my video-camera that was packed away in its carrying case, Rishi and I decided to go out in my car to see if we could get closer to the objects.

As we got into my car, Rishi's job became that of keeping the UFOs in sight as I drove towards the direction which seemed to me most likely to get us closer. As I drove down High Level road towards Sea Point, the UFOs seemed to 'cross' the road in front and above us, flying out towards the ocean. As we reached the end of High Level road, I had to take a right turn to join the lower main road before continuing west.

I followed Rishi's directions and at the end of Sea Point got onto Kloof road, which brought me to higher ground again and less street lights, which made for better observation, although the UFOs were clearly visible even when we were in the brightly lit Sea Point area.

With the mountain on our left and the ocean below us and to the right, we eventually stopped somewhere in the beginnings of Clifton, near one of the less used sets of concrete steps that brings one to the homes of those that live on the steep incline which eventually reaches down to the beach and sea below.

At this point we were almost level with one of the UFO objects. Much in the same way as one can drive past a lamppost, stop the car before reaching it or pull up next to it, we had pulled up 'next' to the UFO, although of course it was still an undefined distance above us and a similarly undefined distance away from us and over the ocean.

Because there were no streetlights half-way down the concrete steps and the house lights below us were mostly off, we had a clear view of the night sky and of the UFOs, which now numbered three.

We had lost one somewhere along the way, or perhaps it was simply hidden by the mountain behind us. In any event, Rishi and I continued to observe the UFOs for about 3 hours. After which we returned home and discussed it with Preshilla and her sister and I also made an audio recording of our experience because I was very tired at this point and I wanted to record all the details I could while they were still fresh in my head, so that even if I later forgot some of them, the tape would be able to jolt my memory.

At about 5 am I headed home, and reaching there I saw two more of the UFO objects directly outside my window some undefined distance away. I went back to Preshilla's apartment to retrieve my video camera, which I had been using there to complete the transfer of some holiday footage onto VHS format, and returned to my own apartment where I proceeded to film the UFOs.

By this point I was very tired (even if still excited), and the video camera of course was not ready to shoot UFO footage.

I had to quickly sort out which tapes I could use to film them and which I was still busy transferring onto VHS format. In all the confusion, the finer points of cinematic photography escaped me, and generally speaking, I set the camera on a tripod, zoomed in at maximum resolution and kept it on one of the two objects for most of the time. My mistake was that I did not film the UFOs in relation to the surroundings very well at all. There are a few seconds of footage where I am zooming in or out that show that indeed the UFOs are in the sky and not just the product of some fancy light-bulb against a dark background, but definitely such shots are in the minority.

In any event, although the video footage does *not* do these objects justice by any standard, it nevertheless shows some interesting things. First of all, the colours which were so readily seen by the naked eye were not as prominent on the video, nor was the intense brightness of the objects. On the other hand, the video footage captured rather well, a phenomenon that although observable to some degree, was much harder to see with the naked eye.

The UFO seemed to radiate some kind of lines of force to three tinier dots of light, which were barely visible with the naked eye, but which appear quite clearly on the fully zoomed image I was filming.

I could see some of these in the viewfinder at the time of filming, so I think the fact that they were harder to see with the naked eye is probably a function of distance rather than of other optical considerations.

The three tinier dots, as well as the main UFO all moved about in a wave-like motion that was not completely regular or rhythmical, but which nevertheless 'flowed' compared to the movement of say a plane or helicopter for example, which could be said to 'cut', 'slice' or 'force' their way through the air. Also, no sounds were ever heard by any of the UFO objects throughout the entire time we saw them.

I had the distinct impression that the three tinier dots were small unmanned probes that were under the control of the larger craft, and I admit to thinking that it looked as if the UFOs were trying to 'clean' the atmosphere with such wavy motions and lines of force that ran between the main UFO and the tinier probes, but of course, I have no idea of what, if anything, these objects were in fact doing.

Whatever the case is, as daylight began to make it's entrance, I left the camera by itself on its tripod and phoned a friend to tell him of what was going on, and during this time, I was lucky, because in broad daylight, with the object still clearly visible in the camera, a couple of seagull flew in front of the UFO, proving that I was not just filming some light in the artificial sky of a studio room.[15]

It was only later, when I watched the complete video footage the next day, that I saw this had happened, because as I said, I was on the phone at the time, as can be heard from the audio component of the video.

I have produced two images of this event, by using a computer to capture the single frames from the full-motion video and the results of this are shown on plate 12.[16] Please keep in mind, that no computer alteration whatever has taken place, computers were merely used to capture the relevant frames from the full-length original video footage in digital format, and in fact, some clarity has been lost in the process. Also, this event occurred right at the end of the video, where the UFO object is considerably smaller/further away than at the beginning of the filming.

Of course, like most UFO photographs, an impartial observer probably can't help wondering if they really are what they pretend to be. Personally I have always found UFO photographs to be particularly unsatisfactory for some reason, and can't help but wonder if instead of a UFO, the thing in the picture is not something else, which may have a simple explanation, or on occasion, if it might not even be a hoax of some kind.

Of course, I can't really do much more than present my story with the accompanying still photograph taken from the video-footage, and let the reader decide for him or herself (and considering their quality, I

---

[15] This actually happened twice, the second time being a single bird.

[16] The computer equipment for the video-capturing was kindly supplied by PCE Electronics in Cape Town, with technical assistance being provided by PCE's Mark Nowitz.

would not be surprised if the tiny white dot in those pictures is taken to be dust on the lens or something equally mundane!) but it may interest you to know that if I am crazy, then at least I am not alone.

I later learned that I had been only one of many that had seen these same objects. They were reported on the radio on at least two occasions I know of and at least one newspaper printed an article on them (on the front page) on the 23rd of April, more than three weeks after my sighting.

Radio 5 FM though beat me to it (although I did not know it at the time and only found out on the 2nd of April) having made a report on the 31st of March that strange lights were moving along the N1 national highway in a southerly direction and had been spotted by several citizens. Among them several policemen, whom at first incredulous, had reluctantly followed two civilians to the highway where they said they could see strange lights in the sky. On arrival, the policemen saw the same thing as the two civilians, and they were shortly joined by their fellow officers in staring upwards into the night at brightly flickering objects that were flying along the main national road of the South African Republic.

The front page of the Cape Metro newspaper of 23 April 1995, had a column on it, by Yvette van Breda, that reported similar lights had been observed by residents of Stillbaai. The same flickering between blue and red was reported,[17] the same ability of splitting into more than one piece, as well as the fact that when accelerating or changing direction, the 'flaring' of light from it seemed to increase. It was correctly reported too, that the objects had first appeared more than two weeks earlier. A local policeman was also quoted as being one of those that had seen the strange lights.

The article also printed the dismissive ideas of one Dr. Dave Laney, an astronomer of the South African Astronomical Observatory.

Laney put the object down as Arcturus, even though he later says it could be Jupiter when asked about its oval shape, or the result of unsteady hands holding the binoculars. What was most obvious

---

[17] The article also mentions yellow, and my girlfriend had seen green too in the flickering, but I believe that what was actually happening was a flickering through the whole range of the spectrum very rapidly, and that occasionally, people with a particularly acute vision, or perhaps whose eyes worked a little differently from another's could see one of the other colours of the rainbow in the flicker.

about Dr. Laney from the article, was that he obviously had not seen the objects himself or taken the trouble to go and investigate them.

How could he do so, you may ask? Well, because like me, one of the citizens quoted in the article has been able to see these lights on a regular basis since they first arrived.

The interesting thing is that the lady quoted in the article makes an observation I have concluded myself too, and that is, that the UFOs have never been as large or prominent as the first time I saw them, nor do they move around as much. They limit themselves to slowly trailing across the sky. The movement is still noticeable but not at a glance, and takes several minutes of observation to notice, nevertheless it is still large enough that it cannot be confused with the natural apparent movement of the stars as the Earth rotates.

I have more than four hours of video tapes of these objects, but since now they tend to simply hover in the air and pretend to be stars, I've stopped filming them.

I originally approached the SABC with my video tape, but received such an unenthusiastic response, that I did not pursue it further.[18]

Maybe it's a reflection of South African society in general, at a time when in this country so many real challenges face the average citizen, UFOs may not be high on the list of priorities, and if the media and authorities are unwilling to take the necessary action to raise public awareness concerning these 'strange lights', why should Joe citizen take it upon himself to do so and risk ridicule and defamation for his trouble too?

But perhaps too, there is some element of this whole UFO phenomena which makes those of us who have experienced it, whether here in South Africa or anywhere else in the world for that matter, feel something along the lines of:

"Well, I don't care who believes it or not. I know what I saw."

And there *is* some kind of a peace in that. A sort of victorious resignation.

---

[18] I was to present them with a copy of the tape, at my own expense I may add, and was informed that they would have to transfer the data "five or six times" during which process information would be lost, and if the quality was not very good after that, they would not use the tape in any way. Considering that I used a normal 8mm video camera to film the UFOs, I do not see how they could ever use the pictures under those circumstances, even if they had been of little green men descending from a flying saucer in broad daylight on the steps of parliament.

On my most recent visit to the South African Astronomical Observatory, I made the acquaintance of Dr. Laney, and in the course of talking with him, I asked his opinion about what I had seen.

At first he postulated that my UFOs could have been helicopters equipped with strange lights. He added that if far enough away no noise would be heard, but in view of the parallax I observed when driving closer to the objects, I am sure that if they had been helicopters equipped with special lights, I would have heard something.

Weather balloons was another possibility quoted by Dr. Laney, although he conceded at least, that some form of lights would have to be added to them to make them look like more credible UFOs.

He admitted though that such material was easier to obtain in his native USA, where students had been known to construct such 'UFOs' for fun.

On my detailing a few more observations to him however, such as the time period involved, the motion of the objects and so on, Dr. Laney made some very useful observations. He pointed out that the 'splitting' of the object can be the result of an optical illusion that is prevalent under certain atmospheric conditions, as is the small 'jumping' motion. I readily agreed with him that this was a definite possibility, since I have seen a similar 'jumping' motion when observing distant survey beacons through a theodolite (particularly if on a hot day), although I must add that the small, quick movement of the UFOs did not appear to me to be of this kind, although the 'splitting' could well have been so.

We considered ball-lightning and a number of other phenomena, during which Dr. Laney said that judging by my description, the objects definitely sounded as though they were artificial in nature.[19]

In the end though, he explained that there are certainly some 'unexplained' phenomena in science, and that whenever observations cannot take place under a certain amount of regularity, science can have difficulty in dealing with such information. He quoted 'sprites' as an example. These are huge pink or orange 'glows' as he described

---

[19] It was interesting for me to note Dr. Laney's statement that one cannot say what or from where such lights may originate, without having a clearer indication of "what the military is doing". A point I agree strongly with him on, but I have to add, that I am absolutely positive that if those UFOs *were* of a military origin, it was the military of some country other than South Africa. I simply do not believe such technology exists here yet.

them to me, that appear occasionally over thunderstorm clouds, and which have only recently become a respectable topic of conversation, since, until a short while ago, they were only seen by the occasional "crazy pilot".

Ultimately, Dr. Laney's position is not dissimilar to my own, while I am convinced that UFOs exist and in my opinion they are artificial in nature and origin, I do not hold that this is the *only* possible explanation for them, just the most likely one.

Dr. Laney on the other hand, although he told me nothing would make him happier than to know UFOs were the result of anti-gravity vehicles, remains sceptical, although not unlikely to change his view if presented with new and solid scientific evidence.

It should also be made clear, that Dr. Laney has never seen anything that resembled what I described to him, and in view of this he was rather open-minded. In fact he gave me his phone number, and we agreed that if I should see these objects again, I was to give him a call. It is my fervent hope, to see these objects again at some point in the future, this time with Dr. Laney present, whom is eminently more qualified than I am to label them artificial vehicles or otherwise.

At any rate, from this point onwards I shall refrain from presenting any more information concerning UFOs and I will simply treat them as factual in nature. Also, although I do not presume to know who controls these objects, in view of the evidence of extra-terrestrial visitations in our past, the possibility that the UFOs are extra-terrestrial in nature, must be at least seriously considered whenever we have them in mind.

Once again, I urge the sceptical reader to do his own research concerning the UFO phenomena and provide a few starting points for interested persons at the end of this chapter.

# UFOs as Extra-terrestrial Crafts

If we assume that at least some of the thousands of UFO phenomena which fall under the category of 'unexplainable' are the result of extra-terrestrial involvement, in fact, even if only *one* of them is, we then have to immediately consider a number of important points related to this idea.

Quite apart from questions such as who are these aliens, what do they look like, what are their intentions and where do they come from, The most important consideration we should concern ourselves with first, is:

What are they using to get here?

And secondly: Is it possible?

Lastly, what are the chances, even if we give them the technology, that they would end up *here*, given the enormity of space.


Back in chapter one I presented a few of the more orthodox ideas concerning interstellar travel, and pretty much concluded along the lines that unless we allow for some science-fiction type of hyperspace capability of an advanced technology, then the only seemingly viable means of travel between the stars would be through the use of ever more specialised nuclear engines, and probably of the fusion rather than fission variety. And controlled fusion reactions of course, are something we are still struggling to learn about, never mind master.

But even if we assume that an extra-terrestrial race has mastered nuclear fusion to the extent that anti-matter ships are as common as cars for us, would this be able to explain the UFO phenomena?

The answer, to this question, it would seem, is a resounding NO.

Because a space-ship that is powered by nuclear fusion, would not behave as the UFOs are reported to.

UFOs seem to have the ability to appear and disappear almost instantaneously, to be almost always associated with luminosity, appear to move without any discernible noise, and most of all their ability to accelerate to fantastic speeds in seconds, are all indicative that what we are dealing with here —if it is an alien craft— is one that is far more advanced than any fusion-thrusting ship we could possibly envision, never mind design.

So where does this leave us? Well, it depends on your point of view. If you are one of those that believes it is written in stone that nothing can ever possibly travel from point A to point B faster than the speed of light, then you would be most likely to conclude that the UFOs must be something *other* than alien spacecraft.

If you believe that a fusion-thrusting engine is the only way anyone or anything can ever travel between the stars, and that it has always been this way and will continue to be forever more, then UFOs cannot possibly be alien craft in your mind.

Of course it still leaves you with the tricky problem of explaining just what UFOs are (which is possibly why so many orthodox scientists can't yet), but at least, you can live in a safe and comfortable universe which follows the generally accepted model of Einstein's relativity theory.

*If* however, you open your mind to the possibility that Einstein's theory is only a part of the puzzle rather than the be all and end all of physics, and if you further note that every discovery that has ever been made about any aspect of science has *always and only* been a better approximation of reality, rather than a perfect definition of that reality, then, you would be allowing for the possibility of a technology that has been developed along different lines from our own while perhaps not being necessarily beyond our reach in the very near future.

I would like to make two observations with regard to this topic. One was made by Arthur C. Clarke, and I must paraphrase, for I do not recall the exact wording:

> When an old, respected and eminent scientists says that something is probably possible, he is usually right.
> When the same old, respected and eminent scientist says that something will never be possible however, he is most probably wrong.

The other is from an unremembered source, but perhaps even more relevant:

> If you'd told Christopher Columbus that one day people would be able to make the journey across the Atlantic in a matter of hours, he would have said:
> "You're crazy. The wind never blows that fast."

We shall examine this idea of machines that can travel between points in space-time faster than light in more detail in chapter 10, but for now, let us, for the sake of argument at least, accept that maybe an alien intelligence older than our own, may indeed have invented such technology a while back already as time goes for the human race on Earth. Let us also, place such a race of beings somewhere in our own Galaxy.

At this point then, we have tentatively answered only one of the three question, and that only by stating that whatever *it is* that the UFOs are using, it's not the best nuclear fusion we can imagine, and so would have to be something else.

For now, we might limit ourselves to suggest "some kind of anti-gravity" but not too loudly, because of course, this goes against the commonly accepted concepts of physics. As to whether such technology is even possible, we shall devote an entire chapter later, so we are left only with: "Why *here*?"

While we should always keep in mind that it may well be that the aliens in question are our stellar neighbours and hence their choice of visiting this particular system is only logical, given the vastness of space this would still be a somewhat unsatisfactory, (although valid) reply.

In fact, considering the evidence of earlier contact by such beings, which we have variously worshipped as 'Gods' in the past, it would seem that if such creatures exist, they have known of our location in space for a very long period of time.

Nevertheless, allow me to introduce a theory which may answer the question "Why here?" more satisfactorily.


Let's start with an alien race, possessing Faster Than Light (FTL) technology which dedicates itself to the exploring of the Galaxy.

Let's assume too, that such a race, having completely colonised their solar system and thus populating all available planets in it, is composed of 10 billion individuals. This is not an unreasonable figure at all, considering that we are rapidly approaching it here on Earth, where we currently have only one planet at our disposal.

Let's say that 0.005% of this alien population is actively involved in the exploration of space, which amounts to some 500,000 people. These half-million beings are divided into groups of ten, and each group is provided with a FTL-capable space-ship. This would amount then, to 50,000 space-exploring ships capable of Faster Than Light travel. Considering a population of 10 billion, 50,000 space ships could be a conservatively low number, especially if we assume that

FTL technology could be as common place for this race, as air-travel is for us.

Each of these ships explores (in a brief manner) two solar systems per year. Since the ships are capable of some sort of 'hyperspace' it is irrelevant (or at least mainly irrelevant) how distant a star is for the purposes of how much time it takes to get there, so let's assume that each ship can explore two new systems per year.

This amounts to some 100,000 systems being explored in the first year, but if we further assume that as the population grows, new planets are settled, and so on, the efforts of space-exploration similarly increases, then this number will become greater with each passing year.

Let us put the increase in the number of exploration ships at a mere 1.56% per year, so that in the second year 50,780 ships are involved in space exploration, in the third year 51,572.168 ships, (the fraction can represent ships not yet complete but under construction as averaged out for the whole period) and so on.

Currently, our population more than triples every 80 years or so, which amounts to an increase of about 1.4% per year, but whereas we cannot continue such expansion indefinitely, a star-spanning race might. Even so, the increase of 1.56% is admittedly a generous one, but let us proceed anyway.

As crew-members of old ships die, they are replaced by new and younger members of the race as the ships periodically re-equip themselves, in effect this also means that a higher percentage of the population eventually becomes involved in space-exploration as time passes, but this too is a real trend.[20]

Now, if this pattern continues for 1000 years, how many star-systems do you think such a race of beings will have explored at least in a basic manner?

The answer is 528,070,858,502. Which considering that our Galaxy is approximately composed (at the highest estimates) of some 500 billion stars, would account for each star in it. Of course, the figures used are generous in a number of ways. No allowance for travel-time wastage is taken into account, nor for the life expectancy of each ship, and the increase of 1.56% is admittedly a high one, nevertheless, the figures illustrate the very useful point, that given such 'ideal'

---

[20] For example, on our planet, not only are new cars being produced every year, but the older ones are passed on to continue useful and active duty for some considerable time still, effectively increasing the number of people actively involved in using cars.

conditions, it would not be impossible for an alien race possessing the required technology to map even the entire Galaxy in as little as 1000 years, which in terms of Galactic time of course is a million times less than the blink of an eye. And in any event, the fact that a much longer time period could have been taken, such as 10,000 years, 100,000 years or even longer, has not been considered. Nor has the possibility that there may be several hundred different alien races, all of which, having achieved such technology, may well be involved in similar exploration programmes.

Given the 500 billion stars or so present in this Galaxy, even if intelligent life has developed only around 1 in every billion stars,[21] that would still amount to some 500 intelligent alien races, most of which may be considerably older and more advance than ours.

But in any event, the aliens possessing FTL travel, need not be on the other side of our Galaxy. All that the above mathematical exercise wants to show, is that if there is such an alien race within 10,000 light years of Earth, which encompasses a circular sector of space 20,000 light years across representing only 4% of the stars in our Galaxy,[22] then given enough time (which by the way need not amount to more than a few thousand years) such beings are almost certain to come across us!

On that last question then, we can comfortably say at least, that given such extraordinary technology as Faster Than Light spaceships, an extra-terrestrial race of beings that exists within 10,000 light years of Earth and is involved in space exploration, is almost certain to have found us if we assume that they have had such technology for at least a few thousand years. But of course, depending on just how near or far to us they are and also on what kind of criteria they have for visiting a particular system, we may be discovered by them as soon as a mere century or two after the inception of FTL technology on their home planet.

---

[21] It is difficult to appreciate such enormous figures, but if we think of a unit of time as small as a second as a base unit, we may begin to get a better understanding of the magnitude of numbers as big as 1 billion. One million seconds is equivalent to a time period of eleven and a half days; one *billion* seconds however, is equivalent to 31.7 *years*.

[22] Actually this figure is based on an equal distribution of stars across the Galaxy, which is untrue, since far more stars are present at the Galaxy's centre, so the figure of 4% is very generous, and perhaps a truer one would be even less than 2%.

Considering that we have no idea of how common or rare intelligent life is in our Galaxy, it would seem wise then, to postulate that if FTL technology can be shown to be a definite possibility, it may be very likely indeed that UFOs have more of an extra-terrestrial connection than we may have previously suspected.

Our problem of course would seem to be that most eminent scientists tell us that FTL travel is an impossibility. But scientists have been consistently wrong when stating that something is 'utterly impossible' as shown by the academics against the Wright brothers, and countless other similar 'impossibilities' which persist even today. Perhaps the most recent 'impossibility' that in raising its head has embarrassed nuclear physicists and their exorbitant budgets the world over, is the 'monster' of Cold Fusion.

An impossibility if ever there was one, according to all the 'scientific' evidence, was the idea of nuclear fusion reactions occurring in a controlled and controllable fashion at room temperature!

And yet through the use of equipment amounting to no more than £90 that is exactly what Professor Martin Fleischmann and Professor Stanley Pons announced, in March of 1989.

They were subsequently vilified by scientists —who said they could not repeat the experiment these two professors said they had achieved with such ease— and the media alike, who took to calling them frauds on the strength of such accusations.

As it turns out, the data collection that seems to be unable to stand up to scrutiny is not that of Fleischmann and Pons, but rather that of those eminent scientists and their institutions that were apparently 'unable' to reproduce it.

Today, the Fleischmann-Pons cold fusion cells have been reproduced with effects as described by these two professors in 92 research organisations in ten countries, yet research funds for cold fusion are only now beginning to be allocated (and in miserly small amounts in view of the beneficial potentials of this new technology) while billions of dollars continue being poured into what may be termed hot fusion, with no comparable gain in terms of technological advancement for the purposes of providing cheap, safe energy.[23]

---

[23] A more complete account of the cold fusion saga is to be found in Richard Milton's *Forbidden Science, Exposing the Secrets of Suppressed Research*. See bibliography for details.

In conclusion then, before stating categorically that Faster Than Light travel is impossible, let's at least try to research the subject a little, and perhaps, in view of what such technology may mean for the human race, let's not stop at 'a little', because as we should by now be learning, the Universe is far from a boring place.

Indeed *very* far from it! Think of it....look at the Sun.

Is that boring? A huge, giant ball of fire floating in space. Who could imagine such a thing would be possible?

If you lived in a cave all your life, do you think you could ever imagine such a thing as a Sun? Even if you lived to be a thousand years old? ***Ever?***

What about a supernova? Or asteroids? Or for that matter, Earth itself along with everything on it! What about life? Life itself.

We live in a Universe so strange and wonderful, that as Shakespeare wrote, it is not so weird that we cannot imagine it, but it is weirder than we *can* imagine!

## Where to find more Information on UFO Phenomena

As usual, the local library should be the first stop whenever researching a subject, as well of course, as the libraries of universities, which are perhaps better stocked in some instances.

Another extremely useful research tool is having Internet access.

Specifically, ongoing discussions on various newsgroups such as alt.alien.visitors, alt.alien.research, and alt.paranet.ufo, are useful places in which to post a message when searching for a resource, be it a book, declassified document, or name of a person or organisation involved in UFO research.

Here too, one can meet and later correspond by e-mail with persons interested in UFO phenomena from around the world, and in doing so, it is only natural that one will come across useful information, although, as always, any information received from the Internet requires that it be double-checked by the honest investigator, since there are a number of people on-line that will not hesitate to type out the most ridiculous nonsense, which by the way can sometimes be funny enough to be a welcome break from the serious research.

On the other hand, more than a few individuals on the Internet are honest and kind enough to give even the most misanthropic among us a fresh and positive outlook on humanity. I have had people mail me documents at their own expense and refuse any kind of compensation for their trouble, simply because they wanted to be helpful, and many others that went to the trouble of finding out a name, title of a book, address of an organisation and so on.

There are several lists of various UFO organisations floating around on the Internet, and some of these are quite extensive, incorporating over a hundred addresses from all over the world.

I include a few of the more well known organisations below:

Citizens Against UFO Secrecy (CAUS)
PO Box 218
Coventry, Connecticut 06238
USA
Publication issued: Just Cause
Founded: 1978

One of the most well known and influential UFO groups. They have had not a little success in wresting declassified (but still sensitive) documents form the US government, sometimes through the use of legal action.

♦ ♦ ♦ ♦ ♦

Fund For UFO Research (FUFOR)
PO Box 277
Mt. Rainier, Maryland 20712
USA
Publication: Quarterly Report
Founded: 1979

FUFOR has raised more than $ 150,000.00 for the scientific investigation of UFO phenomena through contributions from over 2000 members.

♦ ♦ ♦ ♦ ♦

Document Research Services
PO Box 10011
Berkeley, California 94709-5011
USA
They provide declassified Government documents concerning UFOs. Send SASE for list of documents.

♦ ♦ ♦ ♦ ♦

Intercontinental UFO Galactic Spacecraft Research and Analytic Network (ICUFON)
35-40 75th street Suite 4G
Jackson Heights, New York 11372
USA
A non profit organisation trying to convince the UN and the world's governments to establish an official World Authority for UFO Affairs.

♦ ♦ ♦ ♦ ♦

J. Allen Hynek Center for UFO Studies (CUFOS)

2457 West Peterson Avenue
Chicago, Illinois 60659
Publication: The journal of UFO Studies
                International UFO Reporter
Founded: 1973 by Dr. J. Allen Hynek.

Involved in serious research by scientists, academics and volunteers, it has a world-wide network of investigators.
CUFOS retains the world's largest repository of UFO data (apart possibly from the US government) and Dr. Hynek was involved in the UFO phenomena from 1948, when he became Scientific Consultant to the US Air Force. He coined the phrase "close encounters of the third kind" and was technical advisor to Steven Spielberg for the movie of the same name.

♦♦♦♦♦

# 10      Alien Propulsion

Examining those UFO reports that seem most indicative of these objects being extra-terrestrial craft, soon paints a pattern of recurrence. In almost every case, the best of the 'unexplained' UFO reports exhibit certain similarities.

The UFOs in question are silent, they seem capable of extremely rapid acceleration, they are usually seen as bright oval-shaped objects, often giving off brilliance to such extent as to be difficult to observe without some kind of 'haziness' surrounding them.

They also seem to glow brighter when they undergo rapid acceleration, and are occasionally able to appear and disappear from view almost instantaneously.

They are apparently able to evade radar tracking at will, as several reports from air force pilots indicate that they 'lost' the objects on radar (at the same time as did the ground-based installations) but were still able to see them from the cockpit of their jets.

UFOs also seem able to adversely affect any electronic and electromagnetic equipment, and although some of the effects may be automatically induced as a result of the UFO's presence, the degree of the effects do seem to be at least in part under the control of the UFOs. In addition to flight at varying speeds, the UFOs also have the ability to hover motionlessly on the spot, and some capacity to irradiate their immediate surrounding seems to be another of their traits.

All of these abilities tend to suggest that in addition to the almost taboo subject of anti-gravity, one may also do well to investigate electromagnetic phenomena in order to try and gain some clue as to the workings of these objects.

Gravity and electromagnetism though, happen to be two areas in which science is still rather ignorant. What a coincidence then that these two mysterious subjects should both be involved in the UFO phenomena. And considering just how little we really know about these forces, why are scientists so willing to discredit the possibility of anti-gravity machines?

What we know of gravity is very little indeed. Modern science in fact, is limited to describing the effects which gravity has on our external reality, but if the truth be told, just *why* gravity works remains a complete mystery. As for the *how* some people would like us to think we know how it works, but is this really true?

We know certain effects which gravity has on the reality we are able to perceive, but what of those fields of reality we know little or nothing about?

Einstein grasped the way in which gravity works better than any scientist before him, which is why his theory of relativity is still extremely useful. At its most basic level, it was his concept that gravity warped space that earned him his place in history.

Considering that light has no mass, and yet is deflected by the gravity of a massive object, it would seem that this view of Einstein's was indeed a better approximation of reality; certainly so at least, when compared to the earlier Newtonian view of gravity.

One of the best ways to grasp how gravity works, is to envision space as a vast, flat sheet of a very thin and rubbery material which is also very resistant to tearing.

If we drop an object on this sheet of material, it will sink into it, the weight of it stretching the material into a deep 'well'. Depending on the mass of the object, the well will be more or less deep. So for example, while a Moon-sized sphere would create a relatively small 'dimple' on this surface, a Sun-sized sphere would create quite a big and much deeper one.

Such 'dimples' are commonly referred to as *gravity-wells*, and when science-fiction writers —or astronautical engineers for that matter— talk about "climbing out of the Earth's gravity well" what they mean, is that in order to get to the 'flat' surface, a rocket from Earth has to expend enough energy to make it climb completely out of the gravity well created by our planet.

The interesting thing, is that any point on the 'surface' of space which is at the same 'depth' can be reached practically for free once the initial push has been delivered.

So for example, if a rocket does climb out of the gravity well of its planet of origin and reaches the 'flat' surface of space,[1] it can then

---

[1] We could think of 'flat' space as being the datum from which measuring begins, a sort of 'sea-level' for space. Any dimple would result in a 'lowering' of this level, effectively meaning that going into it can be done for free, but climbing out of it requires an expenditure of energy. As far as we know there are no 'mountains' in space, only 'wells'. A 'mountain' in space (if they existed) of course would require an expenditure of energy in order to be 'climbed'.

access any other point in space that is at the same 'flat' level once it receives just one tiny push in the right direction.

If you wanted to go to Alpha Centauri for example, all you need to do is climb out of the Earth's gravity well, then see to it that you climb out of the solar system's well, which can be done almost for free with some clever manoeuvring,[2] and after that, if its headed in the right direction, your probe will continue towards Alpha Centauri until it gets there or slams into something.

Of course it might take a couple of centuries to get there, which could be troublesome if you're planning to make the trip yourself, but, such is life; or at least, so our physicists tell us.

On the other hand, if you don't make it out of the gravity well, then you can stay at the same potential. Since you're half-way up a 'well' or 'tube' the only other points in space at the same potential are defined by a circle, and it is this circle (more or less) that orbiting satellites and spacecraft define. Of course, an orbit can be (and usually is) elliptical in nature, meaning it's oval shaped rather than circular, but keep in mind that this view of space as being 'dimpled' by objects is just an approximation of reality, and not an absolutely accurate depiction of it.

In effect, it is quite likely, that so called 'flat' space, may not even exist at all, because what happens is that we exist in an overlapping pattern of wells.

The Earth for example, is trapped by the Sun's well, which is why it orbits it, but of course, the Earth in turn forms its own well, which traps the Moon, and that in turn forms a well......and so on.

It's fairly safe to assume too, that Galaxies —being concentrations of material— form huge, gently-sloping, troughs in space, and clusters of Galaxies (which do occur and are a known feature of the Universe) form even vaster and gentler-sloping troughs, and so on *ad infinitum*.

So you see, the picture is not quite as clear as a flat sheet of rubbery material, because in truth, this rubbery material is so sensitive, and has so many billions upon billions of 'pebbles' scattered upon it, that

---

[2] By passing the probe close to other planets' gravity wells, it can actually gain momentum as it zips past them and is momentarily grasped by their gravity wells. This will of course also result in a deviation of trajectory, but if all the calculations have been done beforehand, this can work to the advantage of the probe. This type of manoeuvring is known as *gravity assist*. Voyager I & II used this method in order to travel to the outer planets and ultimately out of the Solar System altogether.

it is a veritable rippling of concentric dimples upon dimples upon dimples.

Considering too that each dimple slightly deflects light, one can see why astronomers have some difficulty in accurately measuring the distance to even our closest stellar neighbours.

The idea of two parallel beams of light never crossing is a fallacy, because somewhere along their way, they will pass into a 'dimple' or gravity well that will distort them, so that they will diverge or cross.

Cosmologists ponder on the shape of the Universe, and many postulate that space eventually curves upon itself, and that if we had a powerful enough telescope, what we would see in it is an image of ourselves untold billions of years earlier. As light travels away from us it becomes gently warped, and if the Universe is spherical in shape, eventually, that same ray of light would find its way back to its origin.[3] The process of course would take an incredibly high number of billions upon billions of years, but *if* the Universe is of a 'closed' nature, and *if* it continued to exist for such a long period of time, eventually it would happen; hence the view of our most ancient selves in the theoretical telescope.

The process could be described as being similar to drawing a continuous line with a pencil on the inside of a football. Eventually the pencil line will cross the precise spot it started from, but of course, if this was done with the eyes closed, it would take a long time.

         To date, this is pretty much all that we know about gravity. We can see its effects on the natural world and measure them quite accurately, but that's about it. Or at least, this is the knowledge afforded to the common man, and in the vast majority of cases, to the learned scientist as well. But perhaps, as we shall see a little later, not to *all* the scientists; for there may be a few, that know quite a lot more about how gravity works.

For the moment however, let us shift our attention on electromagnetism, and see if we fare any better there.

As it turns out, the situation with electromagnetism is both better *and* worse than with gravity. For a start, we are able to produce a lot more effects with electricity and magnetism than we can with gravity, but

---

[3] Not quite on the first pass, because all the tiny dimples along the way are sure to have changed its path enough that it would miss, but *eventually*, it might be assumed it would pass through its origin.

on the other hand, despite all its usefulness, we know as little about the origin, the *why* of electromagnetism, as we do that of gravity.

Although more immediately able to produce effects for ourselves in the realm of electromagnetism, the subject remains an incredibly complex and little understood one. No one really knows exactly just what electricity really is, or why it behaves as it does. We talk of positive and negative charges, the flowing of electrons, the build-up of static charge and so on, but these are just words and phrases we have invented to describe an unseen process.

And electricity remains inextricably linked to magnetism, which is perhaps the least understood force of all.

Gravity in comparison appears easy. It's just *there*, and there's not much to discuss about (although of course this may be due to our poor understanding of it).

Magnetic phenomena on the other hand fluctuates. The magnetic Poles of Earth slowly migrate. The magnetic storms caused by solar flares result in adverse weather on Earth. Sunspot activity, still little understood, is the result of some magnetic disturbance.

Compasses have been known to go wild over certain spots of the Earth such as the Bermuda triangle while working perfectly at other times when going over the same route.

Magnetism in short, is the 'magic' of the twentieth century.

Which should come as no surprise when you consider that electricity along with its companion, electromagnetism, are very recent additions to our ever growing scientific database.

But how do UFOs, gravity and electromagnetism all tie together to explain how we may be receiving visits from extra-terrestrial beings? Well, the pattern is a vast one, and its edges are somewhat blurred, but nevertheless, a pattern exists. Let's see if we can't at least define it a little.

# Nikola Tesla : Master of Resonance

No doubt, history has wronged many persons; either by belittling their feats, or exaggerating them, or in the worst cases by ignoring them altogether, but the one man which has been wronged most of all, almost without a doubt, is Nikola Tesla.

Tesla was born in Smiljan, Yugoslavia, but after completing studies as an electrical engineer, he left for the United States, where he would eventually receive citizenship status.

It is not possible, even in a full length book, to do justice to this man and much less so in a few pages. Thus I shall limit myself to simply describing what in my opinion, was the key element of his complex and awesome make-up, which permitted him to be the one person in history whom most truly deserves in its most heroic, poetic and even truly magical sense, the title: *Genius*.

When in his second year of study in 1876 at Gantz in Austria, Tesla made his first encounter with a Gramme machine, he immediately proposed that it could be vastly improved if one could apply alternating current to it rather than direct current.[4]

His lecturer, a Professor Poeschl, was so taken aback by the suggestion that he devoted an entire lecture to tearing apart Tesla's suggestion piece by piece, and finished with the words:

"Mr. Tesla will accomplish great things, but he certainly never will do this. It would be equivalent to converting a steady pulling force like gravity into rotary effort. It is a perpetual motion scheme, an impossible idea."

Prophetic words indeed, not for their correctness, but for their distance from it.

It was in February of 1882 that Tesla finally resolved the problem, and indeed successfully invented the alternating current motor; a problem which had worried him almost constantly for the last six years.

---

[4] The differences between alternating current motors and direct current motors are several, but to put it simply, great energy is lost in a Gramme machine, through sparking at the commutator. More significantly, alternating current can be transmitted along cables of almost any length with only minimal losses, but direct current losses are much greater, making it impracticable to supply electricity to any dwelling beyond a mile or so of the power plant.

Yet, the working model was built only five years later, once Tesla had gained the confidence of J. P. Morgan, the well known millionaire and financier. And here lies the crux of the matter.

Throughout that time, Nikola Tesla kept the design for the alternating current motor in his head.

He did not have so much as a single sketch of it, nor any notes regarding its design written down anywhere. This is further proved by the fact that on his way to America, Tesla lost his luggage, and arrived with literally only the clothes on his back.

More remarkably still, he did not write down any measurements or sketches prior to constructing it for the first time.

He simply required that the relevant materials were cut to the correct size and with the correct shape, giving all instructions verbally only, and he then pieced the parts together.

If this was an isolated case however, it may have gone unnoticed, like so much of Tesla's life, *but this was in fact standard practice for him!*

To say that Tesla was eccentric would probably be an understatement, for he certainly was a complex and in some ways almost fastidious man, nevertheless, from all accounts, he was honest beyond any kind of reproach one might conceive and worked at a pace that none around him could keep up.[5]

One of his peculiarities was that he had a perfect photographic memory and he further believed that this gift was attainable by anyone who tried sufficiently hard, so he expressly forbid any of the people who worked for him to write any measurements down.

Whenever he required someone to construct some piece of machinery, Tesla would sketch it on a piece of paper,[6] and indicate the dimensions of its component parts by word of mouth only. The sketch was then destroyed before he sent the assistant on his way.

---

[5] Incredible as it sounds, it seems that Tesla only slept about 2 hours per night, although in Tesla's biography by John J. O'Neill, it is mentioned that Tesla occasionally stood in his hotel room in a reverie. Attendants could go about their work and Tesla seemed not even to notice their existence, so perhaps although Tesla only slept for 2 hours a night, he took some kind of reprieve during these moments.

[6] Another peculiarity is that no matter how complex the equipment, Tesla's diagrams were never more than an inch across! If he made a small mistake in the sketch, such as a line drawn wrong, instead of making corrections he would start again on a fresh sheet.

This name, Tesla, which should be on a par —if not higher— with that of Albert Einstein, remains unknown to most, despite the fact, that had Tesla not lived, our lives would not only be very different, but also much the poorer for it.

Had Einstein not lived, today we may not be living much worse off than we are, admittedly it would be a curious mixture of 19th Century attitudes with 20th Century technology, but apart from some theoretical physicists, most of us would manage quite well.

If the omission from this planet however had been a man called Nikola Tesla, in all probability, we would only today be extending electrical facilities throughout the Western, or First world countries. Personal computers would almost certainly not exist yet, and in a worst case scenario, the world would still lie in darkness; only the occasional spark from a direct current power plant —which could at best light up one city block at a time— would interrupt the blackness of night on our globe; and that direct current plant, would (in all probability) bear the name of one Thomas Edison.

Edison, the inventor of the light-bulb, is better remembered than Tesla, yet he was an exponent (and fiercely so) of direct current, and when Tesla mentioned to him his ideas concerning alternating current Edison brushed them aside without a second thought.

In fact, even after Tesla constructed and put into operation the first alternating current motors, Edison continued to try and boycott the ideas, with statements to the effect that alternating current was not safe. These two men had an effective feud going between them that lasted most of their adult lives. It's interesting to note, that this begun because of dishonesty on Edison's part —whom is generally thought of as a heroic inventor of the late 19th and early 20th Century— toward the considerably more honourable Tesla, whom is only dimly remembered if at all.

The incident stems from 1885, when Tesla, under the employ of Edison as a junior engineer, suggested some improvements that could be made to the dynamos Edison was using in order to save him a great deal of money and improve the efficiency of the machines.

Edison told Tesla that there was $50,000 in it for him if he could make good on his claims, which Tesla promptly did.

When the time came to pay up however, Edison replied that Tesla did not understand American humour. Tesla resigned immediately.

It was a pattern this man would follow several times in his life. Whenever he was lied to concerning remuneration he simply detached himself completely from that person even if it meant leaving a prominent position for a much worse set of circumstances.

It was this trait of his personality, that ultimately saw Tesla, at the age of 86, die in poverty in the bed of a dingy hotel room. Alone.[7]
He could not tolerate 'small matters' such as finances. Had he bothered to collect, or even allow others to collect for him, but one hundredth of a percent of the royalties which in fairness were owed to him, he would have been an inordinately rich man.
Even despite his complete lack of financial or business sense however, at one point Tesla was relatively rich after he harnessed the power of the Niagara Falls,[8] but due mainly to his peculiar character, he nevertheless died in poverty.

In any event, as I have already stated, I cannot do Tesla any kind of justice here, his life was too complex, his mind even more so, and his achievements, which he attained for the benefit of humanity as a whole, are too vast to even begin to sketch.
Suffice it to say, that regardless of tastes, I believe almost anyone at all, would be completely fascinated by the life of this amazing person, and if you only read ten books in your life, you could do worse than picking one that deals with Tesla's life as one of them.[9]

But what does Tesla have to do with the propulsion of UFOs?
Well, one of the important things to remember about Tesla, is that almost all of his inventions acted on some principle of resonance.
He seemed to be in tune with how any process could be made to function more efficiently, and with greater results, through the principle of resonance.

The Tesla turbine for example, which is more efficient than the conventional turbines we use today, worked on a resonating principle. Why are we not using Tesla turbines today then?

---

[7] Nikola Tesla died on January 7, 1943.

[8] Something which as a child he'd told his father he would one day do.

[9] See bibliography for more details on books about Tesla or dealing with some of his concepts.

The answer would seem to be that it's because when he invented this type of turbine, the other kinds which had been invented earlier were already well entrenched in the world market.

But Tesla had other inventions. Much better ones than the Tesla turbine, and they too are not in use today.

Let's take a look for example, at Tesla's wireless transmission system not only of information, which we might say has been successfully achieved by radio anyway, but of power.

Yes, that is correct. In the last few years before the beginning of the 20th Century, *Nikola Tesla had devised a system for supplying electrical power without wires.*

Surely this can't be right? There must be a catch no?

Maybe it doesn't really work. It's probably too dangerous or something right? I mean, if this wireless system is so great why are we not using it today?

Actually, yes it is right. No there's no catch. No, as far as anyone knows it is not dangerous, and yes it really would be a great system.

As to why it's not in use today...well, we'll get to that.

Let's first have a look at this wireless system though.

In 1897, the first tests of Tesla's wireless system were performed and these were successful, as was reported very modestly in the July 9, 1897 issue of *Electrical Review*.

It is important to note, that Tesla was of the opinion, that any disturbance in what he called the 'electrostatic equilibrium' of the Earth, was capable of being picked up by the correct equipment, and it is with these principles in mind that he built and tested his wireless system transmitters and receivers. In other words, Tesla's means of wireless communication once again used a principle based in resonance and hence is very different from the currently used means of wireless communication.

This first test was between Tesla's laboratory in Houston Street, and a boat 25 miles (40 Km) away, travelling on the Hudson River. The main patents on this system were issued in autumn 1897,[10] shortly after his announcement in the *Electrical Review*.

---

[10] Patent numbers 645,576 and 649,621.

It was in fact, the birth of modern radio, but as John J. O'Neill describes in *Prodigal Genius, The Life of Nikola Tesla*,[11] who has ever heard of Nikola Tesla as being the inventor of the radio?

And yet Tesla published his understanding of the phenomenon we know as radio today, before those whom are normally associated with its discovery did. Also, it is clear from the fact that eventually longer wavelengths were used for radio broadcasting, that his grasp of how radio-waves propagated was infinitely more developed than that of Marconi and Lodge.[12]

Despite this official date of 1897, however, Tesla had already developed the principles of wireless transmission to the demonstrable stage back in 1892. It was in this year that he gave lectures in London and Paris, where he described and displayed the characteristics of his electronic tube. One of the uses of it, as described by Tesla,[13] was as a detector in a radio system.

This piece of equipment however had been already developed in 1890, and there is really little doubt, that Tesla understood the implications of it already by this time, since of course, no one builds something he doesn't at least understand some of the functions of!

At these same 1892 lectures, Tesla also demonstrated lamps which were lit without any wire connections, as well as a motor that also operated without any wire connections.

*Wireless transmission of light and power then, had already been achieved by Tesla back in 1892!*

There are people today, despite at least one or two decent documentaries having been made concerning Tesla, that question whether at these lectures Tesla did not perhaps light the lamps and provide energy to the motor by means of some trick.

---

[11] See bibliography for details of this book.

[12] Guglielmo Marconi is generally known as the man who invented the wireless telegraph, and Sir Oliver Lodge in 1894 built a wireless set two years before Marconi, but both these men were trying to use short wavelengths, which Tesla, as far back as 1892, having already demonstrated wireless reception by means of his electronic tube (developed in 1890), had already understood would not work as well as the longer wavelengths.

[13] As on record in the archives of the Institute of Electrical Engineers; the Royal Society of London; the Physical Society of France and the International Society of Electrical Engineers in Paris.

Apart from the fact that such an accusation is ludicrous simply on the grounds that in addition to being a completely honest individual, Tesla was a scrupulously serious engineer with a scientific purpose; not a charlatan selling his wares from a wagon; here is why such false accusations, or perhaps merely ignorant musings, are way off the mark:

*From 1890 onwards, Tesla permanently used this wireless technique of lighting in his laboratory.*

Whenever light was required at a particular angle, or in a particular place, all that one had to do in Tesla's lab, was move one of the convenient glass tubes in a suitably advantageous position!

Despite this success of the system, Tesla never managed to get his real dream, the World Wireless System, off the ground.[14]

This however was not because the system was impracticable, in fact on the contrary, Tesla's plan, had it been carried to fruition, would have ensured that each of us today could receive all the electrical power we require almost free of charge.

The only costs involved would be to get the initial power transmitters set up, but once that is done, each transmitter, acting in resonating unison with the others would serve to light up the world entire!

Once the transmitters were built, the only other costs would be for their maintenance and upkeep, which of course would be a small fraction of the money being spent nowadays on the maintaining of the millions of miles of electrical cabling that stretch across the globe.

Each home or factory could receive all the power it required by means of a receiving antenna which Tesla envisioned as being mounted on the roof of each dwelling or place of work.

Just imagine what the world would be like if only this one invention of Tesla was to be implemented.

It does sound fantastic, does it not? Unbelievable. An utopian dream that cannot be possibly rooted in reality.

Well, strange, unreal, and yes, even unjust as it sounds, Tesla's World Wireless System is not based on dreams or fantasies. It is as real as a brick thrown through your living room window.

---

[14] It could be argued that the fire on the night of 13 March 1895, which completely gutted Tesla's laboratory may have been one of those events in time that have drastically changed the path of history.

It's practical, it would work, and it would lower the cost of providing energy to those who need it most by factors so large we would have to consider them as magnitudes, rather than percentages.

It is a well know fact that the measure of a country's living standard is governed by the amount of energy available per capita, and Tesla's wireless power transmission would raise not just the energy level per capita of a country or two; it would raise the energy per capita of the world, providing safe, cheap electricity to people that have only heard of it or seen it used by others but to whom this glorious force has been denied.

The country in which I reside at present, South Africa, is a perfect example of the benefits Tesla's technology could bring. The millions of people that live in what are euphemistically referred to as 'townships' could receive electrification at a mere fraction of the cost that more conventional means would require, and in an amount of time so short it would not even be fair to compare it.[15]

The money saved in this manner could then be spent more practically, in the building of schools, homes, hospitals and libraries, all of which in addition to raising the living standards would contribute towards the rapid increase of the education level not only of the illiterate, but of all South Africans.


The obvious question raises it head again and again, if this system is so great, why are we not using it today?

The answer is somewhat complex in its detailed components, but amounts to really only two words: Greed and Stupidity.

And to be fair, stupidity may be the greater part of it.

It is incredibly difficult for us today to understand what life a mere hundred years ago was like. The implicit reason why people like the Wright brothers were not taken seriously even after they had demonstrated their 'flying machines' several times in the presence of numerous witnesses, is not necessarily because the powers that be (academics, 'orthodox' scientists, and politicians) are intentionally evil (all-right, one may wonder about the politicians...) but quite simply, the reason why Orville and Wilbur Wright were ignored and

---

[15] Consider that instead of miles of cabling to be laid down not only across the landscape, but also in each individual 'residence', all that would be required is the construction of several (admittedly large) transmitter/generators, a few similarly large receivers at strategic points throughout the area, and then the populace can be issued with the same kind of electronic tubes Tesla used in his laboratory.

ridiculed, for at least 5 years between 1903 and 1908, is because powered flight by machines heavier than air at the beginning of this century was impossible.

That's it. It was just NOT possible; and anyone with an ounce of sense in their heads knew it.

Try to go back there, to 1908 and realise that powered flight then was not something which may or may not be done, it was just something which could not be done. A hot air balloon could fly of course, any two bit fool knew that, but an aeroplane was just plain nonsense.

Its place was only in works of science fiction[16] and the diseased or fanciful minds of the Wright brothers.

Had some mishap occurred during their public demonstration at Fort Myers, which by the way had been ordered by President Roosevelt, today there might be no such thing as the Concorde, the jumbo jet, and the myriad other types of aeroplanes which exist now.

Tesla had no order from the president to 'show his wares', so to speak. Instead he had a fire that destroyed his laboratory completely, and he lacked altogether a capacity for marketing himself and his ideas to a public that was so far removed from him intellectually that it might be fair to say that he was probably one of the loneliest people that have ever lived.

His concepts, inventions and ideas were also infinitely more complex that a flying machine. After all, such a contraption either flies or it does not, and it is no doubt this simplicity of purpose, easily grasped by anyone, which prompted a prominent politician of the day to order the public demonstration.

But how long would it take to explain even just one of the concepts Tesla envisioned to the same President Roosevelt that showed the world the Wright brothers? And even if such time could have been taken, so much larger, and of such greater importance were Tesla's plans, that even today, more than 50 years after his death and some 100 years since he made his discoveries, we still see them as utopian dreams with no basis in reality.

Tesla, because of personality traits which in no way impaired his genius for producing astounding technology, is already nearly forgotten, and one wonders, how many others that perhaps could have

---

[16] *Bad* science fiction at that, since according to the physicists of the day it would be contravening the known laws of the Universe.

shaped this world into a garden of Eden had we listened to them, has history buried?

But greed too, being a close friend of stupidity, plays its part.

Although a lot of Tesla's work, inventions, ideas and concepts perished with him since he kept few notes (as a result of his astonishing memory) there *are* diaries, notes and concepts which survive him.

Unfortunately, we may never know just exactly what these are, because they were sequestered by the Federal Bureau of Investigation.

Tesla's notes were taken from his safe by the FBI on the day his body was discovered by a chambermaid, and were sealed by the custodian of Alien Property, despite the fact that Tesla had been a naturalised citizen of the USA for over forty years.

Why did the FBI take these notes?

Apparently to inspect them for important secret inventions which may have been of use during WWII, and yet, to this day, their full contents may remain unknown.

Consider too, the billions of dollars of revenue that would be lost by the politicians of the world if Tesla's wireless system of power supply was implemented. Such a system would divert those same billions directly to the people of the world in the form of energy.

In fact, almost as if by Tesla's own concept of resonance, the benefits that would arise from such a set-up, are many times more than what the same amount of money could do if it were spent in conventional technology.

♦ ♦ ♦ ♦ ♦

Why mention Tesla in connection with a book of this nature though?

Because in his later years, when he lacked funding for his inventions, Tesla's claims became even grander than before concerning what he could do once he secured sufficient capital.

Of course some may think these the wanderings of a tortured mind, that having gone from humble beginnings to the greatest heights, and then back down to a humiliating level, tried desperately to relieve its glory by making fantastic claims.

But is such a judgement a fair one?

By all accounts, Tesla was a conservative man, refusing to announce specifications for one of his inventions unless he knew without a doubt that he could achieve his claims.

He always performed many tests concerning his theories before announcing any kind of discovery, and whenever he did announce a discovery, his tone was of a philosophical nature. With almost inhuman modesty, he concentrated more on the beneficial possibilities of the invention rather than on the technical achievements he had made alone, through the exclusive use of his amazing mind.

Furthermore, Tesla claimed —and it was shown repeatedly to all of the people that worked with him on a day to day basis— that he had the ability to build a machine in his imagination only, and then change it's dimensions, proportions or construction to obtain the desired results. This process would take him months at first, during which he would merely be seeding an idea, at which point a flash of inspiration would arrive that told him whether the plan he had in mind was possible or not. Whenever he felt that it was, he proceeded to build his mental models and test each one rigorously in his mind. Only once the mental image worked faultlessly did Tesla then construct it in reality, and unless the parts had been badly machined, as was the case in at least two instances, the real product would work as faultlessly as the imagined one.

This can furthermore be shown by the fact that Tesla built resonating coils and machines in his day and age, without the use of a single diagram,[17] that for us to construct today require a computer to take care of the complex formulae which define the exact distances between loops of the coils as well as their precise dimensions.

It would seem then, that Tesla not only appears to have not possessed the kind of character that would reduce him to boasting in later life, but that he certainly had the ability to create astounding technology *in his mind!*

Throughout his life Tesla seemed not only utterly honest, but also, completely unaware, or at least totally uncaring, about other people's opinions of him, which also further seems to indicate that he would be an unlikely candidate for glorifying his existence by making grandiose, but false, statements.

---

[17] Except, as already stated, those necessary to explain to his attendants the required shape of a part, and giving them verbal instructions only concerning dimensions.

In his lifetime, Tesla achieved the impossible not once or twice, but many times, and in 1938, he made reference to his dynamic theory of gravity, which like most everything concerning this man was once again rooted in the principle of resonance, which, as we have already seen, seems to be a fundamental law of the Universe.

Tesla did not expound on his theory, as was his way until he could concretely have tested for himself the veracity of his facts, but given the genius of the man, we can reasonably assume that whatever its details, it surely was bound to shed new light on that most mysterious of subjects: Gravity.

So much died with Tesla, that it is doubtful whether we will have caught up with him even another hundred years down the line, but considering that this new dynamic theory of gravity was one he more fully developed later in life (although its beginnings, as stated by Tesla himself, were in 1893 and 1894) it is certainly not unlikely that the papers the FBI so promptly sequestered concerned some detailed aspects of it.


It would be wise to keep Nikola Tesla in mind when in a short while we begin to look at the work of Thomas Townsend Brown, and discover that the mythical, utopian dream known as anti-gravity, not only is not so mythical as may seem at first, but also, that it is firmly rooted in aspects of electromagnetic phenomena which are in turn steeped in the principle of resonance.

Keep in mind too, that powered human flight was simply impossible at the beginnings of this century. Not for a lack of technologically advanced equipment of course, but because it contravened the known laws of physics.

That's right. Powered human flight contravened the known laws of physics.

What?.....You say it was early twentieth century physics?

Oh yes...true, true.

Of course we would never make such a mistake *now!*

*Now*, we know the laws of physics in their entirety of course, ha!

And to suggest that anti-gravity machines will exist in the early twenty-first century and once again restructure what we *really* know of physics would be....well, it would be at least as foolish as saying powered human flight was impossible back in 1908.

# The Beginning of UFO Activity

Generally speaking, UFOs seem to have existed in human history since time immemorial, but the modern era of the flying saucers can perhaps be traced back to the sighting made by Kenneth Arnold on 24 June 1947 of nine UFO objects over Mt. Rainier, Washington.

It was this sighting which resulted in the first wide coverage of the UFO phenomena and subsequent exponential growth in interest by the average person into the matter of UFOs.

Shortly afterwards,[18] the much talked about Roswell incident, where apparently a UFO object crashed and was retrieved by the US armed forces, occurred; and the debate as to its veracity is still raging now, almost fifty years later.

Basically, since the end of WWII, UFOs have received more attention from people, but this is by no means the first time during which they appeared 'en masse' so to speak.

In fact, WWII itself seems to have been a catalyst for the great increase in the number of observed UFOs.[19]

To be sure, our improved technology, both from a point of view of detection as well as of communication between countries and people in general, has something to do with this, but nevertheless, it does appear, that certain events of WWII, have resulted in an increased presence of UFO activity.

It is thought by some, that the creation and testing of nuclear weapons is what may have attracted the attention of the UFO's owners, and considering the recent spate of UFO sightings, so close to the renewed nuclear testing by the French, this may be a valid proposition. It is important to keep in mind however, that there were other military developments taking place in these times, which may be just as relevant, or perhaps even more so, than the creation of nuclear weapons.

The US government for example, at around this time (1945-1958) undertook several secret experiments involving equipment of an electromagnetic nature. Just exactly what these experiments entailed is difficult to ascertain, but people like Charles Berlitz and William

---

[18] The Roswell Incident occurred on or near to the date of 7 July 1947.

[19] The term Unidentified Flying Objects, is a relatively recent one, and in WWII, when UFOs were thought to be enemy craft or weaponry, they were referred to as 'foo fighters'.

Moore, whom have co-authored at least two books together involving such covert topics, may have given us at least a clue, in their outstanding book, *The Philadelphia Experiment, Project Invisibility*.[20] The Philadelphia experiment has been, like the Roswell affair, one of those topics that may fall under the classification of urban legend if it were not for the pushy, nosy, and more than a little nervy investigations of people like Berlitz and Moore.

      The young UFO enthusiasts of today, who continue to argue about such topics, would do well to dig up the works of Berlitz concerning Roswell and the Philadelphia Experiment. They may be pleasantly surprised to know, that several years back, exhaustive research complete with documents, personal interviews and photographs of relevant persons, was indeed carried out.

From an investigative point of view, Berlitz's book the *Philadelphia Experiment* has only one drawback. An interview with a scientist that refuses to be identified and his wishes are respected by William Moore, who carried out the interview.

Nevertheless, people connected with the Philadelphia Experiment — that no other investigator had managed to find before— were dug up and once their identities verified (and if willing) they were interviewed.

Old files were found and archives searched and generally speaking, every available lead was followed to its end, whatever that might be.

The conclusion of all this, is that Berlitz and Moore uncovered startling, verified evidence, which confirms that whatever actually happened in the Navy Yard of Philadelphia in 1943, it was something very strange; it involved mainly some aspect of electromagnetism; and it was extremely secret.

It is also fairly certain to assume, that more than a few people perished as a result of whatever took place.

Broadly speaking, the Philadelphia Experiment story states that sometime late in 1943, the Navy ship U.S.S. Eldridge (Destroyer Escort 173) underwent some process, whereby, along with its crew, it was temporarily turned invisible through the use of some electromagnetic or electrogravitic process. The ship apparently suffered some damage but this was not at least of a permanent nature, although strange reports of things disappearing and/or reappearing on

---

[20] See bibliography for details.

it persisted for years later, even after it was sold to the Greek Navy in 1951 and its name changed to *Leon*.[21]

The crew on the other hand, suffered extensively.

Some were apparently lost for good, having simply vanished. Others periodically seemed to undergo episodes of invisibility, particularly if excited, with resultant psychological stresses or dysfunctions. Some of the men quite simply went insane and it seems that most of these eventually perished.

At least a few though, are said to have survived, and after having been discharged as mentally incompetent, went on to lead some kind of existence.

Incredible as this sounds at first, a few things are certain.

1) In 1943, the US government was involved in a Top Secret experiment involving the U.S.S. Eldridge,[22] and in all probability, this was viewed by the crew of the S.S. Furuseth.

2) Most of the surviving people that were thought to have been crewmembers of the U.S.S. Eldridge at the time that Berlitz and Moore wrote *The Philadelphia Experiment*, (1979) were unwilling to discuss this topic, or 'hid' behind the screening of relatives who stated that the person in question did not have anything to say and wished to be left alone concerning this subject.

3) The US Navy lied (and continues to do so) about the whereabouts of these ships at the relevant times, and neither of their logbooks are available, either having been 'lost' or 'destroyed by executive order'. The reason for such destruction remains of a classified and hence unstated nature.[23]

4) At around this time, as is shown by relevant documents available under the freedom of information act, the US was involved in the first stages of a Top Secret research into: "Electrogravitic

---

[21] The latest reports seem to indicate that at least until a few years ago, this ship was still in service under the Greek Navy.

[22] The codename under which this took place was probably *Rainbow*.

[23] The engineering logs and a particularly damning declassfied Action Report however are available, and they do not correspond to the official history of this ship as given by the US Navy.

phenomena".[24] As of about 1958, all reference to such research was completely omitted even from internal memos and (it can safely be assumed) the research continued.

Before —as the 'debunkers' of UFO phenomena are fond of saying— we get caught up in a 'vicious circle of conspiracy theories' though, let's further examine the idea that there is even any basis for claiming that anti-gravity machines are a possibility.

We have already seen that there is video footage shot by NASA's own cameras that would seem to indicate that such technology already exists, but is there in fact any basis for stating that the images captured on that piece of video could be anti-gravity machines?

After all, if we haven't even the faintest of theories that at least indicate anti-gravity may be a possibility, then any idea based on the concept that anti-gravity *is even possible*, would be unsound.

Better still of course, would be a practical demonstration that backs up any such anti-gravity claims as we may come across.

If we *do* have, not only the NASA footage of some Unidentified Flying Objects, but also a theory which at least seems to indicate that anti-gravity is a possibility, we might then begin to take more seriously claims that the US Navy (along with the majority of other Western governments) is actively 'covering up' technology such as that which may be displayed by an artificial construct which is able to perform manoeuvres as the Unidentified Flying Object in the NASA video.

---

[24] See also the end of the video: *Hoagland's Mars Vol II, The UN Briefing (Extended Version)* for further evidence of these documents.

# A New Physics

It is the basis of Relativity theory, that gravity warps space. But it is also true of course, that space and time are inextricably linked.

And Einstein, in his attempts at laying out his Unified Field Theory (UFT) began to view gravity not so much as a force anymore, but just as an observable property of the Universe. He further proposed, that 'matter' as such was only a concentration of energy.

In other words, energy would be the forefather of matter, which if one ascribes to the big bang theory, is difficult to dispute anyway.

Fortunately though, Einstein's own most famous equation, $E=mc^2$ proves that matter is indeed only a concentration of energy, so there is really no question as to whether this basic premise that matter is only 'condensed' energy is valid or not. It's difficult after all to deny an atomic explosion.

Until his death, Einstein stuck to the position that gravity could be mathematically proved to be related to electromagnetism. Even more interesting, is the fact that there are rumours (although of an unprovable nature) that seem to indicate that Einstein may in fact have developed his unified field theory to a conclusion. But in view of what the American government had already done to the Japanese people with the most lethal application of $E=mc^2$ he chose to keep the implications of the unified field theory to himself, apparently even destroying some papers shortly before his death.

There is of course no doubt that Einstein viewed the human race as being as yet unable to comprehend the horrors it could shower upon itself through misuse of certain principles, but it is interesting to see, that he must have conceptualised something even worse than atomic conflagration, because as reported in *Albert Einstein, Creator and Rebel*,[25] he felt obliged '*to use his influence to the utmost to try and save mankind from horrors that, despite Hiroshima and Nagasaki, it did not yet comprehend.*'

Einstein understood, that humanity was simply not ready for certain technology which could be called 'planet-threatening', and what happened to the two Japanese cities was the realisation of his

---

[25] This book was written by Banesh Hoffman and Helen Dukas, whom was Einstein personal secretary and (considering his later years of life) could perhaps be said to have been one of the people who knew Einstein best.

worst fears not so much because of the destruction these two atomic blasts produced on the immediate level, but more importantly, because of what it said about humanity in general, and how it would behave.

Personally, I have no doubt that having seen the results of the atomic explosions of Hiroshima and Nagasaki, had Einstein later discovered a new theory of the Universe which however had the potential to be used with consequences that would be even more devastating than atomic explosions, he would most likely have destroyed such findings.

Whatever the case may be concerning the unified field theory's completion by Einstein however, the point is that if gravity could be shown to have any connection to electromagnetic phenomena, then anti-gravity would most definitely no longer be a mere possibility; it would have to be a virtual certainty.

◆ ◆ ◆ ◆ ◆

Most people know that an electric current produces a magnetic field; the powerful electromagnets of industry and particle accelerators being the ultimate implication of this. Similarly, an electric current can be generated by use of a magnetic field, as is shown by electrical generators. Similarly, if it could be shown that gravity is somehow connected to either electrical or magnetic fields, then several interesting propositions arise.

First of all, given the relationship between electric and magnetic fields, it is unlikely that gravity would concern itself with only one or the other. In fact, our view of separating magnetism from electricity is only one of our convenient ways of reducing something to smaller parts for the purposes of understanding it, but it would be more accurate to talk about *electromagnetism*, rather than electric or magnetic fields.

Now; we know that gravity affects light by bending it, and in the case of a black hole, light is affected to the extent that it cannot escape, but light of course is an electromagnetic phenomenon, and although the relationship is tenuous, this already shows that there is *some* connection between gravity and electromagnetism.

This should already be evident anyway, since ultimately, absolutely everything in the Universe is somehow connected.

The important point, is whether this connection between electromagnetism and gravity is of a nature which can produce useful results for us, such as a local nullification of gravity.

How so?

Well, let's assume that we start off with a natural magnetic field we want to nullify.

One way of doing it would be to use an electric current to create an *artificial* magnetic field which in turn cancels the natural magnetic field by interference.

Similarly, an electric current could be cancelled out using the same principle; and the interfering electric current could of course be created by means of a magnetic field.

Intellectually at least, it's not a big step to see that if electromagnetism and gravity are related in a useful manner, it may be possible to create a similarly nullifying gravitic field which in turn could be produced by electromagnetic means.

The problems then seem to be two. Firstly can it be shown that indeed there is such a connection, and secondly, if there is, can it be taken advantage of in a useful manner. It is worth pointing out however, that if the connection between gravity and electromagnetism exists, then it is almost certain that such a situation could be put to many fruitful and practical endeavours.

It is also interesting to point out, that what are perhaps the two greatest minds that ever lived, those of Nikola Tesla and Albert Einstein, both seemed to feel that such a connection existed.

Einstein believed it emphatically for the last 39 years of his life, and a mere two years prior to his death stated that he had highly convincing results concerning the theory that there was indeed a mathematical proof to show the connection between electromagnetism and gravity.

Tesla on the other hand, not having had the possibility to test his ideas concretely, and because of his conservatism concerning details, never said as much. But the implication is clear from what little he did say on the subject, and when we keep in mind that Tesla spent his life dealing with electromagnetic phenomena of forms so unique that some of them would appear to have been lost with him, it becomes obvious that his dynamic theory of gravity must also have had its roots in some aspect of electromagnetism.

The 'final' version of the Unified Field Theory as it was left to us by Einstein comprises sixteen very complex quantities represented by tensor equations (which are already an advanced form of mathematics in themselves). It is interesting to note that the little sense scientists have been able to make of these indicates that while a gravitic field can exist without an electromagnetic one, the reverse is not the case. In other words, a pure electromagnetic field must be accompanied by some sort of gravitic field.

But Einstein himself also made it clear that his equations were not necessarily in their final form, which in a way seems to add weight to the 'rumour' that he felt the human race was not yet ready for an even more advanced concept of the Universe than his already avant-garde theory of relativity.

Apart from theoretical considerations however, there is always the more practical application of such theories into an experimental format, which of course occasionally bypass the more abstract or intellectual aspects of a theory to produce very convincing results.

To put it plainly, it's a lot easier to produce a magnetic field by using electricity than it is to define the relationship between electric and magnetic fields in a mathematical manner. Might it not then be the same with respect to the production of an artificial gravity field?

According to the experiments performed by Thomas Townsend Brown, this would indeed seem to be the case.

# T. T. Brown : Master of Antigravity

Thomas Townsend Brown was born in Ohio in 1905, and like Tesla remains an obscure yet brilliant scientist.

He may not have been as prolific an inventor as Tesla, but what Thomas T. Brown discovered is certainly on a par with Tesla's wireless power distribution and perhaps of even greater value to the human race ultimately.

Brown was interested in space travel from an early age, which at the time was of course a subject of not little ridicule considering that the Wright brothers had only just begun to convince the world of the possibility of flight in 1908. In addition, Thomas was interested in electronics, with which he experimented freely and often.

Brown had the advantage of being born in a relatively prominent family, and so did not altogether lack the means to fund even his youthful experiments, to be sure though, like any teenager, he no doubt had to 'sweat' a little for his money.

The first discovery Brown made, was through the use of a Coolidge tube, which produces X-rays. He had set up the apparatus in a delicate balance to find out if the X-rays produced had any sort of measurable thrust which might come in handy one day for the propelling of space vehicles.

Of course, there was no such result forthcoming as a result of the X-rays, which as we know today produce negligible thrust under such conditions, but Brown did notice that whenever the apparatus was used, the Coolidge tube itself seemed to undergo some sort of slight movement. Further investigation revealed that it was not in fact the X-rays which produced this movement, but rather the high voltage required to create them.

Excited by his find, Brown decided to conduct further experiments of an altogether different nature, and these culminated in the construction of a rather unimpressive looking piece of apparatus which Brown called a 'gravitor'.

This device consisted of a Bakelite case about a foot long and having a cross section of some 4 inches square (30 cm long by 5cm deep and 5cm wide) which when placed on a scale and connected to a 100 Kilovolt power source produced an unusual effect. Depending on the polarity of the apparatus, the case would either gain or loose about 1 percent of its weight.

Aside from a few reports of his discovery having appeared in newspapers though, no one seemed to take notice. Why this is so is not exactly clear, but perhaps the fact that eminent professors working with electromagnetism are not partial to taking advice from a boy that has yet to complete highschool, might have had something to do with it.

Despite some two years of trying to convince his teachers that his experiments into electrogravity had value, it was only in 1924, at the age of nineteen or twenty, that Brown managed to have one of his professor at the Denison University of Granville Ohio (where he had just recently transferred) take note of his work.

This man, Dr Paul Alfred Biefeld, whom was professor of physics and astronomy at the University, not only listened to Brown, but together with him begun a series of experiments which developed into a principle which has since come to be known as the *Biefeld-Brown Effect*. This effect basically was that already discovered by Brown by himself and consisted of the tendency of motion towards the positive pole by highly charged electrical condensers.

For the next twenty years, Brown worked in several capacities, the most interesting one being that which he held with the US Naval Research Laboratory from 1939 to 1943.

Brown's position with the NRL was a prominent one. He was in charge of several highly qualified researchers and an expense budget of about $50 million. Furthermore, his field of expertise was related to electromagnetic work in connection with Navy ships, as shown by his rank of officer in charge of magnetic and acoustic minesweeping, which involved working with magnetic degaussers.

In December of 1943, a scant two months after the time during which the tragic Philadelphia Experiment took place, Brown suffered a nervous breakdown, and was shortly afterwards discharged from service under the advisement of naval physicians.

In 1944, having recovered from his nervous collapse, Brown became employed as a radar consultant for Lockheed-Vega Aircraft Corporation in California, but eventually moved to Hawaii to continue his research into his 'gravitor' apparatus, although he now preferred to use the term *dielectric stress*[26] rather than gravity when discussing it.

It was in Hawaii too, that Brown became a consultant to the Pearl harbour Navy Yard, after having impressed Admiral Arthur W. Radford,[27] then Commander in Chief of the US Pacific Fleet, with his new and improved 'gravitor' apparatus.

But only in 1952 however, and after having moved to Cleveland, did Brown begin to 'market' his 'gravitor' somewhat more aggressively.

Although having a rather meek countenance, he had finally decided to try and impress his achievements to the military, to whom (he felt sure) he could no doubt make sale of his idea if it was presented to them in the proper manner.

One of his presentations, performed in 1953 under laboratory conditions, was the successful flight of saucer shaped objects 2 feet in diameter (60 cm). These discs were attached by a 10 foot wire (through which the direct current flowed to them) to a central pole which was free to rotate. Being supplied with a direct current potential of 50,000 volts with continuous input of 50 watts, the discs achieved a top speed of 17 feet per second (5.18 metres per second).

His next demonstration surpassed even this feat, flying discs 3 feet in diameter around a 50 foot diameter circuit. This test, however, was immediately classified due to the spectacular performance of the discs, so no other details are available.

Despite these astounding results, scientists present at the demonstrations attributed the flight of these discs to 'electric wind'.

Apart from the fact that this idea is quite ridiculous, since the energy required to produce an 'electric wind' of sufficient strength to move 2 foot discs at a speed of 17 feet per second is quite astronomical, it is also completely false, since Brown's discs were also flown in a high vacuum a few years later, and in fact functioned more efficiently without the presence of an atmosphere.

---

[26] A dielectric is a material that can store electrical energy or charge without normally passing it on to nearby materials. There are principally two types of these, those that can be charged slowly and those that can be charged rapidly. The ones which can be charged and discharged rapidly (at rates of several thousand times per second) at very high potential, are the ones Brown used mainly, along with high voltage direct current to produce the Biefeld-Brown effect.

[27] Radford later became Chairman of the Joint Chiefs of Staff under President Eisenhower from 1953 to 1957.

By 1955 Brown realised that although his experiments had been successful, no money would be forthcoming from the government, so he went to demonstrate his discovery in Europe.

So successful were Brown's demonstrations in France at the research laboratories of SNCASO (La Société Nationale de Construction Aeronautique Sud Ouest) that immediately plans were drawn up to build a large vacuum chamber with a 500,000 volt power supply for further tests.[28]

But SNCASO was in the process of merging with a larger company, Sud Est, and the new president, showing complete lack of interest in such 'far-out' ideas concerning propulsion, cancelled all research concerning electrogravitic efforts.

Dejected if not defeated, Brown returned to the United States where in 1956 he became chief researcher for the Whitehall-Rand Project, which was an investigation into antigravity and was being directed personally by Agnew Bahnson.[29]

Bahnson had used his own funds to construct a well equipped laboratory for Brown's research into antigravity phenomena. It looked as though Brown's ideas concerning this topic would finally receive the attention they deserved. Unfortunately, Bahnson suddenly died in the crash of his private aeroplane.

The remaining Bahnson heirs however were not interested in antigravity research and once again Brown found himself out in the cold.

At the time Brown was interviewed for the book *The Philadelphia Experiment*, in 1979, he was in semi-retirement in California and quietly continuing to pursue his ideas, being involved in a project which had the support and assistance of the University of California and the Ames Research Centre of NASA.

The nature of the project was to try and determine if there was (as Brown believed) a connection between the Earth's gravitational field, and rock electricity (petroelectricity).

The outcome of such research is unknown to this author nor is it known whether the results are available to the public yet (or ever),

---

[28] At this time, speeds of several hundred miles per hour for the discs, using voltages of between 100,000 to 200,000 volts, were envisioned as being easily feasible.

[29] Bahnson was president of the Bahnson Company, located in Winston-Salem North Carolina.

however had Brown been able to show that petroelectricity is a result of some kind of induction by the Earth's gravitational field,[30] then it would have been obvious that there is a practically applicable connection between electromagnetism and gravity, as Brown's flying dielectric discs had in any event already demonstrated beyond any doubt.[31]

---

[30] In this respect, the work done by Dr. Larry Lloyd Babcock concerning the Basic Electromagnetic Identity, comes particularly to mind and may be instrumental in redefining our understanding of gravity. A synopsis of this theory is available from Synergy Information Systems 3661 N. Campbell Avenue Suite 388 Tucson Arizona 85719-1527, Tel (520) 888-9345.

[31] Some of the more relevant patents awarded to Thomas Townsend Brown concerning his experiments with anti-gravity are the following:

300,311     T.T. Brown   Nov. 15, 1928 A Method of, and an Apparatus or Machine for Producing Force or Motion.
1,974,483   T.T. Brown   Sep. 25, 1934 Electrostatic Motor.
2,949,550   T.T. Brown   Aug. 16, 1960 Electrokinetic Apparatus.
3,022,430   T.T. Brown   Feb. 20, 1962 Electrokinetic Generator.
3,187,206   T.T. Brown   June 1, 1965 Electrokinetic Apparatus.
3,296,491   T.T. Brown   Jan. 3, 1967 Method and Apparatus for Producing Ions and Electrically Charged Aerosols.
3,518,462   T.T. Brown   June 30, 1970 Fluid Flow Control System.

These patents were all awarded in the USA, and can be found in public libraries of that country. Brown was also awarded some patents from outside of his native country.

# The Reality of a Hyperspace Drive

Based on the work of Nikola Tesla, Albert Einstein and most importantly of all, Thomas Townsend Brown, it would seem then, that an antigravity machine is not only a possibility, but that crude versions of it were being manufactured by Brown in the 1950s.

There is however, an aspect of gravity which we have so far barely mentioned, the implications of which —as related to an antigravity vehicle— are of the utmost importance.

This property of gravity is directly related to *gravitational redshift*; a suitably ominous sounding phrase which can be easily explained in simpler words. Gravity has a tendency to weaken light, and this of course results in a shift towards the red end of the spectrum, since in the narrow band of visible light, that which is coloured red has the slowest frequency.

Gravitational redshift has been confirmed by scientific observation, and as such forms one of the pillars which hold up Einstein's general theory of relativity, which until the time when the concept of antigravity is added to it, remains the best theory of gravitation to date.

The important point to keep in mind however, is that *gravitational redshift leads directly to the premise that gravity slows time!*

This is not just a fanciful idea, it's part of that section of the general theory of relativity which is responsible for numerous and important advancements in physics and our understanding of the Universe.

But if gravity in addition to warping space also slows down time, then a vehicle which is able to create gravity fields might well be able to travel in such a way that it might effectively be the equivalent of what is known as a hyperspace engine in science fiction novels.

The exact details of just how such a vehicle would operate, both from a mathematical and practical point of view cannot easily be demonstrated unequivocally without a great deal of research, but it does not take a rocket scientist to see that since gravity warps space, and considering that space is intimately linked with time, obviously a machine which creates gravity fields must affect time too in some way. And furthermore, Einstein's theory indicates that what would be achieved by gravity control is a *reduction of time*.

If very powerful gravity fields could be created however, and shaped to produce the desired effects, then what one would effectively be doing is warping space to the extent that the gravity well or 'tunnel' formed might project the spaceship into a realm of space which is very distant indeed from the original location of the spaceship.

It is such a principle that is behind the notion that wormholes and black holes may be some kind of a gateway to different sectors of space if not to a different plane of existence altogether.

It is also interesting to note, that the idea of wormholes being used as a theoretical means of space travel between stars is one that finds itself rather comfortable in the mind of what is currently thought to be the greatest theoretical physicist alive today; the world famous Stephen Hawking.

Ultimately then, the notion of fantastic 'hyperspace drives' is not one that finds its roots in the realms of fantasy, but rather something which may be well within our reach if the necessary research into electrogravitics is performed not only by black-budgeted, top secret government departments, but by all the scientists of the world.


If research into electrogravitics were to take place perhaps in a manner not dissimilar to the way in which investigation into the nature of matter have been undertaken, and with a similar influx of cash as has been placed into the manufacture of particle accelerators, it is my belief that we would not only see antigravity vehicles in a very short time —a foregone conclusion considering the achievements of Townsend Brown almost half a century ago— but also that interstellar travel would become a practical and affordable possibility within the lifetimes of most people alive today.

Similarly, the use of gravity control would open up the rest of our solar system to exploration and —more importantly— colonisation.

The transport as well as erection of heavy materials for this purpose would be achieved at a fraction of what it presently costs to put a similar amount of equipment in Earth orbit, never mind the shipping of it to one of our neighbouring planets.

In addition, the discovery and use of antigravity would no doubt open up the doors of countless other paths which in themselves would be of further benefit in the advancement of humanity towards an ideal future where problems such as hunger, disease and poverty are but old nightmares of a reality we can leave behind with considerable more ease than might be suspected.

# 11  The Cover-Up

Once again, whole books can be (and have been) written concerning the topic of government concealment of facts pertinent to the UFO phenomena.

Although an in-depth study is not permitted here due to space restrictions as well as the lengthy nature of any such investigation, the topic can nevertheless be at least touched upon.

It is important to understand however, that in order to provide incontrovertible proof of government cover-ups one needs to have access to certain classified information, which of course is not easily achieved since the very reason that it is classified is to restrict public access to it.

Secondly, even in the event that such documentation is produced, it can always be declared false by the concerned and relevant military or government departments.

Additionally, one of the most effective misinformation techniques is to provide the 'enemy' with false but juicy data which stems from the very department being investigated. Once the story has been accepted as relevant by the proponents of a conspiracy theory, that very same data can effectively be proved to be false by reference to some carefully placed details which indeed seem to be the work of forgers or hoaxers. The famous Majestic-12 document may be a case in point. Supposedly a classified document which proved the US government —with the knowledge of President Eisenhower— was involved in investigation of UFOs with a view that they were extra-terrestrial in nature, it seems to have been effectively labelled as a fake since the president's signature (Truman's) at the end of appendix A appears to be a forgery.

Interestingly enough though, the document contains such details, and the people it mentions as supposed members of MJ-12 have such affiliations (of a very pertinent nature!) that it seems highly unlikely that it could have been fabricated by anyone other than the very same people who now proudly present the MJ-12 document as a 'typical example of the kind of "proof" paranoid conspiracy theorists produce'; namely, the US military. Secondly, if the original MJ-12 document was real, then by supplying a fake addendum A, the powers that be succeeded in covering up their leak.

In view of these facts a complete study of government involvement which provides undeniable proof would be a time consuming affair. Timothy Good's *Above Top Secret*, and Charles Berlitz and William Moore's *The Roswell Incident*, may thus remain two of the most well documented books of such a nature for some time; especially since the official documents received through the freedom of information act and reproduced in *Above Top Secret*, already prove unequivocally that the FBI has lied about its involvement with matters concerning the UFO phenomena.[1]

At any rate, the fact that the US government lies to its citizens on a regular basis and at times with severe, disastrous or even fatal results for certain individuals, is something which is well known and has been recently highlighted by the number of documentaries over the last few years concerning themselves with the experiments with radioactive substances the US government undertook (with fatal consequences) on its own citizens and without their knowledge.

Nor is the US government the only one to perform such inhumane experiments, or to hide from its citizens information of a 'sensitive' nature, as the British experiments with anthrax as a biological weapon at the time of WW II clearly indicate.

The assassination of President Kennedy is similarly fraught with inconsistencies if one is to believe the official explanation for it. It does not take much to figure out that Lee Harvey Oswald could not possibly have fired the three shots which supposedly killed Kennedy. Any half-competent marksman will attest to the fact that the feat supposedly carried out by Oswald (given the rifle used, distance from the target and positioning with respect to it) is unattainable.

---

[1] See pg. 471, of the book *Above Top Secret* where a 1973 letter from the FBI states no involvement into the UFO phenomena has ever taken place on the part of the Bureau. Several other documents from the FBI in the same book however make it plain that some involvement did indeed take place prior to 1973 (see for example pg. 523 and 537). And in any event, the 1976 release by the FBI of over 1,100 pages of UFO related documents further proves that the contents of the 1973 letter was as bold-faced a lie as one can muster. Of similar interest is the letter reproduced on page 463 by the Department of National Defence of Canada, written in 1972, where it is admitted by that office that concerning UFO incidents, "It has not been the practice to allow the general public to study these files." Furthermore that although no apparent threat is forthcoming from these UFOs, they seem to exhibit a "...unique scientific or advance technology that could possibly contribute to scientific or technical research."

Just as damning of course, is the original footage of the assassination as was shown and became popular in the Oliver Stone movie, *JFK*. It is clear from the way the bullets impact on Kennedy, that he was shot from the front as well, and from a height approximating his own.

Although such things may not be immediately obvious to a person that has little knowledge of ballistics or marksmanship, any person even only somewhat knowledgeable of these topics could have already clearly seen that something was amiss. It is very revealing (and not a little disturbing) that the official government explanation for Kennedy's assassination blatantly ignores such obvious evidence.

And yet, what has the American public been able to do about such injustice? It is very doubtful that any level-headed person that has been shown what evidence there is concerning the Kennedy assassination would arrive at the same conclusion of the official reports unless he or she had vested interests in doing so; yet, those same official reports persist right up to the present.

The case with respect to the UFO phenomena is not dissimilar to the Kennedy affair. In this instance though, the governments of the world have had a useful ace up their sleeve: the natural tendency of the average person to view reports of 'little green men' as something which exists in a sphere of reality that has little to do with day to day existence.

As much as possible, UFO reports are ridiculed as well as silenced. Emphasis is given to those UFO phenomena which most obviously are unlikely to be evidence of any extra-terrestrial craft, but instead are explainable as some other event. And where there *is* a security leak, a systematic campaign of misinformation is the course taken by way of remedy.

Another example may be the case involving the notorious Bob Lazar, who claims he was involved in a reverse engineering project in which, using captured extra-terrestrial craft, the US government was attempting to learn how to use and manufacture similar vehicles.

It is interesting to note, that Lazar was denounced by the relevant departments he mentions as 'never having been employed by them', and yet Lazar seems to have produced ID cards and tax return forms that clearly indicate he indeed *was* employed by these government departments. If these ID documents are forgeries, why is Lazar not being taken to task on it?

And if they are *not* forgeries, why aren't more people listening to him?

Whether Bob Lazar is telling the truth or not with respect to his claims is not as important for the purposes of this chapter, as the fact that he seems to have effectively been 'neutered' by a campaign of misinformation carried out in large part by government officials, or at least by people employed by the government.[2]

It is usual for us to take in only a part of what is presented to us, and considering we are already conditioned to generally assume that UFOs belong in the same box as Elvis apparitions, how seriously would we take a person that claims to have worked on just such UFOs for the US government?

Especially when there are reports going around to the effect that such a person never actually worked for the government, as well as the fact that he has been found guilty of something or other that is connected to prostitution.

None of us really has the time to go and investigate the details of Bob Lazar's life, and even if we *did* have such time, it is doubtful we would use it in such a way, considering that the whole topic is somewhat 'fishy'.

At the end of it, what it boils down to is a case of a few individuals shouting "Yes you are!", and the US government (along with others) shouting back "No we're not!".

But do we need solid evidence of government cover-ups in order to show that at least some of the unexplained UFOs are very likely to in fact be extra-terrestrial crafts?

Although it would certainly make matters easier if the roof of the Pentagon building were to blow off in a storm and in doing so reveal to the world hundreds of flying saucers and pickled alien bodies preserved under spirit, it is by no means absolutely necessary.

---

[2] Lazar has also had some problems with the law concerning some charges related to the production of amphetamines (these were thrown out of court after a preliminary court hearing though) and involvement with the writing or producing of a computer program for a woman involved in prostitution while apparently also taking a percentage of the money she received from such activities. Lazar pleaded guilty to this charge in some plea-bargaining deal and received a 6 months probation sentence. It is of note however that his passport was apparently taken away from him and never returned by the authorities and (again, apparently at least) it seems that no explanation for this has been given.

Once again we can use a simple thought experiment to arrive at an answer which can in fact be quite conclusive.

Keeping in mind such things as the US radioactive testing it conducted on its own citizens; the convenient coincidence of 'undesirables' being drafted more regularly for the Vietnam war; the agreement between America and Japan at the close of WW II, not to prosecute Japanese war criminals who tested (among other things) biological weapons on prisoners of war, as long as America could be made privy to the test results; the Kennedy assassinations and other equally disturbing facts concerning the US government,[3] let's make a few assumptions concerning the affair of wireless power transmissions, UFOs and anti-gravity technology.

Let us first of all begin with the premise that none of these things are at all real. Tesla was wrong, UFOs are all naturally explicable phenomena, the result of drug induced (mass?) hysteria or hoaxes carried out by bored citizens, and anti-gravity is a nice topic for science fiction writers but has really nothing to do with real life.
If we start with this premise then we have a decidedly difficult task before us, because we have to explain away thousands of unexplainable UFO sightings, and I am only referring to those sightings reported by such creditable witness as military or commercial aircraft pilots of such diverse countries as the USA, UK, Belgium and France to name a few, which incidentally often also have corroborating evidence from radar operators on the ground as well as eyewitnesses.
Additionally we also have to explain away the flying discs of Thomas Townsend Brown, and the well documented fact that Tesla lit his laboratory without the use of any wires as well as gave demonstrations (again documented) where he had motors running without any wiring attached to them from which a supply of power might come.
We are also left with the somewhat easier task (but somewhat more disturbing one) of explaining away the deaths of people like doctor

---

[3] Although perhaps not in the same class for sheer brutality, one may also want to keep in mind that Stan McDaniel, who conducted a two year study of NASA's policy concerning the Face on Mars, points out that the same shape-from-shading techniques Mark Carlotto used on the Face on Mars are also used by NASA itself to determine the shape of other objects photographed by space probes, including some from the Viking missions, and in view of this, NASA's continual denial that the Face is anything other than a "trick of the light" or "natural rock formation" is unethical.

Jessup[4] and of explaining away the numerous 'cover-up' theories some of which, like the Roswell Incident, have been surprisingly well documented and no doubt involve a number of high ranking officers as part of the 'hoax'.

An indication of just what kind of mammoth task this position would entail, might be found in the results of a recent survey performed by Scripps Howard news Service and Ohio University.

The survey, which was reported in the *Arizona Republic* of Saturday, July 8, 1995, took place on June 12 to 16 of 1995, included only adult Americans from all walks of life and from all 50 states, and involved the asking of the following question to 1,006 persons:

"*Some Americans feel that flying saucers are real and that the government is hiding the truth about them from us. Do you think this is very likely, somewhat likely or unlikely?*"

The results were as follows:

43% of respondents said this was unlikely
31% of respondents said this was somewhat likely
19% of respondents said this was very likely
 7% of respondents said they were uncertain

Effectively this means that some 50% of adult Americans believe that UFOs are real. Before one goes off half-cocked with a comment on the gullibility of Americans in general, which in any event may or may not have some merit, it might be useful to point out that similar surveys performed in European countries yield similar results.

In fact, there are areas even in 'primitive' Africa, where the native population has been convinced for a long time of the existence of what can only be described as UFOs.

In view of the fact that a sizeable portion of the planet's population seems to believe that at least some UFOs are the result of extra-terrestrial visitation, it becomes difficult to explain away all of these people, some of whom arc perfectly capable scientists, engineers, and (mostly ex) military persons.

---

[4] See *The Philadelphia Experiment* pg. 89-91 and also *Without a Trace*, pg. 170-172

Some of these same people have gone on to form organisations which are petitioning not only the governments of their respective countries, but also the United Nations, to release all information pertaining to UFOs.

We would also have to explain why so many government documents, released to the public through the freedom of information act, although censored by black marker still make clear references to foo-fighters, UFOs, flying saucers, flying discs, or fastwalkers.

If, however, we start from a different premise, the opposite one in fact, then we find that we encounter no such problems whatsoever.

If we assume that indeed some UFOs are the result of extra-terrestrial visitation by anti-gravity capable craft, that anti-gravity is a possibility, and that Tesla wireless power transmission is also a reality, then the picture suddenly makes a lot more sense.

In fact, even if only one of those three things is true it still makes ample sense. Of course, the concept of UFOs being extra-terrestrial in nature would automatically make anti-gravity a necessity, as has been outlined in chapter 10 already, and although not as directly linked, if anti-gravity was found to be indeed a possibility, then this would similarly add a lot of weight to the idea that some UFOs are extra-terrestrial vehicles.

Let us now in fact consider what the world would be like if antigravity machines where to be researched and produced openly.

Considering the advances already made by T. T. Brown almost single-handedly, it is not inconceivable, that if fully fledged and government sponsored research were undertaken, in a short space of time, any government would be the proud owner of the secret of antigravity.

One of the most interesting things about antigravity machines would seem to be their relative ease of construction. Obviously many millions of dollars would be needed to ensure that such vehicles could be built in a 'safe' configuration for use by organisations, and eventually, it is hoped, private individuals; but the basic design of antigravity vehicles is not that difficult as Brown's flying discs, which were little more than crude antigravity machines, clearly showed.

The main problem would seem to be the installing of a sufficiently small yet powerful generator of energy which would be able to provide the craft with the required oscillations of high frequency voltage needed to produce the desired gravity effect.

We shall return to this point a little later, but for now let's assume that such power plants might be manufactured with relative ease.

What would planet Earth become if antigravity vehicles were to become as common as jetliners?

Well, for a start we would no longer need jetliners. The already built airports would in fact be able to handle larger volumes of antigravity vehicles than the current number of planes they routinely process in view of the simpler landing and take off procedures inherent in a machine that can control its own personal gravity.

Secondly, the maintenance of these machines would in the worst case only amount to a comparable amount to the one currently being spent on the upkeep of jetliners.

The transport of goods would gradually shift to antigravity cargo ships quite rapidly because of the much lower shipping times when compared to travel by land, air or sea, and this in effect would make the demand for antigravity machines rise, which would eventually result in their being adapted for civilian use, in a similar way to the transition which in a scant twenty years, computers made from the sterile rooms of megacorporations to the desk of every other kid who just had to have the ultimate games machine.

Of course, in a world with antigravity machines, cars, trucks and boats would die a rapid death. The only survivors might be sailing yachts, which would serve much the same purpose they serve today: strenuous luxury.

Without cars and trucks of course there is no need for roads. In fact, antigravity ships would be able to land quite safely on any reasonably level surface, thus making the need for landing pads more of an aesthetic consideration than a practical one.

Of course, without roads and trucks and cars, the petroleum industry would collapse.

Crude oil would become severely limited in its applications, and apart from this being economically bad news for oil rich countries, it would also improve the current level of toxicity in the air as a result of the burning of so much billions of tons of fuel every week, month and year.

Smog would be a thing of the past. Consequently acid rain would similarly be eliminated. Additionally, antigravity machines could be used to facilitate and greatly reduce the cost of major engineering projects.

It may be impossible to build a Great Pyramid with our current technology, but an antigravity machine would make such a task certainly feasible if we managed to figure out how to keep the thing level to such precise standards once it's built.

Still concentrating solely on Earth, what would it really mean to have antigravity machines around?

It would ideally mean an almost complete stop to pollution, and perhaps more importantly a raising in living standard for every single human being on this planet.

If we also add such concepts as Tesla's wireless power transmission to the equation, then we suddenly have truly utopian conditions.

The billions and billions of dollars currently being spent on the building of vehicles which use fossil fuels, the construction and maintenance of roads, and the electrification and upkeep of present day power grids would all be used for the construction of antigravity machines and wireless power transmitters and receivers.

The savings would amount to an almost incalculable quantity of resources, certainly enough to clothe, feed, educate and adequately shelter every human being on this planet in comfortable conditions.

People all over the world would be able to enjoy television, electricity, adequate supply of food and clean water, as well as have the opportunity to become educated enough to properly care for the continued and prosperous existence of this planet and their fellow humans.

So why would anyone want to keep such technology hidden?

As it happens, we have so far only looked at the positive side of things, but the truth of the matter is that —as always— antigravity, or any other technology for that matter, has two sides to it.

The picture painted so far would hold true only if the beneficial sides of antigravity are considered, but the same technology can of course be used for crimes of such a nature as to make a few atom bombs going off in densely populated areas look like something which would be infinitely preferable.

Consider that an antigravity machine, as already sketched out in the previous chapter, is by its very nature also an item capable of hyperdimensional travel.

By shaping a gravity 'tunnel' or small wormhole from one point of space-time to another, an antigravity vehicle can then travel down that tunnel without being crushed by the enormous gravity distortions because of course it can control the gravity around itself.

By intensifying the 'density' of the gravity 'tunnel' to very high levels, a great deal of space-time can be warped to the extent that very large distances can be travelled very quickly, since not only is space-time warped to favour such passage, but time too is altered to be greatly reduced if not altogether nullified.

This of course is so far merely a theory expressed in layman's terms, but if antigravity exists, then the way it has been described here is a very good approximation of how it will work. Einstein's definition of the Universe, which so far continues to prove the best model ever conceived, clearly indicates this.

A vehicle capable of almost literally teleporting from one point to another can itself be used for evil ends. Secondly, another inherent property of antigravity is not only radar but even visual invisibility, since all one has to do is bend the electromagnetic light waves around the antigravity machine.[5]

There may be different levels of intensity of this invisibility, which may also explain why UFOs seem to be able to blink in and out of sight and radar at will.

With teleportation and invisibility comes the frightening prospect of weapons of war which are scary indeed, and that is only if we use but the tiniest part of our imagination.

In a similar way to how Mars was destroyed, any half-bit madman with an antigravity machine could also of course resort to the old trick of catching a suitably sized asteroid in a gravity field strong enough to alter its course and direct it towards the offending ground dwellers.

Nor would reaching such asteroids prove to be any kind of problem, because another of the properties of antigravity machines is that they would open up solar system travel in a way that has no parallel in human exploration.

This would hold true regardless of whether the hyperdimensional travel capacities of antigravity were a possibility or not.

In addition to the enormous dangers directly linked to antigravity though, we should also keep in mind, that a complete restructuring of human society would take place, with all the attendant 'teething' problems.

---

[5] By controlling the gravity field immediately surrounding the craft.

In a sense it would be as if each person had hand-grenade sized atom bombs. There is no easy way of tracking and keeping tab of antigravity vehicles, as is evidenced by the fact that although supposedly in possession of extra-terrestrial technology as a result of the Roswell crash, there are not many rumours going around of the US military capturing alien craft every other day.

Each individual would have to be self-regulating to a level which may approach the old Wild West. And we all know what a neat, lawful society existed back then, don't we.

The difference in the case of antigravity machines is that instead of being responsible for the deaths of a few dozen people at most, a lone desperado could conceivably be the cause of the deaths of millions if not billions.

A society with widespread use of antigravity has to exist in a level of harmony which is akin to the highly idealistic and lofty aims of the Human Federation in Star Trek programs.

For any degree of safety to be achieved, we would all first have to learn to hold nothing more sacred than each other's lives. Not our religion, not our ideals, not our beliefs, but *each other's lives!*

That means we have to become the kind of person that is willing to step all over our own personal beliefs in order to save and protect the life of a person whom is diametrically opposed to our views.

The Moslem would have to renounce the Koran if in doing so he could save the life of a Jew. Similarly, the Jew must refute the Torah and Old Testament if in doing so he can save the life of a Moslem.

The Christian fundamentalist must throw down his Jesus and his Bible to save an Atheist, and this in turn must give praise to an almighty God to save the religious zealot.

Do we have such a situation on Earth? Far from it!

In view of this then, it is perhaps not altogether a bad thing that governments around the world have suppressed certain technology, but as I have explained at the end of part I, we are now rapidly approaching a nexus.

We are running out of time.

We need to begin introducing these advanced technologies, despite the risk, because the longer we wait, the greater that risk becomes, and the less time we are left with to allow the world to make the gradual change to a better way of life.

The time when such ideas begin to filter through to humanity may not (in fact cannot) be too far away, so it is imperative, that we all strive towards not only reading ourselves for such change, but begin to live in this new way from day to day.

Smiling at a stranger; providing good service; being honest in all you do. Striving to become more than you are in a healthy, positive, friendly and co-operative way is far more important than just being 'good public relations'.

Smile from your heart to that rude, difficult customer. Smile and try to understand him deeply. Ease his day. Be a good friend to him.

Because one day he may be driving an antigravity machine, and you don't want him to be in a bad mood in your city if he ever does.

# 12  Loose Ends

## 1. Government Misinformation Agents

The fourth group of individuals which become associated with the UFO subculture, incredible as it may sound at first, are government agents[1] whose only purpose is to spread rumours, confusion and lies among the UFO enthusiasts.

These people normally operate as individuals and each usually has a distinctive style of operation, although of course, their ultimate aim is the same in each case, and that is, to make useful trading of data between UFO investigators as difficult as possible.

Although such a statement smacks of delusional paranoia, once again, interested readers are invited to research this phenomena for themselves. It is not as hard as you may think. All you really need is Internet access and some spare time.

Internet access enables a person to do several things depending on the type of account chosen, but basically these boil down to just five.

1)  The ability to send and receive electronic mail.
2)  The ability to browse through the almost infinite data stored on the myriad servers around the world through World Wide Web or a similar program. Keywords can be entered when searching for information on a particular subject and a list of sites will be shown. Selecting a site will then link a user with that particular location where he can then peruse the relevant information.
3)  The opportunity to read articles posted by users from around the world in newsgroups. Newsgroups have different names to show the kind of topic they primarily deal with. For example a newsgroup dealing with the topic of extra-terrestrials, UFOs and similar phenomena is called alt.alien.research.
4)  The ability to download or upload files, articles and pictures from the World Wide Web servers, ftp sites or the newsgroups.
5)  The ability to have Internet Relay Chat sessions, during which a person can 'talk' in real-time with other users around the world

---

[1] Legally speaking a better term may be 'alleged' government agents, but the simple fact is that no organisation other than governmental ones would benefit from the spread of misinformation.

by typing what they want to say on their screens. In this manner one can communicate with several people at once or just with one person at the time.

There may be other ways to ferret out government operatives on the Internet, but becoming a regular on one of the newgroups which deal with UFO phenomena is probably the easiest way to convince yourself that indeed there is a Big Brother, and it would seem he works for Uncle Sam.

On first joining a newsgroup that deals with UFO phenomena you will read several articles, some interesting, some funny, some just plain silly and so on.

If you decide to follow these articles by posting your own comments and observations for a while, pretty soon you will begin to receive electronic mail from some of the people in the newsgroup and of course, answers to your replies and comments will also (primarily) be posted in the newsgroup by the other contributors.

In this way, in a relatively short space of time, you will become 'one of the group'.

However, it would be wise to just read the articles at first without posting any replies. This will give you the opportunity to see that there are several types of people that are 'regulars' on a newsgroup of this nature. Firstly there are 'stragglers'; people that come and go and are not regular contributors to the newsgroup; these can generally be ignored although of course it will take you a while to know just who is a regular and who is a straggler.

Among the people who post frequently, you will find several types of persons, these can generally be described as follows:

1) Pro UFOs— Believe that some UFOs are the product of extra-terrestrial intelligence.

2) Agnostic Contributors— Although these usually are not against the idea of UFOs being of extra-terrestrial origin they would like more information/data/proof before they become convinced.

3) Sceptics— Do not believe UFOs are extra-terrestrial in nature but are nevertheless interested in them and are willing to change their mind if presented with more proof/evidence.

4) Noise-makers— Do not make useful or even interesting contributions. These people basically take up bandwidth space for no clear reason.

5) Weirdos— Seriously strange people that believe that we are about to be invaded by Sauroid aliens that are approaching Earth 'as we speak' from behind Jupiter in a hollowed out comet.

6) Jokers— These usually poke fun and make jokes. Although almost everyone falls into this category at one time or another, there are people who only post joke-articles, and one can usually sense where their sympathies lie in the pro/against UFOs debate by taking note of who or what they poke fun at.

7) Government Employees— These are the Net-spies that spread misinformation. They can use any of the other personality types as cover and can and do switch between them, but some of course just stick with one type.

In a relatively short time, you will notice that there are certain persons whom seem to post an incredible number of articles. Not only in the one newsgroup you are in, but also in other related ones.

But which of these are misinformation agents and which are bored students with plenty of time on their hands?

The pattern goes something like this:

1) You are welcomed to the group by someone that agrees that intelligent life must exist elsewhere in the Universe other than on Earth.

2) On asking a few questions (perhaps specific to a recent UFO sighting or incident you may have heard about) the person indicates that it was a hoax, that it was Venus (or a meteorite, or an aeroplane...etc.). In many cases of course this *is* the case, but in this instance I am referring to an incident that is clearly *not* something that can be explained so easily. The Mexico City sightings would be a case in point. Viewed by professional astronomers that to this day are unable to explain them away as any naturally occurring phenomena, it is clear that someone that said this sighting was 'Venus shining brightly during a solar eclipse' either does not have all the facts, is ignorant of the event, or is purposefully trying to be misleading.

3) On keeping a careful eye on such a person you notice that he spends *all* his time trying to 'debunk' all and any UFO phenomena. There is of course nothing wrong with this if it's done in an honest manner, but when the person resorts to falsity in order to justify their wild claims, then something begins to feel suspicious.

4) When confronted with bold facts concerning such falsity, or the incorrectness of the claims made, the person ignores those articles but continues to post his theory unaltered, particularly to newcomers to the group.

5) The number of postings made by this person are so numerous as to suggest that he/she spends the better part of each day reading and posting articles and even over a span of time of several months the number of postings remains consistently high.

It would seem by now, that either we're dealing with an eccentric millionaire that can afford to spend his days in front of a machine, or that what we have here is a person that is being paid to do this.

How interesting to note then, that none of these eccentric millionaires are pro-UFOs. Especially strange when you consider that the majority of people that contribute to such groups of course are UFO enthusiasts and 'believers'.

So persistent are some of these people in their two-faced and indefatigable attempts at the swaying of newcomers, that some people have even contemplated that these supposed misinformation agents are in fact pro-UFOs. You see, some newcomers that were originally sceptical of UFOs eventually become pro-UFOs at least in part due to the obvious lies and deceit of the misinformation agents!

But how do you, or I, or anyone know that indeed there are such things as government employees that keep tabs on people that want to share information concerning UFOs on the Internet?

First of all, keep in mind that it is a known fact that certain branches of the Federal Bureau of Investigation do indeed have agents scouring the Internet. Generally speaking though, Joe-citizen is told that such persons are there in order to find (and hopefully punish) people that are producing pornographic material involving minors or children and/or who are using the Internet to try and solicit minors and/or children. There is nothing wrong of course with this kind of government involvement. Personally I am all for a free and unregulated Internet myself, but there is no place for paedophiles in my world either.

What it does show though, is that government officials most definitely *are* on the Internet.

Secondly, here is another disturbing piece of news concerning the sharing of information between individuals: the US government wants to ban the use of an encryption program that renders electronic mail unreadable unless you possess the correct software and codes.

Why? Because the program is virtually uncrackable.

But so what? Aren't individuals allowed to have true privacy and confidentiality? Not if the US government has its way. And how badly does it want to have its way? Pretty badly.

The man who wrote the program has been arrested because the encryption program was placed on a public access bulletin board and people outside the US downloaded it and began using it.

This, according to the US government, constituted "...selling arms to foreign countries". So the man went to jail.[2]

Note of course that this is a shareware program, meaning it can be used for no charge and without the need to pay any kind of remuneration or royalties by non-commercial users.

Note too that the man that was arrested was not responsible for the program being in the public access bulletin board. He only wrote it, not distributed it to the world.

There can only be one reason why a simple thing like an encryption program warrants being labelled as "munitions", making the export of it a Federal Crime.[3]

Namely, the US government wants to be able to read anyone's electronic mail at will.

But as usual, don't take my word for it concerning the alleged presence of Net-spies; I may just be one more of a growing number of paranoid delusionals. Buy a computer with a modem if you don't already own one, get an Internet account and see for yourself.

Some relevant newsgroups are listed below.

alt.alien.research
alt.alien.visitors
alt.paranet.ufo
alt.ufo.reports

Although not strictly related to UFOs, the following two newsgroups are also useful to know about. Some serious researchers currently

---

[2] See pg. 35 of PC POWER magazine, issue 9 September 1994 and pg. 20 and 22 of SA Computer Buyer Magazine, Vol 3 No. 9, October 1995.

[3] Apparently simply writing the program seems to constitute a crime!

investigating electrogravitic phenomena may occasionally be found here.

sci.physics.electromag
sci.physics.research

## 2. A Personal Example of Internet Misinformation

Although I believe I have indeed come across at least one or two people that I would expect to be government misinformation agents judging by their attitudes, one of the strangest pieces of misinformation came to me from a newsgroup that unlike one of the UFO related ones, we would not normally associate with net-spies: sci.archaeology.

Having posted a few questions on this newsgroup, I received private e-mail replies from a certain individual who professed to be an archaeologist. This person assured me not only that Herodotus never wrote that the Great Pyramid was described to him as having an artificial lake under it, but also that the Vyse markings were nothing as what was described by Sitchin.[4]

He further added that he was absolutely positive of this last point because he "had access" to the Vyse markings himself and had personally "deciphered them".

It needs to be pointed out that the only way one can see copies of the Vyse markings is either from some photographic evidence that may exist, or, short of having access to the Great Pyramid's chambers themselves, by referring to the archives of the British Museum in London.

No photocopies of the Vyse markings, or indeed of any of the Vyse notes are permitted, and the only way for a researcher to find out how these markings looked like is for them to see firsthand the copies being stored in the archives.

A person has to go through three security checks just to enter the rooms in which these documents are kept and only laptop computers or pencils are allowed in the room, no ball-point pens are permitted for fear that some ink may accidentally get on one of the ancient tomes.

How then a resident of the United States of America can possibly have access to copies of the Vyse markings is unclear, nor was any light shed on this topic by the man in question when I asked

---

[4] In the context of this topic, it should be remembered that while the Internet is a great research tool, none of the information found there should simply be taken as valid. The main aid of Internet access for research purposes is to provide clues to an interested investigator; but like a good police inspector, it is up to the individual to make sure that the clues are real and meaningful.

about it on more than one occasion. But even assuming that the archives had been visited on a trip to London at some date in the past, then the person whom saw fit to e-mail me would still have to explain the gross error he made concerning what Herodotus did or did not say in his writings.

It is easily confirmed that Herodotus wrote about an artificial lake being located under the Great Pyramid by perusing any one of the official translations of his Histories, as I have taken pain to point out in footnote number 4 of chapter three.

The point is that this would be an unseemly oversight for a qualified archaeologist, and thus the case is only one of two, either the misinformation is the result of gross ignorance (the poster was not a qualified archaeologist, was a trickster out for kicks etc.) or the mistake was deliberate and sent with the intention to mislead.

I would find the first explanation to be somewhat unsatisfactory in view of the fact that judging from his grammar, syntax and eloquence, the writer of these (several) posts was obviously a person whom possessed an above average level of education. On the other hand, the idea of misinformation agents being so widespread seems to me to also be somewhat unlikely.

In this particular instance, I feel the answer may lie somewhere between theses two extremes: the sad case of an Egyptologist perhaps, trying to hold on to his coveted theories even by having to resort to a few 'white lies'.[5]

---

[5] In the last few days before going to press, while on IRC in a UFO group on the Internet, a person that had been taking up a lot of space and time with ridiculous claims, made threats of intercepting my transfers of (anti-gravity related) text files to other members of the group. Ignoring him, I continued to distribute the file to my net-friends. Shortly afterwards the IRC program I was using began to behave in a manner I have never seen before, and which cannot be attributed to some kind of malfunction, but was obviously the result of some sort of 'net-bomb' if not a virus. Although no permanent damage to the system seems to have resulted, I was forced to disconnect and reboot the system in order to resume control. Happily, the various computer whizzes of that group have since managed to effectively ban that persistent character. While of course this cannot seriously be considered as proof of a misinformation attempt, the fact that the character was perfectly happy to continue wasting everyone's time with irrelevancies until the point was reached when some useful data was being exchanged, is at least disturbing.

## 3. Rumours (?)

Retired US Army Lieutenant Colonel Thomas E. Bearden,[6] among others, has written a number of books of a rather startling nature.
Broadly speaking they can be classified into three sections, although Bearden touches on numerous releated topics.

1)   Electromagnetic Biological Weapons.
2)   Scalar Electromagnetic Weapons.
3)   Free Energy Systems.[7]

Even more broadly speaking, the concept of electromagnetic weapons with the kind of properties defined by Bearden have resulted in the coining of a new phrase; Psychotronic Weapons.
Whether Bearden is completely correct in all of his assessments concerning the existence and testing of such weapons (in particular by the Soviets) remains to be seen, but it is certainly doubtful that Bearden is making *all* of this up, especially since if one accepts the concepts outlined in the last two chapters, some of these weapons would undoubtedly be a reality.


The ideas concerning electromagnetic weaponry as detailed by Thomas Bearden are somewhat complex, but we can perhaps mention just a few of the more easily defined possibilities if we remember Tesla's wireless power transmission system and add what may or may not be some conjecture to it.
          In order to grasp some of the theories of psychotronic weaponry, we should assume that Tesla's wireless power transmission works by a principle of sending energy into 'hyperspace' where it then travels almost instantaneously to its destination. Here, a suitable receptor then converts the energy back into normal space, and this becomes available for use.
If the person using such technology is of a benign nature we can assume that such energy would be used for similarly benign purposes, but the world being what it is, if there are negative possibilities we

---

[6] Many of the Bearden's books are available through the Tesla Book Company, see the bibliography for their address.

[7] The use of the words 'free energy' are somewhat of an exaggeration since there are some costs involved in the maintenance and construction of these machines, but they are minimal when compared with the energy output supposedly generated by them.

should at least be made aware of them before unleashing such technology on an unsuspecting world. And if this principle of wireless power transmission does indeed work as described in this section, then the negative implications are of the utmost severity.

The most obvious negative use is that a chamber could be designed that would instantly 'hyperspace' vast amounts of energy to a specified location with all the attendant explosive results.

But this is merely the first application of such technology in its crudest form as far as psychotronic weaponry is concerned, and Bearden goes on to describe some truly frightening possibilities.

For example, it may be possible to transmit the energy of a single explosion in this way to several different locations which have receivers set on the same frequency, in rapid sequence. In this way a single atomic explosion could be used to attack any number of targets. And since the receivers contain no explosives themselves and can apparently be as small as a normal transistor radio, it would be very difficult indeed to know where any number of them may be located. No wonder that some of the more paranoid conspiracy theorists (or is it just more informed?) have been going around saying that the Russians are responsible for the implanting of cheap receivers in household TV sets.

Even without getting into the somewhat 'New Age' topic of human auras, it is still a generally accepted and understood medical fact that the human body is an electrical generator, and that the fields it produces are measurable.

More importantly, the fields generated by a person suffering from a disease or other imbalance, would differ from those of a healthy person. And since the symptoms of say a cancer victim are different from those of a person that has ingested poison, the electrical fields would also be different from one patient to another.

In fact the difference should be evident even between similar types of diseases if the equipment measuring the electrical fields is sensitive enough. A brain tumour for example would produce a field that is different from the one of a leukaemia sufferer.[8]

---

[8] Recent research involving study of the Kirilian aura have already proved this basic principle since the gradual change of auric fields from the fingertips of patients were photographed at regular intervals after they had taken medication (tranquillisers or stimulants mainly) and these showed a consistent pattern of change depending on the drug taken.

A biological psychotronic weapon would be able to send out electromagnetic pulses to specific locations which would then scramble the field of a healthy person to resemble the one of a sick person.

Enter germless warfare. Which although 'bug free', nonetheless results in the deaths of those unfortunate enough to have their personal fields altered.

Obviously it is also possible to reverse the process, so that a sick person's field can be made to vibrate at the correct and healthy frequency. Theoretically, such a process may result in the almost complete redundancy of the multi-billion dollar industry of medicinal drug production.

Since the germs, viruses or cancer cells that make a person sick would be unable to exist if subjected to a healthy field, what effectively would occur is the killing off of such parasites without any harmful side effects to the patient.

Holistic medicine would have its ultimate revenge! Who could have imagined a system that would effectively cure a person of any disease simply by 'cleaning their aura'?

Certain surgical procedures would of course remain unavoidable, but healing after the operation could still be speeded up by use of the same 'aura cleaner'.

Although not as devastating as some of the other psychotronic weapons, the biological implications somehow make one's hair stand on end.[9] If we ascribe to these principles for example, it is perfectly possible not only to kill off vast number of people by apparently 'natural' means, but it's also possible to conduct a more subtle kind of warfare.

By making all the people living around nuclear reactors in an enemy country more susceptible to certain types of cancer, the owner of psychotronic weaponry can influence how nuclear reactors are viewed by the population of that country.

The few examples mentioned here by the way are only the most obvious ones and far from the worst that are apparently possible if Bearden is to be believed.

Interestingly enough (and slightly terrifyingly so) the Soviets have most undoubtedly been involved in large scale experiments with what

---

[9] One only needs to know the title of one of Bearden's books for the idea of 'incurable lab diseases' to sink in: *Aids: Biological Warfare*.

the paranoid subculture of conspiracy theorists has come to know by the acronym ELF.[10]

Extremely Low Frequency vibrations are said to have mind altering or even controlling properties, and while this may not have been proven in the overt laboratories of science, no real scientist would state that such a claim is wholly unfounded.

The Schumann resonance, which is the frequency at which the whole planet vibrates with electrical energy, is the same one found in the human brain wave range of frequencies, and intense or prolonged exposure to electromagnetic pulses of similar frequency seem to result in nausea, headaches, mood swings and 'ringing' in the ears.

Interestingly enough again, we should keep in mind that safety standards for non-ionising radiation are far more stringent in Russia than they are in Europe or the USA, and one can't help wonder if this is perhaps as a result of the deeper insights into such phenomena that the Ex-U.S.S.R. may have compared to the rest of the world.

Although with similarly catastrophic possibilities as the other psychotronic weaponry just mentioned, Bearden also writes about the concept of receiving what for all intents and purposes could be construed to be free energy.

The idea behind an instrument that will provide almost unlimited energy for negligible cost is one that has hounded modern man in this century in a way not dissimilar to that of powered human flight in the last. And who's to say that the close of the twentieth century will not, like that of the nineteenth, herald a completely new and astounding science.

---

[10] In the 70s Canada in particular, but the UK as well, had some periods of very bad reception on all their radio and TV receivers (if memory serves me correctly). When investigated, the triangulation of the origin of the ELFs which produced this effect around the world pointed to a location in the U.S.S.R.. The explanation given was that some experimentation in low frequency vibrations was being undertaken although details were never disclosed and the testing seemed to have stopped shortly after the complaints. Although I remember reading this in at least two different places, one of which also quoted newspaper reports on the subject, it is one of those topics I did not have the time to investigate further and unfortunately, the original sources are lost in the depths of my memory.

The principle behind a 'free energy' machine is relatively simple to explain, although the implementation of it in practical terms may be a somewhat more difficult matter.

A useful analogy reflecting the principle of such a system —and one I shall adopt here— was presented in a documentary produced by The Royal Atlantis Film which concerned itself with such topics as UFOs as being extra-terrestrial in nature as well as the ideas behind the concept of antigravity.

There are a great deal of very serious physicists that believe that the vacuum of space contains incredible amounts of energy. So vast in fact, that if the energy that is theoretically found in a volume of space as large as one cubic centimetre (0.0610 cubic inches) were to be released, it would obliterated Earth if not the entire solar system.

The reason why we do not experience such incalculable amounts of power in our everyday lives, is easily understood when we consider a tiny bird, say a sparrow, sitting comfortably on a high tension cable.

Unbeknownst to the little bird, vast quantities of power are flowing right under his tiny feet, but he's perfectly safe as long as he remains at the same electrical potential. Should the unfortunate bird in some way manage to rest one of its claws on the wire at the same time that its other foot made contact with the ground (perhaps by means of a small chain) we can rest assured at least, that the little guy would not suffer, because —to quote the original version of this analogy— "He would be fried to a greasy spot."

In a similar way, we remain perfectly safe and unaware of the veritable maelstrom of energy which surrounds us because we are effectively sitting in it.

In our case, the analogy may be more akin to a fish that on hearing that the sea is a most powerful force, goes around asking "Where is the sea?".

If however we could build some kind of device that has the ability to create even just the tiniest potential difference and thus have access to the unlimited energy of the Universe, even the most inefficient system would still be able to provide Humanity with all the energy it might ever need.[11]

---

[11] The three part documentary of the Royal Atlantis Film (Titled Mysteries of the Unknown) mentioned that even a level of efficiency in conversion as low as ten to the power of negative eighty would still provide more than enough energy for all human purposes.

In my personal opinion, there is another very good indication that this theory of unlimited energy existing all around us is a good one.

The Universe.

Although this is a personal belief and as such its weight may be negligible for some readers, it is my opinion that the whole 'Big Bang' experience that created our Universe, was nothing more than the tiniest 'rip' in the fabric of space time.

This tiniest 'rip' or potential difference allowed an enormous amount of still expanding energy to come into being in 'our' state of physical existence.

Over the aeons this energy spillout formed the present planets, stars and galaxies and in fact our whole 'Universe'.

Exactly what made this tiny rip, fluctuation or disturbance in space-time is open to speculation, but the answer "God" would certainly seem to be as good (and in many ways better) as "A storm in hyperdimensional space".

Of course the wonder-machines of Thomas Bearden have to remain in the realm of rumours for now, but it would be unwise to dismiss them completely. In fact, a little thought concerning the topic of antigravity will lead an investigator to the conclusion that Bearden's claims, outlandish as they may seem, are almost a certain possibility is we assume that antigravity itself is possible.

And this of course, would also further explain why such technology has been suppressed so thoroughly and for so long, as well as why only a gradual introduction of it, accompanied by extensive social re-education can hope to result in its being used solely for beneficial purposes.

# 4. Net Rumours

I include here three different articles culled from public information locations on the Internet. I make no claim to either believe or disbelieve their contents, as I have not personally investigated them, but I would like to point out that they are certainly interesting and particularly the first and last articles would seem to merit further research.
No corrections of grammar or syntax have been performed on these posts; partly in order to preserve their original 'flavour', but also so as not to accidentally interpret a misspelled word or badly construed phrase to mean something more than what it intended.

### Article No.1 (received by e-mail from an Internet friend)

Copy and Distribute Freely CENTER OF ATTENTION July 26, 1995
2221 Bowers Ave.,
Santa Clara,
Ca. 95051 (408) 241-7981 and FAX

The Bi-Weekly News and Comment of the Ascension Movement
(excerpts from ) ISSUE 17

The Looming End to The UFO Cover - Up Part 2 By Richard J. Boylan, Ph.D.

Astronaut Gordon Cooper meanwhile has been cooperating with Irish
film producer Jackie Dunn, who is working with a Canadian film production company to have a major UFO documentary film ready by summer, 1995.
Dr. Steven Greer, Director of CSETI (the Center for the Study of Extraterrestrial Intelligence), is reported to be cooperating with Jackie Dunn on this film.
        Another film due out soon shows dramatic camera footage shot at Clear Lake, CA, showing a UFO coming over Mt. Konocti and proceeding cross above the lake, then stopping and shooting straight up out of sight.
        In 1993, Laurence Rockefeller, long an interested follower of the UFO phenomenon, assisted by his "point man" on UFO matters, Cdr. Scott Jones, Ph. D., USN-ONI (Ret.), former career intelligence officer, began the White House Initiative to get the current Administration to reveal to the public promptly, yet responsibly, what the Government knows about UFOs and ET visitation.  Rockefeller and Jones met with Clinton White House Science Advisor Dr. John Gibbons in March, 1993,

and presented a "Matrix of UFO Belief" analysis monograph. A follow - up visit by Mr. Rockefeller took place with Dr. Gibbons at the White House Feb.4, 1994.
Thereafter, it is reliably reported that President Clinton agreed to the White House Initiative.

Subsequently, White House Science Advisor Gibbons has been asked to find out everything about UFOs. A recent Freedom of Information Act requested directed to Dr. Gibbons yielded a cache of UFO - related correspondence between Dr. Gibbons and Laurence Rockefeller, Cdr. Scott Jones, U.S. Air Force officials, Former French Government UFO investigations office (GEPAN) official, Dr. Jacques Vallee, and former Defense Secretary Melvin Laird. Secretary Laird is the official who also authored a letter (I have seen) to incoming Clinton Administration Defense Secretary Les Aspin, offering to brief him about UFO matters.

Dr. Jacques Vallee, in April, 1995, briefed Dr. Gibbons on his knowledge of the UFO phenomenon, and on what President Clinton does not have access to about UFOs. (Just like President Carter was denied access to UFO information which the CIA had, by then - CIA Director George Bush; so also President Clinton is not satisfied with the briefings he has gotten from the CIA, because they conflict with the information which his Science Advisor has acquired.)

Dr. Steven Greer of CSETI has also provided two White House briefings to Science Advisor Dr. Gibbons on his UFO findings.
Dr. Greer has been making international contacts with European governmental leaders to attempt to achieve consensus on an international joint statement acknowledging UFO and ET visitation reality. Dr. Greer engages in this project in concert with the BSW Foundation and Marie Galbraith, a good friend of Laurence Rockefeller, and presumably with the support of Prince Hans - Adam of Liechtenstein, a longtime advocate and patron of the release of UFO information. (Prince Hans-Adam and Las Vegas entrepreneur Bob Bigelow co-financed the 1992 Abduction Study Conference at MIT.)

On June 5, 1995 at a Bay area lecture, Dr. Steven Greer revealed further findings. He has received leaked information that the Air Force, through its North American Air Defense Command (NORAD) facility deep inside Cheyenne Mountain, Colorado, has t racked an average of 500 "fastwalkers" (UFOs) entering the Earth's atmosphere from deep space annually.

Dr. Greer has talked to a fellow medical doctor, the nephew of a pilot with General Jimmy Doolittle, who was commissioned by President Harry Truman to report on Foo-Fighters during World War II.
(Foo-Fighters were levitating lighted globes, perhaps a meter in diameter, which silently paced Allied and Nazi fighters and bombers during WW II, and were most likely of extraterrestrial origin.)

Two separate Intelligence sources have told Dr. Greer that unaccountable units in covert compartments of military agencies are directing Black Budget weapons systems at UFOs, and have downed two UFOs in the past two years. One source, a physicist in the Office of Naval Intelligence, said that this shooting - down of nonhostile UFOs is done by out - of control, arrogant "cowboys", using "unimaginable technologies", because they (UFOs) are there. Dr. Greer went on to note that key members of the government, military, intelligence and international leaders are "out of the loop" on detailed UFO information, and not being informed about hostile actions directed against these Visitors.

Former NATO (SHAPE) Intelligence Sergeant Robert Dean has revealed the contents of an extremely - classified SHAPE 1964 study of UFOs and ETs, which stated, among other things, that UFOs are real extraterrestrial spacecraft, that at least 8 - 12 ET races are visiting Earth, That their intentions do not seem warlike, and that it appears that they have been studying the Earth for a very long time.

Mr. Dean has assembled 20 astronauts, former intelligence officers, generals and admirals who are willing to testify to a Congressional Committee about what they know about UFOs, provided that they are released from their National Security oaths. The videotaped depositions of sworn key witnesses have been taken by a prestigious Washington, D.C. law firm, and stored in its safe, awaiting public hearings.

Mr. Dean is working with CSETI'S Dr. Greer, in concert with former astronaut Gordon Cooper, another high - ranking military officer, and a General, to plan the release of UFO information to which they are privy.

Dr. Greer and Sgt. Major Dean are part of a Coalition which has been putting together the best evidence of UFO/ET reality. The evidence includes not only military/autopsies in ET corpses, but also fighter pilots, generals, astronauts and cosmonauts who have witnessed UFOs close-up, as well as UFO metal samples and ET tissue samples.

The Coalition's plan is to take their Briefing Document and evidence to world leaders, the U.N., scientific academies, and religious leaders for a pre - briefing. Then the Coalition will make a Public Disclosure within the next 12 months.

Dr. Greer reports that the White House, the Joint Chiefs of Staff at the Pentagon, and the United Nations are being enlisted to assist, and no one has said that this cannot come out.

There is an unconfirmed report that a Presidential Release form his National Security Oath is under consideration for astronaut Gordon Cooper, to enable him to tell what he knows about UFOs.

On Friday, May 5, 1995 British TV producer Ray Santilli held a bidding and press conference at the London Museum to announce the first presentation (20 minutes) of excerpts from the 14 reels of 16mm military -

intelligence film he has come into possession of, which include scenes from various UFO crash retrieval operations by classified Army and Air Force special units, along with scenes of autopsies in progress on several extraterrestrial corpses.

Mr. Santilli states that he was given this film by Jack Barrett, an 82 year old Army photographer, who says that he kept back a personal copy of the classified footage he shot at the Rosewell UFO crash retrieval operation in July, 1947. But Christopher Cary, a business associate of Mr. Santilli, feels that account does not fit, and that there are indications that the films came by way of Intelligence agencies "leaks". Furthermore, supposed time - discrepancies in the 20 - minute segment shown May 5, (coiled telephone cord, hand held camera and possibly zoom lens), may be inconsistent with that segment being of the Rosewell 1947 UFO crash.

Furthermore, the six - fingered, 5-6 foot tall ET shown in the autopsy film presented on May 5 is not the same race as the four - fingered, 3 1/2 foot tall ET shown in a different film reel. Such anomalies suggest that films may actually be of different UFO crash - retrieval operations, some at a later date that Rosewell.

These inconsistencies may point to a sophisticated negative - misinformation campaign by American intelligence compartments to divert attention form mushrooming civilian evidence about a 1947 UFO crash at Rosewell, by foisting onto an unsuspecting Mr. S antilli and the world public some hoaxed film. Alternatively, the inconsistencies may point to an even more sophisticated "positive - misinformation" campaign, in which actual military - intelligence footage of a variety of UFO crash retrievals of various places and dates are cobbled together under the rubric of "Rosewell", so that it will require discernment to distinguish the Rosewell scenes from the other UFO sites and times. This latter strategy would permit fearful disbelieves to reassure themsel ves that Roswell "didn't happen", by pointing to scenes in certain film sequences inconsistent with investigated Rosewell facts.

Attending the May 5, 1995 Santilli press conference were: BBC - TV, BBC Radio, NC, a reporter form the American TV documentary "Sightings", and various British newspapers, including the Sunday People, The Sunday Express, etc. The Times of London had already run a story about the upcoming presentation of the alien autopsy film.

According to the NBC crew in attendance, the 20 - minute excerpt depicted a grainy image of an autopsy taking place in a poorly – lit space. This segment is reportedly the second of three autopsies which are shown on various film reels. There are three humans shown in a hospital - like environment. Two people are wearing "radiation - type" (biohazard-isolation) white suits, while a third person is behind a glass partition, looking on.

The two in isolation suits are working on a humanoid body shown on a black slab. The ET body is humanoid in appearance, and has a height of 5 - 6 feet. Her head is larger than a human's, and is shaped differently. She has large, black eyes. The torso is extremely skinny. Extremities have six fingers and six toes. One of her legs is charred, (presumably form the crash). There is no cranial or axillary hair. The body appears muscular. There is no rib cage. Ears, nose and mouth are extremely small and rudimentary by human comparison. There is no navel. Female genitalia are noted, but there are no apparent breasts or nipples. One researcher feels that this corpse may be a hybrid of ET - human origins, (because of the genitalia, [not ET], an d the absence of breasts/nipples, [not human].)

Surgical instruments, a clock, a telephone, and a warning sign, "Danger", were seen. One of the two isolation - suited workers is shown taking notes. On the heading of the notes, the name "Dr. Bronk" [Detlev Bronk, M.D. of Johns Hopkins University Medical School?] reportedly can be made out. Incisions are made. One incision goes up the thorax, and the heart and some intestines are removed. The pathologist removes what appears to be a geometric crystallization from the abdomen. The film shows f urther incisions on the head to remove skin samples, and the skull is opened. When the pathologist cuts into the head, no skull was in evidence, but rather a jelly - like, plastic - like material reportedly was found.

Another film segment, seen earlier by Phillip Mantle, an official with the British UFO Research Association (BUFORA), shows an autopsy involving an extraterrestrial who is 3 1/2 to 4 foot tall, and who has four fingers at the end of his (?) upper limbs.

Reactions from viewers of the May 5 film preview include: "The footage is certainly compelling", (NBC reporter); "I'll be saying, 'Here it is; make your own mind up.'" (Sunday People newspaper).

The same day as the press conference, BBC Radio broadcast a report on it.

A copy of the film is reportedly being presented for examination by a pathologist from the British Home Office, (the equivalent of the U.S. State Department). Mr. Santilli is pledged to give the Kodax Company some film to analyze (as being of 1947 vint age). BBC - Television is producing a TV documentary on the UFO crash retrieval film which Mr. Santilli has, which will be broadcast in August. A clip from these saucer crash site recovery operations films is being aired by the BBC in June.

A much lengthier screening from these 14 reels of UFO crash retrieval and ET autopsy operations will be shown on August 19 at the BUFORA International Conference at the University of Hallam, England. This screening was announced on CNN in late April, four months in advance!

A prestigious three - day international conference was held at the Sheraton Washington in Washington D.C., May 27 - 29, on the subject of what should be the Earth's proper response, when it comes out in the open that we are being visited by cultures f rom elsewhere. The "When Cosmic Cultures Meet" International Conference featured presentations by scientists, academics, governmental leaders, research professionals, military officers, journalists and religious spokespersons.  This major, world - class conference, revealing the solid acceptance by political, academic, scientific and journalism figures of the realism of preparing for extraterrestrial contact, provided a number of compelling statements and revelations.

Arlington Institute National Security specialist John L. Petersen compared the current shift in society and culture involving dramatic breakthroughs in energy sources, ET contact, and technology, to the shift from the Middle Ages to the Enlightenment.

Anthropologist - journalist Michael Hesseman likened ET contact to a second Copernican Revolution.  He also reported that Soviet KGB UFO files have now become public, revealing, for example, that in 1989 a UFO hovered two hours over a Soviet nuclear w eapons storage facility, until finally a MIG fighter came and caused its departure.

Harvard psychiatrist Dr. John Mack presented arresting videotapes of his interviewing of Southern Africa fourth graders who witnessed a UFO set down at the edge of their recess playground.  Then several Ets emerged, one approaching within three meters ( nine feet) of one schoolgirl as others watched.

Washington Post journalist Ruth Montgomery related how she had received multiple reports about UFO reality from various military officers with whom she had spoken.

The producer of Mexico's "Sixty Minutes" television documentary program, Jamine Maussan, showed extensive videotapes of UFOs over Mexican population centers.  These videotapes show structured craft, flotillas of UFOs, and vertical column of UFOs inside a translucent plasma? field.

The most arresting footage is a nighttime shot of a UFO hovering near the ground, and later, a "Praying - Mantis" - type extraterrestrial, illuminated, is seen turning towards the camera from perhaps one block away.

Sumerian culture expert Zecharia Sitchin presented evidence from ancient Sumerian tablets for previous contact between extraterrestrial and modern humans' ancestors.

The convener of the Cosmic Cultures Conference, Dr. Scott Jones, plans to make videotapes of the Conference available to key governmental figures, so that they can be aware that representatives of the scientific, academic, professional and journalistic communities are well - informed and accepting of UFO and extraterrestrial reality.

A source in contact with CIA's UFO Desk within the Directorate for Science and Technology reports that President Clinton has already made several pilot videos of proposed alternate announcements about UFOs and ET - presence/visitation.

In view of multiple pressures compelling disclosure of UFO reality, and the exigencies of Presidential election year politics, Congressman Phil Gramm and Senator Bob Dole had perhaps best be prepared for the possibility of the Democratic incumbent launching an unprecedented kind of "October surprise".  In this vein, one UFO investigator, Dan Smith, has entered into discussions with the Dole Presidential Campaign staff, offering them his informational services, so that Mr. Dole can minimize his risk of being blind - sided by the UFO Cover - Up/Disclosure Issue, which may become the sleeper issue of the 1996 Presidential Campaign.

In an apparent effort to avoid being taken by surprise by the politics of UFO disclosures, Republican National Chairman Hailee Barbour and other RNC officials reportedly plan to contact various astronauts to learn from them what they know about UFOs.

Multiple    previously   -   reliable   sources   expect   that   an announcement of UFO reality will occur in the Fall of 1995, possibly in October.  These sources also expect that a formalized meeting of human and ET representatives will occur in 1996.  Subsequent events in 1997, including a predicted announcement of the forthcoming establishment of the first openly - declared, official ET community on Earth, will make it virtually impossible by 1996 for anyone to deny extraterrestrial reality."

Richard J. Boylan, Ph. D.

You can reach Dr. Richard J. Boylan at:
2826 O Street, Suite 2,  Sacramento, CA  95816
(916) 455-0120

## Article No.2 (Text on WWW)

UN General Assembly decision 33/426, 1978

Establishment of an agency or a department of the United Nations for undertaking, co-ordinating and disseminating the results of research into unidentified flying objects and related phenomena.

At its 87th plenary meeting, on 18 December 1978, the General Assembly, on the recommendation of the Special Political Committee adopted the following text as representing the consensus of the members of the Assembly:

"1. The General Assembly has taken note of the statements made, and draft resolutions submitted, by Grenada at the thirty-second and thirty-third sessions of the General Assembly regarding unidentified flying objects and related phenomena.

"2. the General Assembly invites interested Member States to take appropriate steps to coordinate on a national level scientific research and investigation into extraterrestrial life, including unidentified flying objects, and to inform the Secretary-General of the observations, research and evaluation of such activities.

"3. The General Assembly requests the Secretary-general to transmit the statements of the delegation of Grenada and the relevant documentation to the Committee on the Peaceful Uses of Outer Space, so that it may consider them at its session in 1979.

"4. The Committee on the Peaceful Uses of Outer Space will permit Grenada, upon its request, to present its views to the Committee at its session in 1979. the committee's deliberation will be included in its report which will be considered by the General Assembly at its thirty-fourth session."

**Article No. 3 (taken from alt.alien.visitors)**

In reply to a sceptic commenting on this subject, poster 1 makes his comments. These are followed by the reply of poster 2.
Poster 3 apparently did some research concerning the identity of the scientists concerned, and poster number 4 adds a last note on the subject of scientists being detained by the National Security Agency.

>.....................< Comments from contributor No. 1
>>................<< Comments from contributor No. 2
>>>............<<< Comments from contributor No. 3
>>>>........<<<< Comments from contributor No. 4

**From:** (personal e-mail address of poster removed by author)
**Subject:** Re: Scientists detained by NSA after discovering gravitational anomaly in same city as NSA headquarters.
**Summary:** (personal name of poster removed by author)
**Date:** Thu, 24 Aug 1995 05:56:27 GMT

> .....if the poster elaborates, you, personally, would have to verify it before you could believe it anyway. You certainly wouldn't trust anyone else, so ultimately it is up to each one of us individually to verify whatever to our own individual levels of satisfaction. You always provide commentary first, before you even ask a question. This smacks of laziness and an axe to grind. <

>> Excellent points. I tried to verify the existance of these guys Gerald Ollman and Robert Wayne via the Internet WEB page for the University of Maryland [ http://www.umd.edu ]. Unfortunately, I couldn't access the Phone Books section because my web browser hasen't implemented the CSO protocol. However, I couldn't find  these guys anywhere, nor the department of Geophysics (they may just not be online). A search of the Libraries at U of M turned up no papers written by either of them. Maybe they are graduate students. Who knows. However, my search wasn't really intensive, I just used the easiest thing at my finger tips. That's all the effort I'm going to put in now. <<

>>> I was able to find the following:
University of Maryland, Eastern Shore
TULL, ROBERT WAYNE
office_phone: (410) 651-6091
office_location: J.T. WILLIAM HALL (EXECUTIVE OFFICE) B210
department: COMPTROLLERS OFFICE
title: ACCOUNTANT III UM

That was the result of a search for "Robert Wayne".
No matches at all for "Gerald Ollman".
Not that I have any particular interest in this thread, just happen to be using a web browser that works, so there it is. <<<

>>>> I just got an email response from the source of the article. He said he heard this from an associate of these professors at the university, and that there was no media coverage. Whatever that's worth.<<<<

## 5. Net Data

On posting a request for information concerning antigravity research or information to the newsgroup sci.physics.electromag I received a reply by a helpful individual who e-mailed me a long text file containing a vast number of references, some where in the form of Internet addresses, titles of books, or numbered technical papers, but amazingly, the names and addresses of a few research groups supposedly involved in antigravity research or technology where also included. Due to the length of the text file (233 Kb) it is not possible for me to reproduce it fully here, but I have compiled a small selection of entries below, and I hope this would prove useful to anyone interested in researching this topic further.

### Internet Sites

Elektromagnum web site by David Jonsson:
http://nucleus.ibg.uu.se/elektromagnum

KeelyNet:
http://www.protree.com/KeelyNet/

Los Alamos National Lab Physics E-Print Archive:
http://xxx.lanl.gov/

Center for Gravitational Physics and Geometry:
http://vishnu.nirvana.phys.psu.edu/

Embry-Riddle Aeronautical University's Aerospace Virtual Library:
http://macwww.db.erau.edu/www_virtual_lib/aerospace.html

Jack R. Hunt Memorial Library (aerospace):
http://amelia.db.erau.edu/

American Institute of Aeronautics & Astronautics (AIAA) home page:
http://www-leland.stanford.edu/group/aiaa/national

NASA Langley Research Center Library:
http://blearg.larc.nasa.gov/library/larc-lib.html

NASA Scientific and Technical Information:
http://www.sti.nasa.gov/STI-homepage.html
Levesque's (laurent@ee.umanitoba.ca) web site:

http://www.ee.umanitoba.ca/~laurent

UFOs and the New Physics:
http://www.hia.com/hia/pcr/ufo.html

Interstellar Propulsion Society:
http://www.digimark.net/ips/

## Patents

Patents for anti-gravity devices and systems have been issued to
Brown, Hooper, and Wallace.
A less well known one is by Hooper -- patent No. 3,610,971.

Also:
   US patent No. 1,006,786, Piggott.
   US Patent No. 5142861, Nonlinear Electromagnetic Propulsion System,
            R.L. Schlicher et al. 1992.
   US Patent No. 2,886,976, System For Converting Rotary Motion Into
            Unidirectional Motion, Dean, Norman L.,
            (Dean Drive).
   US Patent No. 4,631,971, Apparatus For Developing A Propulsive Force,
            Thornson, Brandson R.
   US Patent No. 5,024,112, Gyroscopic Apparatus, Kidd, Alexander D.
            (Aka Sandy).

## Journals, Papers and Publications

I have recently come into possession of a paper on magneto-gravitics and
field resonance systems, presented by A.C. Holt from NASA Johnson Space
Center to the American Institute of Aeronautics and Astronautics' 16th Joint
Propulsion Conference, June 30-July 2, 1980.
Holt presents a project using an already existing system known as the
Coherent Field and Energy Resonance System (CoFERS) [probably located
at Los Alamos Labs' High Magnetic Field Research Laboratory].
CoFERS utilizes a toroidal-shaped energy guide with megagauss magnetic
field sources located along radius vectors equally spaced around the toroid.
CoFERS is shaped like a thick flying disc.
Holt goes on to say: "By converying an object's normal space-time energy
pattern to an energy pattern which differs substantially from the normal
pattern, the gravitational forces acting on the object are changed. The object's
new pattern interacts with the surrounding space-time and virtual energy
patterns, such that the interactive forces are substantially altered. The

alteration of the characteristics of the continuous field of force results in the apparent motion of the object *through space-time*."

[...] "Since the gravitational forces acting on the propulsion system can be quickly altered to achieve the desired motion, the *spacecraft* can make right-angle turns at very-high velocities without adversely affecting the crew or system elements. The effective gravitational field the *spacecraft/aircraft* experiences can be nearly simultaneously reoriented at a 90-degree angle, resulting in a smooth continuous motion as far as the occupants are concerned." [ ... ] "The gravimagnetic system is perhaps best suited for use in and around ... a large mass such as the Earth."

"While the gravimagnetic system is likely to be the first field-dependent propulsion system developed, the field resonance system will **bring stellar and galactic travel out of the realm of science fiction**. The field resonance system artificially generates an energy pattern which precisely matches or resonates with a virtual pattern associated with a distant space-time point. According to the model, if a fundamental or precise resonance is established, (using hydromagnetic wave fine-tuning techniques), the spacecraft will be very strongly and equally repelled by surrounding virtual patterns. At the same time, through the virtual many-dimensional structure of space-time, a very strong attraction with the virtual pattern of a distant space-time point will exist. ...this combination of very strong forces will result in the translocation of the spacecraft from its initial position through the many-dimensional virtual structure to the distant space-time point. [ ... ] "A space-time 'jump' already appears to be supported by astrophysical research."

Should you wish the entirety of this report, "Prospects for a Breakthrough in Field Dependent Propulsion" by A.C. Holt, you can order it from:
AeroPlus Dispatch,
1722 Gilbreth Road,
Burlingame, CA 94010;
phone: (800)-662-AERO.
It'll cost $11.50. AeroPlus is a sales agency for AIAA documents.
- Rich Boylan

---

There is a reprint of an article that appeared in "Interavia, Volume XI - No. 5, 1956" a March 23, 1956 article titled "Towards Flight without Stress or Strain...or Weight" This article has a photograph of T.T.Brown holding one of his flying disks, and another photograph of the flying disk by itself. There is some info on the opperation of the electrokinetic apparatus.

"Anti-Gravity Electronics", H. Aspden, Electroncis & Wireless World, Jan 1 1989, Vol 95 No 1635

"The Latest Antigravity Gossip", Rock & Ice, Nov 1 1994 No 64
"Propulsion by Gyro", Eric Laithwaite, Space, Sep 1989 Vol 5 No 5

In an attempt to reveal the strange, hidden properties of gyroscopes, Professor Eric Laithwaite explains the physics behind the idea that a propulsion system could be built using gyros.

"Negative Mass in General Relativity", H. Bondi, Reviews of Modern Physics, Vol 29, July 1957, pp 423-428

"Looking for New Gravitational Forces with Antiprotons", M.M. Nieto and B.E. Bonner, Proceedings RAND Workshop on Anti Proton Science and Technology, World Scientific, Singapore, 1988 pp 328-341

Scott, W.B.   "Black World engineers, scientists encourage using highly classified technology for civil applications."
Aviation Week & Space Technology, March 9, 1992, pp. 66,67.

Brown, T.T.  "How I Control Gravity."  Science and Invention Magazine, August 1929. Reprinted in Psychic Observer 37(1) pp.14 - 18.

Burridge, G.  "Another Step Towards Antigravity."
The American Mercury 86(6) (1958):77 - 82.

Sigma, Rho, "Ether Technology: A Rational Approach to Gravity Control."
Lakemont, GA: CSA Printing & Bindery, 1977, p. 44-49,
quoteing a letter from T. Townsend Brown dated February 14, 1973.

Intel.  "Towards Flight Without Stress or Strain...Or Weight."
Intervia Magazine 11(5) (1956):373-374

Rose, M.  "The Flying Saucer: The Application of the Biefield-Brown Effect to the Solution of the Problems of Space Navigation."
University for Social reesearch, April 8, 1952.

LaViolette, P.A.  "An Introduction to Subquantum Kinetics: Part Journal of General Systems, Special Issue on Systems Thinking in Physics" 11(1985):295-328.

LaViolette, P.A. "Beyond the Big Bang: Ancient Myth and the Science of Continuous Creation." Rochester, VT:Inner Traditions Intl., 1994.

LaViolette, P.A.  "A Theory of Electrogravtics." Electric Spacecraft Journal, Issue 8, 1993, pp. 33 - 36.

LaViolette, P.A. "A Tesla Wave Physics for a Free Energy Universe."
Proceedings of the 1990 International Tesla Symposium,
Colorado Springs, CO: International Tesla Society, 1991, pp. 5.1 - 5.19.

Aviation Studies (International) Ltd. "Electrogravitic Systems:
An Examination of Electrostatic Motion, Dynaimc Counterbary and
Barycentric Control." Report GRG 013/56 by Aviation Studies, Special
Weapons Study Unit, London, February 1956.
(Library of Congress No.3,1401,00034,5879; Call No. TL565.A9).

LaViolette, P. "Electrogravitics: Back to the Future." Electric Spacecraft
Journal, Issue 4, 1992, pp. 23 - 28.

LaViolette, P. "Electrogravtics: An Energy-Efficient Means of Spacecraft
Propulsion." Explore 3 (1991): 76 - 79; idea No. 100159
submitted to NASA's 1990 Space Exploration Outreach Program.

Aviation Studies (International) Ltd. "The Gravitics Situation".
prepared by Gravity Rand Ltd. - a divison of Aviation Studies,
London, December 1956.

Northrup Studying Sonic Boom Remedy." Aviation Week & Space
Technology, Jan. 22, 1968, p.21.

Rhodes, L. "Ex-NASA Expert Says Stealth Uses Parts from UFO."
Arkansas Democrat, Little Rock, AR., April 9, 1990.

NEXUS Magazine Volume 2, Number 17
P.O. Box 177
Kempton, IL 60946
Phone: (815) 252-6464
Fax: (815) 253-6300

Extraordinary Science
Volume VI. Issue 2
ISSN 1043-3716

SPACE ENERGY NEWS
The Space Energy Association is dedicated to the pioneering work of several
entists and inventors, including Nikola Telsa, Viktor Schauberger, T Henry
Moray, T.T. Brown, Alfred Hubbard, T.J.J. Erwin Saxi, Hans Coler and
others.
PO Box 11422
Clearwater FL 34616
Suscription info:$35, 4 issu/yr.,
Editors: Donald A. Kelly See, Michael Marino
Look up a paper published by Miguel Alcubierre in Classical and Quantum
Gravity 11 (1994) pp. L73-L77. It's titled "The Warp-Drive: Hyper-Fast

Travel within General Relativity". If you can follow a lot of math (or at least the gist of it) it is fascinating.

---

ELECTRIC SPACECRAFT JOURNAL          $24 yr.
US P.O. BOX 18387                    $29 yr. Canada/Mexico
Asheville, North Carolina 28814      $39 yr. other countries
I highly recommend this semi-pro publication.
Has articles on energy anomalies, Tesla, unconventional
hobby projects, unconventional physics, etc.

---

AN EXPERIMENT TO TEST THE GRAVITATIONAL AHARONOV-BOHM EFFECT
Ho, Vu B.   Morgan, Michael J.   Monash University, Clayton, Victoria, Australia 1994 8 PAGES, Australian Journal of Physics
(ISSN 0004-9506) vol. 47, no. 3 1994 p. 245-252 HTN-95-92507
The gravitational Aharonov-Bohm (AB) effect is examined in the weak-field approximation to general relativity. In analogy with the electromagnetic AB effect, we find that a gravitoelectromagnetic 4-vector potential gives rise to interference effects. A matter wave interferometry experiment, based on a modification of the gravity-induced quantum interference experiment of Colella, Overhauser and Werner (COW), is proposed to explicitly test the gravitoelectric version of the AB effect in a uniform gravitational field.
CASI Accession Number: A95-87327

---

Electric Spacecraft Journal, 73 Sunlight Dr, Leicester, NC 28748,
704.683.0313 Voice / 704.683.3511 FAX / 615.952.5638 BBS
Published since 1991, Quarterly, Subscription Rate: $24/Yr
(Only U.S. publication dedicated to Space Drives R&D)

---

Stine, G. Harry, "Detesters, Phasers, and Dean Drives,"
ANALOG, Jun 1976, pp. 68-8O
All those interested in advanced propulsion concepts should check out Jane's Defence Weekly, 10 June 1995. An article discusses anti-gravity schemes and shows drawings of sauceroid vehicles from British Aerospace among others. Area 51 is mentioned, as well as an unclassified paper done for the USAF by Science Applications International Corp. in 1990. The subject was [Electric Propulsion], a[n] euphemism for anti-gravity according to Jane's.

## Books

The paper entitled the "U.S. Antigravity Squadron" paper appears with others in the book:
ELECTROGRAVITICS SYSTEMS: Reports on a New Propulsion Methodology
Edited by Thomas Valone (Washington, D.C.: Integrity Research Institute, 1994); ISBN 0-9641070-0-7.

In addition to this paper, this book also includes the following:
1) The 1956 paper "Electrogravitics Systems" (prepared by the Special Weapons Study Unit of Aviation Studies Ltd., a UK-based aviation industry intelligence firm). It was declassified from a confidential status some time prior to 1985 and entered the public domain as a result of a request I placed through the Wright-Patterson Air Force Base Technical Library.
2) The 1956 paper "The Gravitics Situation" (prepared by Gravity Rand Ltd., a division of Aviation Studies Ltd. This includes six appendices with papers by various authors including the text from T. Townsend Brown's 1929 gravitor patent.
3) A paper by Banesh Hoffman entitled "Negative Mass as a Gravitational Source of Energy in the Quasistellar Radio Sources.
4) A collection of diagrams copied from various patents by T. Townsend Brown.
You may order a copy from:
Starburst Publications, 1176 Hedgewood Lane,
Schenectady NY 12309, USA

Also available from Starburst Publications is the book
"Subquantum Kinetics: The Alchemy of Creation"
(ISBN 0-9642025-0-6).
Subquantum kinetics is a new approach to microphysical theory that utilizes concepts from the fields of nonlinear chemical kinetics, irreversible thermodynamics, and general system theory, replacing the current mechanistic foundation of physics with a reaction-kinetic model. This new approach resolves a number of problems that plague classical and modern physics also may provide some insights into the electrogravitic connection that Brown was researching. In particular, chapter 9 gives some background information on Townsend Brown's electrogravitics.

Another very interesting research on anti-gravity is done (and still going on) by the Japanese prof. Shinishi SEIKE. He published his findings in the book " The Principles of Ultra Relativity ".
For his highly mathematical (no nonsense) book write to:
Shinichi SEIKE

G Research Institute
Box 33
UWAJIMA/Ehime (798)
JAPAN

---

For other 'unusual books':
Health Research
8349 Lafayette Street
Mokehumne Hill, CA 95245

Borderland Research Foundation
PO Box 429
Garberville, CA 95440-0429

---

A book 'Ether Technology' (and others):
International Tesla Society
330-A West Uintah Street - Suite 215
Colorado Springs, CO 80905-1095

---

The Anti-Gravity Handbook (revised ed.)
Compiled by D. Hatcher Childress
Published by - Adventures Unlimited Press
303 Main St., Kempton, Illnois 60949 USA
Tel. (815) 253-6390
ISBN: 0-932813-20-8
Pub date: 1993  (First edition was in 1985)

---

Scott, W.B.  "Inside the Stealth Bomber"  Tab/Aero Books: New
York, 1991.

---

ANTIGRAVITY:  The Dream made Reality
[The Story of John R. R. Searl]
by John A. Thomas Jr.
Published by Direct International Science Consortium
13 Blackburn, Low Strand, Grahame Park Estate, London
NW95NG England
Available in the USA through John A. Thomas, Jr.
373 Rock Beach Rd.
Rochester, NY  14617-1316

---

INTRODUCTION TO EXPERIMENTAL GRAVITONICS
Abstract of book by S.M.Poliakov and O.S.Poliakov

The experiments part covers the following subjects:

1. Light-beam curvature and optical-radiation frequency shift is created and investigated in an artificial nonhomogeneous gravitational field. A new gravitational effect, named "quadrature" frequency shift in the curved light beam is predicted and calculated.
2. Magnetostriction is at last explained as a secondary gravitational effect. An equation derived for magnetostriction permits to calculate the magnetostriction curve.
3. The propagation velocity of gravitational radiation (generated by a laboratory source) was measured for "quadrupole" - 9x10E20 cm/s or squared light velocity.
4. It was demonstrated that gravitation is only one of NONLINEAR-MECHANICS EFFECT, that can be created in mechanical system or in ferromagnetic.

The book was published at the author's expense in 1991. Most powerful experimental result described in this book is more than 1200 gramms of pulsed G-force. Several mechanical systems and systems using ferrites are detailed here.

Second edition in English is ready for copy process (disket's text). Editors and investors are interested in joint project for publication can get direct contact with Dr. Poliakov by address:

Moscow area,
141120, FRIAZINO,
60-let str., 1-167.
Phone 7-095-4658822.

Alexander V. Frolov
P.O.Box 37, St.-Petersburg, 193024, Russia.
E-mail: alex@frolov.spb.su

---

Conception of Edmund Whittaker ( papers of 1903 -1904 ) is developed by T.E.Bearden in his book "Gravitobiology", published by Tesla Book Co., P.O.Box 121873, Chula Vista, CA 91912, USA.

---

AUTHOR: Terletskii, IAkov Petrovich, 1912-
         Paradoksy teorii otnositelnosti. Russian/English
TITLE: Paradoxes in the theory of relativity, by Yakov P. Terletskii. With a
         foreword by Banesh Hoffmann.
PUBL.: New York, Plenum Press, 1968
NOTES: Translation of Paradoksy teorii otnositelnosti.

---

AUTHOR: Thomas E. Bearden
TITLE: Excalibur Briefing
COPYRIGHT DATE: 1980, 1988
PUBLISHER: Strawberry Hill Press/A Walnut Hill Book

ISBN# 0-89407-060-6
PURCHASED FROM: Tesla Book Co. or Fry's INC. INQ.

COMMENTS : According to the front and rear covers this book explains paranormal phenomena and the interaction of mind and matter. There are 4 chapters plus a glossary and bibliography. 332 pages, 42 photographs, and 40 illustrations.
Chapter one is called... A Sampling of Specific Paranormal Phenomena. Some of the subjects in this chapter are...Remote Viewing The Moray Radiant Energy Device, Thought Photography, Pavlita's Psychotronic Generators, UFO's, Kirlian Photograpy, Psychic Surgery.
Chapter Two is called...A Theoretical Background for Understanding PT, UFO's and PSI Phenomena...Some of the subjects are....Unexplained Mysteries of Physics, Two Slit Experiment, Radionics, Biofields and Maverick Worlds.
Chapter Three is called...New Military Applications of PSI Research... Some of the subjects covered are...Background to Psychotronic Research in the U.S and the U.S.S.R., Radiation of the U.S. Embassy, Hyperspace Howitzer operation, Virtual States and Hyperspaces, Feynman diagrams The Neurophone, Soviet Woodpecker signals The last chapter covers Soviet Phase Conjugate Directed Energy Weapons (Weapons that use time reversed Electromagnetic Waves) The Glossary is about 30 pages long and is very useful.

---

AUTHOR: George Trinkhaus
TITLE: Tesla: The Lost Inventions
COPYRIGHT DATE: 1988
PUBLISHER: High Voltage Press
ISBN# N/A
PURCHASED FROM: Lindsay Publications
COMMENTS: Paper, 33 Pages, 42 Illustrations.
Describes Tesla's lost inventions in plain, easy to understand English. According to the author, patents are hard to understand. In the illustrations he shows the patent number. Some of the inventions include...
> Disk Turbine Rotary Engine
> Magnifying Transmitter
> Transport
> Free Energy Receiver

---

AUTHOR: Margaret Cheney
TITLE: Tesla: Man out of Time
COPYRIGHT DATE: 1981
PUBLISHER: Laurel Book by Dell Publishing Co.
ISBN# 0-440-39077-X
PURCHASED FROM: Tesla Book Co. or Lindsay Publications

COMMENTS: Paperback 320 Pages, 8 Pages of Rare Photographs
Good biography of Tesla. 30  Chapters plus Reference Notes.
Chapter 29 deals with Tesla's papers and what may have happened to them
after he died.

---

I have one reference for you.
The book is called "Suppressed Inventions and other Discoveries". It's an
anthology edited by Jonathon Eisen. Authors include: Dr. Brian O'Leary,
Christopher Bird, Jeanne Manning, Barry Lynes, and others. As well as
Townsend Brown, the inventers/doctors (as well as inventions the book also
covers various cancer treatments which have had research suppressed) who
are discussed include Naessons, RifeHoxsey, Gerson, Tesla, Brown, Reich
and others. The book covers free energy, various "unfree" though different
motive technologies, cancer cures which have worked but not seen
approval by the AMA, Roswell, the Mars face, and as a delight to conspiracy
buffs, there are also chapters on how various Government bodies have
suppressed these technologies, as well as how the AMA came to be all
powerful in the field of suppressing alternate treatments.
The book is published by:
Auckland Institute of Technology Press
Private Bag 92006
Auckland, New Zealand
ISBN No. 0-9583334-7-5

---

Fer-de-Lance by T.E. Beardon
Tesla Book Company
P.O. Box 121873
Chula Vista, CA  91912  USA

---

New book by Matt Visser. Published 1995
"Lorentzian Wormholes---from Einstein to Hawking",
by Matt Visser  (Washington University in St. Louis).
American Institute of Physics Press (Woodbury, New York).
ISBN 1-56396-394-9
412 pages (including index and 38 figures); hardback;
US$59.00 (US$47.20 for APS/AIP members).
To order---Voice: 1-800-809-2247; FAX: 1-802-864-7626.
Table of contents:

## Other Sources

Anomalous Info Nexus
S P A C E   D R I V E S
PO Box  228
Kingston Springs, TN
Introductory Reading List             U S A    37O82-O228

---

Anomalous Info Nexus, 615.952.5638, 3/12/24/96/14.4 Kbps v.32bis, for Space Drive Info, Files, and graphics.

---

Leading Edge Research Group
(Leading Edge Journal #77 12/94)
P.O. Box 7530 Ste 58
Yelm, Washington  98597 USA
About electric dipole precession. The article "Electricity" in Britannica includes a resonance equation for dipole precession in dielectrics. It was identical in form to the one used in magnetic resonance, except for the obvious differences in units. Dielectric precession (resonance) frequencies were in the optical range.

Brown didn't use resonance; but he did use a steady frequency. His frequency, too, would damp out if it were discontinued. Greater results than Brown's could probably be achieved with lasers. But I doubt you'll find a better description of dielectric dipole resonance. The Britannica article gives the mathematics.

# Epilogue

A new imaging technique being developed at NASA's Ames Research Centre may one day soon have some important implications for the pictures of the Face on Mars.

The technique is known as Super-Resolution Surface Modeling, and some examples of its amazing capacity to improve the quality of low-resolution pictures are shown on the World Wide Web at:

**http://ic-www.arc.nasa.gov/ic/projects/bayes-group/group/super-res/**

where more information on the program and its designers can also be found.

The team of technicians developing this new algorithm is led by Dr. Peter Cheeseman whom I e-mailed in connection with his program, which at this stage is not yet available commercially and has been used only on some of the Martian features.

My questions concerned themselves mainly with whether the images of the Mars Face could be improved by using this new techniques.

I reprint below the reply I received from Bob Kanefsky, whom along with John Stutz is also involved in the project.

Date: Mon, 13 Nov 95 17:22:04 PST
From: (E-mail adress deleted. Information on Bob Kanefsky, John
        Stutz and Dr. Peter Cheeseman can all be found at the World
        Wide Web site given above.)

Our super-resolution algorithm needs a fairly large number of images to make any significant improvement in resolution. We've been able to get subtle improvements by combining four images; fewer than that couldn't be expected to help much. Also, we only have a 2d super-resolution algorithm, and it requires images that were taken from approximately the same viewing angle and with the same lighting angle.

As you know, there are only two images of the feature you're writing about, plus two more much lower-resolution images that wouldn't help. The two images are from significantly different viewing angles, which is how stereo images have been produced by others. Our algorithm isn't designed to handle that, and in fact I tried registering them once and couldn't get them registered using a 2d transform. Again, even if I could, the improvement from combining them would have been very slight.

As for availability, I'm in the process of translating the program to C, but it's still experimental, and it will be some time before it's commercially available. When it is, NASA usually releases software in the U.S. a year before making it available internationally, since the development is paid for with U.S. tax money.

Bob Kanefsky
(Caelum Research Corporation)
NASA Ames Research Center
Computational Sciences Division

As Kanefski's reply makes clear, at this stage, the program being developed by his team is of little use in reinterpreting the pictures of the Martian Face and nearby pyramids, but with the rapid increase in technology which is constantly taking place, it may be that in a few years the two available images may be enough to get a much more detailed three dimensional perspective of these structures.

Although it is doubtful whether such developments in computer technology will take place before one of the several planned probe missions to Mars takes place, the technique would still come in handy for double checking any future images of the Martian Face which NASA takes in the near future.

If nothing else, if the images of the Face produced by such an algorithm using the 1976 Viking frames were to closely resemble those taken by a near-future orbiter mission to Mars, this would at least reassure the conspiracy theorists that the US is not in fact busy deceiving the world (at least with respect to these structures).

Several interesting developments have recently come to light with respect to research concerning antigravity systems and related technologies. The same documentary by Royal Atlantis Film mentioned in the last chapter, showed some of the results achieved by one John Hutchison[1] in the field of antigravity research.

---

[1] The spelling of the name is phonetic, as no credits were given at the end of this documentary, although some may have been presented in the first part (I only saw and taped half of part 2 and the whole of part 3). The reason I mention this documentary, is because it contained personages which are rather well known, such as Rupert Sheldrake and a number of prominent physicists, although the presentation was rather subdued and documentation by way of written names of interviewees almost non-existent. I believe the Royal Atlantis Film to be a British based company, but I may be

Using two Tesla coils as the primary sources of field generation, Hutchison aims the created fields at a target placed between them. The video tape evidence showed metal objects which must have weighed in the region of a few kilograms float vertically upwards rather rapidly once the fields were activated. It also demonstrated that it was possible to render metals malleable and apparently even liquid by electromagnetic vibratory means, without any raise in temperature. The footage showed a metal spoon inside a chamber 'trembling' as though it was being molten by the electromagnetic fields it was subjected to.

Hutchinson was said to be a lone researcher and inventor in Canada, and is therefore (presumably) not connected with any government agencies. If this is the case, the finding of this man and his lab, and the precise documenting of his achievements may be the first step towards having the world realize that antigravity is quite real.


With respect to some of the ideas of Retired Lt. Col. Thomas E. Bearden concerning the travel of energy beams through some kind of hyperspace, an article printed in *New Scientist*[2] may be particularly interesting and worthy of keeping in mind.

Appropriately titled *Faster than the Speed of Light*, it goes on to describe how some physicists have been effectively breaking that most sacred of barriers; lightspeed.

            In quantum mechanics, the phenomena known as quantum tunnelling is more or less the everyday equivalent of throwing a tennis ball at a wall and instead of having it bounce back, go straight through it as if the barrier did not exist.

Such unlikely occurrences as a ball going through a solid wall are if not *de rigoure* at least commonplace in the field of quantum phenomena, which in itself should already be telling physicists that there are some flaws in our current models of how the Universe works; but at any rate, the interesting thing about quantum tunnelling, is that, as was already suggested as far back as 1932 by L. MacColl of Bell Labs, when a particle behaves in such a strange manner, it seems that it tunnels through the barrier without any appreciable delay.

---

wrong about this. The documentary was called *Mysteries of the Unknown* and had the subtitle *UFO*.

[2] New Scientist April 1, 1995, (No.1971 ISSN 0262 4079) Pgs.26 to 30.

The problem with quantum mechanics, is that due to our inability to correctly appreciate the true structure of matter at its most fundamental level, more than one model exists to try and explain the various phenomena. While in some cases the analogy of matter being composed of tiny particles works fine, for other situations we have to resort to thinking of it as waves, the size of the wave depending on the probability of finding the particle along any point in its length. In this case, there are several ways of measuring the supposed speed of such a wave-packet of matter; group velocity, (related to the way the peak of the wave-packet moves) phase velocity (the movement of a single oscillation within the packet) and energy velocity (the speed at which energy is transported) being three of them; and just to keep thing even more confusing, each of these ways of measuring the speed can yield quite different results.

It is necessary to resort to thinking of the particle as a wave-packet though, because if we consider it as being a tiny billiard ball, then we can only explain its tunnelling ability by assuming it borrows energy from its surroundings in order to do so, because otherwise the particle would have negative kinetic energy while it is in the process of tunnelling. But kinetic energy is normally taken to be related to the square of the speed, and if the energy is negative, then the speed would become the square root of a negative number,
which of course doesn't make any sense. And yet, if we assume that the particle indeed does borrow energy from its surroundings, that in itself seems even weirder than having a speed that is the square root of a negative number! Where does the particle borrow the energy from? And even more shockingly, *how does it know when to borrow it?* One might as well go around stating that particles have a whole social structure based on principles of banking similar to our own, and that some of them go into deficit in order to 'hyperspace' through walls just so they can baffle physicists.

In 1955 Eugene Wigner and L. Eisenbud at Princeton University came to the conclusion that under certain conditions, tunnelling particles could travel faster than light. Their calculations, which were based on the peak of the wave-packet, suggested that during the actual process of tunnelling time would become "saturated", or in plainer language, reach a maximum point that would stay the same no matter how thick the barrier that was being tunnelled through.

In effect, this would mean that a tunnelling particle would take the same amount of time to tunnel through a barrier a few atomic layers thick or one that was (for argument's sake) a couple of kilometres thick.

The important thing to remember, is that as the barrier thickness increases, the likelihood of quantum tunnelling correspondingly decreases, but nevertheless, those few particles that tunnelled would do so without any regard for that 'ultimate' speed limit, the speed of light.

More recently, Wigner and Eisenbud's calculations have been confirmed by actual experimentation. The first to do this in some degree were Stephen Chu and Stephen Wong at AT&T Bell Labs in New Jersey. Although the first to have measured superluminal[3] velocities, their work was largely ignored. A pattern that by now we should be starting to recognize. It is not as a result of malice or willful stupidity that their work was ignored, but more likely simply because since everyone knew (and to a large extent still 'knows') that nothing can travel faster than light, the announcement of superluminal velocity having being achieved did not quite sink in.

In fact, similar tests by Anedio Ranfagni's group at the National Institute for Research into Electromagnetic Waves in Florence produced results that showed a speed that was slower than that of light, so it seemed that the status quo remained unchanged.

It took Günter Nimtz and his team from the University of Cologne, performing the same test as the Italians but using thicker barriers, to produce results that were truly faster than light. When their results were presented in 1992, the Florence team repeated the experiments using thicker barriers and sure enough, they too recorded superluminal velocities. Since then, others have reproduced the results not necessarily using the same experimental setup as Nimtz, but in any case involving the principle of quantum tunnelling. In 1993, Raymond Chiao along with Aephraim Steinberg and Paul Kwiat at the University of California, Berkley, managed to get light photons to travel at 1.7 times the speed of light. Astonishing as it may seem, Chiao still maintains that despite this achievement, because the number of photons that tunnel cannot be predetermined since they are

---

[3] Faster than light.

probabilistic in nature,[4] "...it would not be possible to send any useful information."

Chiao seems to be under the impression that we cannot transmit useful information by these means of light which travels at superluminal velocities. And yet, even a layman like myself, can see that probabilistic as they may be, over a long period of time, the number of photons which arrive at any given second will average out to a given value X. Since quantum phenomena can occur in the order of several million events per second, then it should be obvious that if one operates the equipment for say three seconds, then some photons will definitely tunnel. Similarly, if the equipment is then switched off for three seconds, no photons will arrive.

By alternatively switching the equipment on and off, and having the receiver interpret any photons which arrive in any three second interval as being a dash and any 'blank' three second period as being a dot, it would then be possible to have an effective Morse code which can be transmitted at speeds faster than light.

Obviously it may take quite a bit of refining of the equipment used, which is probably not currently geared to oscillate on and off in the manner described, but with the rapid advances in technology being made every day, I find Chiao's own statement somewhat blinkered.

In any event, the process I described is just the crudest alternative, since with present day computers, the dash-dot transmission could just as easily be interpreted as binary code, the digital language of computers, which in addition could also control the rapid oscillation between 'on' and 'off' of the equipment, and hence make the transmission of data at superluminal velocities an achievable and useful reality.

At any rate, Chiao's statement was proved wrong when in October of 1994, at a special meeting in Snowbird Utah, Nimtz announced that his team had sent Mozart's 40th Symphony through 12 centimetres of space at 4.7 times the speed of light![5]

Even more astounding, despite this amazing achievement by the Cologne team (who played the recording of the transmitted symphony

---

[4] In other words, it cannot be predicted exactly how many or at which times these tunneling photons will actually arrive at the detector.

[5] Nimtz's work was also featured on the BBC 2's Horizon program on 2nd December 1996 at 8.00pm.

to the audience of physicists) Chiao remained of the opinion that this was not a real transmission of a signal.

Chiao's objection is that because the distance is so short, one can still predict what the music will sound like before it arrives, but in principle there is nothing that prevents Nimtz's team (given the correct equipment) to repeat the experiment using a distance much longer than 12 centimetres and hence send a signal faster than light which would arrive much sooner than one that was travelling merely at lightspeed; and this of course would result in the discrepancy being large enough that one could not predict what the signal is like before it arrives.

But so what if you can guess what the next note is likely to be or not? And this question highlights the problem of blinkered scientists perhaps best of all.

You see, the special theory of relativity of Einstein included a principle of causality which stated that no signal can be transmitted faster than light, the reason for this is that (the theory goes) if you could do that, then all sort of impossibilities concerned with time would theoretically be possible, such as the receiving of a signal before it has actually been sent.

The point is that this idea of violating Einstein's causality principle so scares physicists that they would rather ignore phenomena that appears to violate the special theory of relativity than attempt a revision of it. It is an approach of which I am sure Einstein himself would disapprove most strongly.

In any event, although I do not possess the mathematics required to prove it empirically, I do not believe that travelling faster than light would allow for any impossibilities to occur.

It seems to me, (and from here on I want to warn the reader that these are purely my thoughts on the matter and hence they may want to ignore them if they so wish) that physicists have taken Einstein's idea of time and space being inextricably linked in the wrong context.

I do agree that time and space *are* linked of course, but at this point I would like to go back to the analogy I used earlier, of time being like ripples in a pond that is spreading out from every minutest particle of matter in the Universe (and from every ray of energy too by the way). If we imagine time as being that, and we also imagine space, physical distance, to somehow be enmeshed into this very 'wave' of expanding time, then I believe we would be approaching the matter from a point of view that may be more useful in explaining certain phenomena such as antigravity and quantum tunnelling particles.

Lastly, we may want to remember that this whole fabric is 'ruled' by gravity. That is, gravity shapes this fabric of space-time with an effect on both the physical space as well as the time factor.

Now, keeping this image in mind, let us assume that we could create a very strong gravity field at a particular point in space. This would result in a very deep 'hole' resulting in the fabric of space-time and it is my belief that if the gravity field is perhaps vibrated at a particular frequency, or perhaps set into a resonating frequency with either itself, or the 'frequency' of space or time (whatever those may be) then a certain critical point is reached at which a 'break' or 'tunnel' is created in the fabric of space time.

There are two ways of viewing such a 'tunnel' and both seem equally valid. The first one, and perhaps the most commonly known one thanks to its use by science fiction writers, is that by opening such a 'tunnel' we have effectively opened the door to an alternate Universe where the laws of physics don't apply. If we were to enter such a dimension, time may pass differently for us, or distances could be greatly reduced, or both, effectively placing the entire Universe at our fingertips, since moving one inch in one second in *that* Universe, would be the equivalent of traversing 10 billion light years (or some large number anyway) in the blink of an eye in ours.

By moving an inch and then re-opening the portal by the same use of a gravity field as before, we would then be able to traverse vast distances in space.

The other idea is one I have formulated myself and perhaps because of it I tend to prefer it.

One of the questions that has haunted me from as early as I can remember, is that of Infinity. I was a small child when on asking      (I believe it was to my mother) where did space end, I first heard about Infinity, and the concept has pretty much been nagging at me ever since.

My immediate reaction was to ask a question that is so primordial its roots surely go back to the first protoscimian that had any tendencies to evolve millions of years later into what we now call physicists,[6] and that is: "But what is there after Infinity ends?"

---

[6] No offence to physicists intended, they happen to be one of my favourite primates as far as primates on this planet go.

My mind could not grasp this concept of Infinity and so it naturally wanted to place a barrier beyond which not even nothingness could exist. As absurd as this concept sounds, I believe that although it's a somewhat imperfect analogy, it helps to explain my thoughts concerning effects such as those shown by the quantum tunnelling particles as well as a theoretical hyperspace or antigravity engine.

On opening a 'tunnel' in space-time by the means described above of vibrating a gravity field at some particular (and most likely resonating) frequency, what we have in effect done is opened a portal onto the 'end' of Infinity. That is, if there was anything to look into we would be looking into that place where 'even nothingness ceases to exist'. Since a place where even nothingness cannot exist, cannot, for our purposes, actually be manifest, what happens is that we are thrown back into our own Universe; in which nothingness, along with all the other stuff in it, *can* exist. The interesting side effect is that by having breached the 'skin' so to speak of our 'football' from the inside, and since not even nothingness can exist on the outside, we get a kind of *'free pass to anywhere inside the football'* concession.

It's almost as if a long suffering God, on being somewhat exasperated (and perhaps a little pleased too) at anything that seems to have discovered the limits of the expanding football and that on doing so tries to go beyond it by 'tunnelling', nonchalantly drops us back inside the thing, but with the concession of putting us wherever we want in it.

By breaching onto a place that cannot exist (as it is the limit of the Universe) one has effectively placed within his reach any point in that Universe.

There may be of course, certain constraints to travel time or length. For example, the total distance we can travel in such a manner may be a function of the size of the breach, which in turn would be a function of the strength of the gravity field. And time too may be affected more or less by the size of the breach, but I feel that effectively, the idea of a 'time saturation' is quite correct and the main factor affected would be distance, time remaining a constant or varying by only the smallest of amounts regardless of distance travelled.

There is one more aspect of this idea of 'tunnelling' that is most fascinating if somewhat bizarre. One of the claims that Bob Lazar made concerning his alleged involvement with the site S-4 at the Top

Secret Area-51 facility also known as Groom Lake, is the use by the US military of what he calls Element 115.

According to Lazar, this material (the name derives from its atomic number) occurs naturally and cannot be synthesized, furthermore, he stated that the research group at S-4 had about 500 pounds of it and he claims it must have either been given to them by extra-terrestrial beings, or found aboard one of the nine flying saucers he says are being kept at Area-51, because Element 115 does not exist here on Earth.[7]

Although Lazar's claims seem rather startling, one should keep in mind that if the US military indeed is reverse engineering extra-terrestrial craft, any claims made by someone who decided to speak up would indeed sound ridiculous. Secondly, although what Lazar says cannot really be proved true or correct, no one (to my knowledge) has yet proved that he is lying either, interestingly enough though, it seems that indeed there are certain irregularities concerning Lazar's official papers, that is, they do seem to be either missing or at least not easily found or occasionally even different in

nature from what one would expect. Given Lazar's attitude however, it seems somewhat unlikely that all of these irregularities are due to complete dishonesty on his part; considering the offers he's been made, I think he could have got a lot more publicity than what he has in effect received if this was his ultimate aim.

While Lazar says he does not know how the US came into possession of the alien craft, given their generally pristine condition, he feels they may have been 'donated' or perhaps loaned by extra-terrestrials interested in trying to get Humanity 'off the ground', so to speak.

At any rate, the interesting thing about Lazar is that if his claims seem absurd, the physics behind the supposed alien spaceships makes somewhat more sense than one normally expects from crank

---

[7] It is generally accepted that elements with atomic numbers ranging from something like 114 to 118 should be more stable than the ones with atomic numbers of say 110, which have incredibly short half-lives. It needs to be pointed out though, that the properties of such elements cannot really be decided on with impunity, because even using computer generated models of their internal structure, the field of subatomic phenomena, which is the realm of the hazy quantum mechanics, is far from properly understood. Lazar's claim that Element 115 cannot be synthesized on Earth seems to be at least sound, because the amount of matter that can be generated by present day particle accelerators is in the order of a couple of atoms a day, and according to him, the spaceships take about 223 grams or so in order to work.

'physicists' trying to make a quick buck as a result of some wild claims.

Several points about Lazar's story struck me, and I have listed the main ones below in no particular order of importance.

1) The craft operate by means of a device capable of creating gravity fields.

2) The field is created by a process of amplifying the natural gravity wave of Element 115. This seems interesting in particular, because amplification is generally a result of a resonating process, something I hold to be not only a basic principle of the Universe, but one which I feel must be present in some way for the operation of any kind of antigravity device.

3) Lazar claims that the atomic gravity wave of the super-heavy Element 115 begins to extend outside of the atomic structure itself, and it is this wave, that when tapped off in very small quantities and amplified, results in the creation of an artificial gravity field.

4) Furthermore, the creation of a gravity field for the purposes of travelling from one point to another in space, judging from Lazar's description of certain experiments he supposedly conducted or was present at while at S-4, definitely resulted in a 'saturation of time' effect. One of his claims, is that a burning candle introduced in such a field became effectively 'frozen', that is, the wax would not run, nor the flame flicker. The whole thing seemingly 'stuck' in an infinitesimally small slice of 'frozen' or 'saturated' time.

5) Another experiment Lazar talks about it the effective curving of light by gravity fields to such an extent that a black spot was created in the middle of the field. Such an idea ties in rather well with the ideas behind events such as the Philadelphia Experiment.

6) Lazar claims that these ships travel through space by focussing their gravity field towards the point they want to travel to and then amplifying the gravity fields which emanate (as I understand it) from three points under the ship. This effectively folds space he says, so that the ship can travel practically instantly from one point to the next, without really any appreciable time passing for the travellers, nor the observers. He furthermore states that it is his opinion that the travel time from say Zeta Reticuli[8] to Earth is less

---

[8] Although apparently unwilling to talk about it at first because he says he had no first-hand experience of these events, Lazar says he read top secret military papers that stated that some of the aliens came from Zeta-4 (fourth planet out from the Star

than days, and in fact hinted at the possibility it may happen in a very short time indeed. This notion of course fits perfectly with the idea of a saturation in time being formed when creating a tunnel in space-time.

If we return to the concept of space-time outlined a few pages ago as an enmeshing of the expanding ripple of time with the fabric of space, and we keep in mind some of the points mentioned, then we can see that by quantum tunnelling or travel by antigravity drives, one can effectively 'jump' ahead of the ripples of time that start from a given origin.

For example, envision the ripples of time as expanding outward from yourself sitting inside a spaceship with antigravity engines.
By using your ship, you can open a portal and jump in front of the expanding time-ripple, and here is where physicists have a problem, because in so doing, you seem to have arrived before you left. That is, if you now sit there and wait for the time ripple to 'catch up' with you, while looking towards your point of departure with a powerful telescope, you will soon see yourself in the telescope moving around in your ship as you prepare to make the jump which you have already concluded.

This is because while the light travelling to you is limited by the speed of some $3 \times 10^8$ metres per second (lightspeed) you made the trip instantly. But the illusion is an apparent one and not a real one, while the light given off by you is still travelling in space, you have already arrived where you are. Just because you can still see yourself way back there at your point of origin, does not mean that you actually *are* there as well. You are limited to existing in one point in space only at the time as far physicality goes anyways, so the idea that a machine that travels faster than light would allow time travel is a fallacy.
If you were now to jump back to your origin prior to the time ripple having caught up with you, you would not meet yourself, because no matter how quickly you do it, some minimal fraction of a second must pass between jumps (the portal has to be closed for you to arrive) and

---

designated Zeta) in the Reticulum system, which is approximately 32 to 34 light years away. He also claims that a basic description of these aliens was mentioned in the report, and it is probably from here that what is now taken to be the 'common' or garden variety 'grey' alien comes from.

this is enough to ensure that you have effectively left the origin so that you cannot bump into yourself. It's a bit like running to and fro between nearby rooms. No matter how fast you do it, you will still be only in one room at the time.

Similarly, the theory that you can jump to a point where our most ancient light waves are still busy passing through, say 10 billion light years away and hence travel 'back in time' is again illusory.

While theoretically one may be able to jump say 10 billion light years away, it would take rather a powerful telescope to see our tiny speck of planet Earth wouldn't you say?

Although with advances in science arising from the implementation of gravity-producing devices it is probable that telescopes that are able to view planetary details at several light years distance will be possible,[9] and while these may certainly be used to look at our planet as it was found, say for example, in the 14th century, we would still be limited to being observers of our history, and not actually able to in any way affect or alter it.

Cause and effect would remain largely unchanged from a practical point of view, hence, the physicists do not have to worry about faster than light travel (whether by ships or by light itself) resulting in total chaos. It is not necessary to scrap relativity theory entirely, it was a useful tool and continues to be one, but it *is* certainly necessary to revise it!

As a last point with respect to quantum tunnelling, may it not be, that instead of having to have troublesome things like particles that borrow energy from the surroundings whenever they want to tunnel through a barrier, what is actually happening, is that a few wave-packets, which just happen to be in resonance with respect to their internal atomic gravity wave, are actually opening tiny gravity holes and effectively doing what an antigravity spaceship would do, they are tunnelling, or folding space, as you prefer, and appearing again on the other side.

---

[9] By stationing several gravity fields in a particular location at some distance from Earth (let's say 500 light years away) it may be possible to warp space in that locality in such a way that light waves passing through that sector of space are focussed onto another point, behind this *gravity lens* where an appropriate telescope can then view the concentrated light that passes through it. The larger the gravity lens, the more detail the telescope should be able to pick up since a larger quantity of light (originating from the object being observed) will have passed through the sector of space being warped by the gravity devices.

After all, considering the few billion billion 'entities' present in a tiny flash of light it is not unreasonable to assume that a few of them might be in natural resonance[10] with each other enough to amplify the atomic gravity wave of a single photon or two to the extent that it can travel through space-time by the same principles of faster than light travel discussed here. Furthermore, since their direction is controlled by the experimenters, their reappearance in space-time is predeterminable, if not perhaps precisely, at least enough so that we can detect them, as Nimtz and others have been doing; and if this theory on the *why it happens* of quantum tunnelling is correct (which it may or may not be) then by future refinements of the equipment used which might allow the studying of how many photons are actually tunnelling with each 'jump', we may be able to make a correlation between the size of the tunnel's 'opening' and the distance this allows a traversing photon or particle to travel through.

◆ ◆ ◆ ◆ ◆

Another development which may have some very important implications, is the theory by several UFO groups, paranormal investigators and the like that we have been informed for some time now, through the use of media such as television programs, books and (perhaps most importantly) films, with tidbits of information that are slowly conditioning us for an announcement sometime in the near-future concerning our contact with extraterrestrial species.
It is of interest to note however, that as far back as 1950 or so, when the very first books on the UFO phenomena were just coming out, authors of books on the topic and the few people that could be said to have been UFO investigators, were making similar claims, so perhaps this is merely an echo of those ideas.

Even so, the number of people today that are interested in UFOs, the possibility of antigravity vehicles, contact with extraterrestrials and the alleged abduction phenomena, number not in the thousands but in the millions, and several hundred of them at that, many of which are not only convinced that extraterrestrial contact has already been achieved (despite its being kept under wraps) but that in

---

[10] Also, given the fact that these particles undergo the same sort of 'treatment' by being shot through the same experimental apparatus, we may also expect the number of particles or wave-packets in resonance with one another to be higher than normal.

addition are perfectly able and professional scientists, engineers, doctors, pilots, lawmen or military persons.

While certainly I have never been an advocate of the adage that there is truth in quantity, and hence just because there are millions of believers this does not necessarily make what they believe true, I do believe that there is truth in quality. And some of the persons that believe in UFOs as attributable to extraterrestrial contact are most definitely in the category of excellent as far as quality goes; not only from a human point of view (after all some children believe in Santa Claus, and from a human point of view some of these children are no doubt of a high order, but this a real Father Christmas does not make), but more importantly, from a logical, or objective point of view when it comes to observation, the laws of physics, or scientific study and analysis of phenomena that while marginal, can still be processed in a logical way with the means and information we *do* have at hand.

The recent spate of movies concerning themselves with a new type of extraterrestrial contact (as opposed to perhaps the 'classical' view presented in films such as *Predator*, and the three *Alien* movies), may indeed be an indication that such a process of soft conditioning is in fact taking place; whether by the subtle influences of governmental powers (which seems doubtful) or the more likely process of natural induction that can be expected when such a large number of people are talking, writing and thinking about the UFO phenomena and related events.

A few examples of movies which may be in this category could be *Stargate*, *Species*, and the soon to be released *Men in Black*. While movies such as *Waterworld*, and *Rapa-Nui*, although their connection to extraterrestrials is not obvious, may in fact be helping to shape a new idea in our minds of the kind of catastrophe which the human race may have endured thousands of years ago as a result of alien involvement.

As for those that feel that no government would be able to effectively keep hidden such an important development as extraterrestrial contact on the scale suggested by Lazar and others, since the inevitable leak would surface sooner or later, I would remind them that investigations by individuals into such topics as the Roswell Incident, the release of information by Lazar and such occurrences as the release of the Majestic-12 documents (if marred by the fake addendum A) would all in effect constitute leaks, which can only be repaired by a rather extensive campaign of misinformation designed to ridicule or 'expose' as fraudulent whatever truth there in fact is in such stories.

Model designer John Andrews of Testor Corporation ruffled some feathers in high places in 1986 when he designed the plastic airplane model kit called F-19, because it was uncomfortably close in shape, size, and more importantly, science, behind the Air Force's latest, and most secret, stealth bomber plane.

In an eerie echo of what has become the standard embarrassment story concerning US military cover-ups we find the production of a relatively new computer game called XCom. In it, you control quick-response teams of special forces agents that react to radar sightings of UFO craft manned by hostile aliens bent on abduction and experimentation of humans. After shooting down the UFOs with either conventional fighter planes or reverse engineered UFO craft of your own, the response teams quickly land near the area to eliminate any surviving aliens and capture new technology and materials found in the downed UFO. These items they then have to submit to their scientists back in their subterranean bases for eventual understanding and inception into the escalating secret war.

Interestingly enough, the material required to power the alien crafts is called Element-115 and the humans have one bit of technology which seems to be unknown to the aliens: laser weapons.

Although the game appeared only after Lazar had made his voice heard, it is still a little uncomfortable to consider that the game designers made the assumption that well equipped, secret underground bases manned by a small number of people could conceivably exist.

But perhaps I am letting my imagination run way ahead of itself. Mind you, there *is* that NASA footage of what appears to be a UFO being shot at by orbital gauss guns...

◆ ◆ ◆ ◆ ◆

And finally: Stephen Hawking, which is considered to be possibly the finest physicist alive today, and who certainly does not suffer from any kind of reputation as a crank scientist, has recently publically stated that he has changed his mind with respect to antigravity, and he now believes that it is possible.

It has been a cause of concern to me on occasion to wonder how far I should allow myself to speculate on some of the topics I have discussed in this book, and generally I have taken the approach of presenting factual evidence where this is available, including further references when possible, and then constructing a theory or idea based on what I believe to be a logical analysis of the evidence presented or at least referred to. Where I have then extrapolated somewhat more than usual, I have normally taken the proviso of at least warning the reader that I was doing so. Despite this, being aware of my only too human shortcomings, I fully expect that I may have made a mistake here or there in some of my assumptions or conclusions, although I generally hold the main concepts to be sound. Furthermore I also expect that some people will no doubt consider my theories to be outlandish and crazy, but with this latest admission by Hawking concerning antigravity, it is a cause of personal satisfaction for me to know that if I *am* crazy, then at least I am in good company.

# What Now?

One of the emotions I felt when researching the various topics I touch upon, is intense frustration at the way in which certain technologies have been effectively suppressed either wilfully or by the blind lumberings of a society too bulky to take advantage of new developments. To this end, I decided to form a research group that would investigate such missed opportunities, not for any personal benefit which may arise from such a venture, but simply because I find it impossible to sit idly by while these things float past me and the rest of the world.

Sadly, what any research group needs most of, apart from imagination and co-operation, is money. The cost of building a Tesla coil of useful dimensions is quite high since the parts have to be precisely engineered, similarly, the finding of people such a Paul Cook for the purposes of seeing if something more can be done with respect to bringing his eye-training technology to the people of the world, can be expensive as well as time consuming.

Where I a Bill Gates, I would be able to follow up such claims of wondrous technology with impunity, but Bill, I'm afraid, has considerably more funds at his disposal than I do (although he seems too busy developing erratic operating systems to put his money to some good use) so I am reduced to what I find to be the somewhat distasteful task of asking for funds from the public at large and perhaps being also fortunate enough to secure the support of some individual or organisation that would be willing to financially sponsor specific projects.

On page 443 you will find a form that you can fill in and post alongside your yearly membership fees, that will result in your receiving a by-yearly newsletter containing details of any advances made either in terms of actual research or of investigations into topics of general relevance to the subjects covered in this book, such as what the eventual findings of the still sealed chamber found in the Great Pyramid will be when it is eventually opened, as well as a more detailed and thorough investigation concerning the Vyse markings, which I will be undertaking as time permits when I am next in the United Kingdom.

Hopefully, as the organisation grows it will be possible to increase the frequency of these newsletters, but it must be kept in mind that at first, our priority must be the actual investigation, research and development of new technology, and that only once definite and demonstrable advances in this regard have been made, will it be our main objective to publicise these findings.

It is not the intention of this group to monopolise any new technology, and generous provisions will be made to ensure that access is made available to all interested parties.

Although of course it will not be against the principle of the organisation to involve itself into profitable ventures, this will be done solely in order to further increase the researching capacities and the power of making the findings of these public.

At this stage of course I cannot guarantee that given X amount of money and Y amount of time the Coherent Light group will be able to produce such and such technology. The aim of research after all is to first discover what is known, then what may be possible, and only finally of pushing and stretching the reality envelope by creating something new, and at this stage it can be said that we are still undertaking the preliminary investigations of any research which may eventually take place.

What I *can* guarantee though is that any money Coherent Light receives will be spent on aspects of investigation, research, or the making public of any findings by whatever means, be they newsletters to members, articles, books or any other means deemed appropriate for the spreading of news.

While it may be necessary to eventually give some sort of remuneration to people involved in full time research, (even scientists have to eat after all) the personal gain of such people or of its founders or members is *not* Coherent Light's objective.

# Coherent Light cc

Name:_____

Address:_____

_____

_____

City:_____

Postal Code:_____

Tel:_____

Fax:_____

E-mail:_____

Signature: _____

(Parent or guardian to sign for persons under the age of 18)

Please find enclosed my cheque/money order to the value of R 65.00 (or £ 12) which includes my payment for 2 newsletters for the year 1996, as well as all postage and packaging fees.

## NOTE:

1)  Persons wishing for confirmation of their inclusion in our database should include a self-addressed, stamped envelope, or if from overseas include an additional R 2.00 (£ 0.50) for postage, in which case they will receive a postcard confirming the arrival of their subscription.
2)  Further donations, either by way of funds, technical expertise or materials are most welcome, and interested persons can either write to the same address as below or send E-mail to: russellp@iafrica.com.
3)  Please include self-addressed stamped envelopes if merely requesting information.
4)  Please send cheques/money orders in Rands or Pounds Sterling only.
5)  All foreign orders (from outside South Africa) add an additional R 5.00 (£ 1.50) for extra postage.

Post your subscription fees along with this completed page (or reasonable facsimile thereof) to:

**ADDRESS REMOVED** - This section "What Now?" is reproduced only for posterity. The research group no longer exists in its original form and any further information or research can be done by first familiarising yourself with the author's website at:

**www.GFilotto.com**

# **Acknowledgements**

For all the Viking Orbiter images which were the starting point for Plates 3, 4, 5, 6, 7, 8 and 9, the cover picture of this book and the unaltered images found on Plates 1 and 2, my thanks to the National Space Science Data Center, the Viking Orbiter Experiment Team Leader, Dr. Michael H. Carr and the Planetary Data System.

For the assistance in the video-capturing of the frames found on Plate 12, I am indebted to PCE Electronics and Mark Nowitz.

Footnote 8 from page 207 of the book *The Great Pyramid Decoded* by Peter Lemesurier was reproduced in chapter three with the kind permission of the publishers, Element Books Limited of Shaftesbury, Dorset.

The sketch map of Mars found on Plate 11 and figures 5, 11, 12, 16, 17 and 18 in chapter three were done by Rahman Cassiem who despite time constraints got all the drawings to me on time.

The photograph found on Plate 10 as well as the colour inset on Plate 9 were supplied and reproduced in this book with the kind permission of photographer Gerald Cubitt.

Gerald is one of those rare persons one is sometimes lucky enough to meet under the oddest of circumstances (such as the writing of this book was for me). Not only do I recommend him to anyone that requires a photograph, for he is an outstanding master of his art, but he is also perhaps one of the most interesting people I have ever met, his travels having taken him on many and varied adventures all around the world.

Gerald can be reached at his Cape Town number: (021) 238-627 or fax: (021) 238301.

Yes, unlike all the other people I mention in this book, this *is* a shameless plug for someone I admire, but let it be known that first of all, he did not ask for it, and given his modesty, I do not believe he ever would do anything of the sort; and secondly, considering not only the number of books he's taken pictures for, but the fact that it's his photograph that was used for the etching of the rhino found on the ten Rand note, I doubt he needs it. Yes, the ten Rand note.

I *did* say he was modest didn't I?

1995

2014

### About the Author

Giuseppe Filotto was born in Turin, Italy and in addition to his native country has lived in Nigeria, Botswana and the United Kingdom. Astronomy, space travel and the physics related to them have interested him from childhood and continue to fascinate him.

He enjoys travelling to new places and believes that true philosophy cannot be studied in a classroom but must be lived, thought of and experienced in everyday life in order to be meaningful.

For the last five years he has lived in Cape Town, South Africa, where he received his Civil Engineering Diploma. Together with Russell Petersen he owns and manages a computer hardware firm that specialises in the supply and installation of computer equipment.

### The Computer Imager

Russell Petersen has had a symbiotic relationship with computers since the time he learned that pushing a button on the plastic keyboard of his (now extinct) 80 Kb memory Spectravideo computer would produce a corresponding character on the TV screen.

Since those halcyon days, he has continued to follow the natural evolution of PCs by acquiring the latest available technology at every opportunity and maddening those around him by refusing to ever actually keep his machines bolted shut, since this gets in the way of his tinkering.

# Bibliography

This bibliography is by no means complete. The research I undertook for this book spanned a period of close to three years and for a good deal of the year prior to these I was not in the habit of taking down the titles of books I read on the various subjects. During this period, I was busy writing a science fiction novel on this topic, and it is only when I decided to write a non-fiction book on the subject that I began to take note of where I had come across a particular piece of information. Additionally, numerous articles from scientific magazines (such as Popular Science, Astronomy, Sky & Telescope, etc.) which pertain mainly to some aspect of Mars, its environment or its moons, have been omitted from this bibliography.

The books or articles whose titles are written in bold are recommended reading for those requiring more information on a particular subject but not wanting to wade aimlessly through all of the texts mentioned.

Selecting just one of the books in bold for each topic should be more than enough for most people to familiarise themselves with the general concepts concerned. Those titles preceded by this symbol: Ξ are highly recommended and could be considered as optional companion volumes to this book.

**ALIEN ORIGINS OF THE HUMAN RACE & RELATED TOPICS**

Ξ  **Above Top Secret, The world-wide UFO Cover-up**
        **Timothy Good**
        **Sidgwick & Jackson Ltd. 1987**
        **ISBN 0-283-99496-7**
**NB: This book should still be available from the Tesla Book Company (see later section of this bibliography under Nikola Tesla).**

Flying Saucers Have Landed
        Desmond Leslie
        Neville Spearman Ltd. 1970
        (First published in 1953)
        SBN 85435-180-9

Chariots of the Gods?
>    Erich Von Däniken
>    Corgi Books 1971
>    (First published in German in 1969)

The Gold of the Gods
>    Erich Von Däniken
>    Corgi Books 1974
>    (First published in German in 1972)

Guardians of the Universe
>    Ronald Story
>    New English Library 1980
>    ISBN 0-450-04446-7

**The Roswell Incident**
>    **Charles Berlitz and William Moore**
>    **Granada Publishing Limited 1980**
>    **ISBN 0-246-11384-7**

Spacemen in the Ancient East
>    W. Raymond Drake
>    (no publication date given, but written after 1966)
>    Neville Spearman
>    (No ISBN No. provided)

The 12th Planet, book one of the Earth Chronicles.
>    Zecharia Sitchin
>    Avon Books 1976
>    ISBN 0-380-39362-X

The Stairway to Heaven, the second book of the Earth Chronicles.
>    Zecharia Sitchin
>    Avon Books 1983
>    (First published in 1980)
>    ISBN 0-380-63339-6

The Wars of Gods and Men, the third book of the Earth Chronicles.
    Zecharia Sitchin
    Avon Books 1985
    ISBN 0-380-89585-4

## ASTRONOMY, PLANETS & RELATED TOPICS

Asteroids
    Edited by Tom Gehrels
    The University of Arizona Press, Tucson Arizona 1979
    ISBN 0-8165-0695-7

Astronomia
    Mario Cavedon
    Arnoldo Mondadori Editore 1980
    (No ISBN No. provided)

Astronomy & Astrophysics (Subvolume 2a)
    Edited by K. Schaifers & H.H. Voigt, © by Springer- Verlag
Berlin-Heidelberg
    Landolt-Börnstein 1981
    (No ISBN No. provided)

Astrophysical Quantities
    C. W. Allen (Prof. of Astronomy, London University)
    The Athlone Press, University of London 1964
    (First published in 1955)
    (No ISBN No. provided)

Atlas of the Planets
    Paul Doherty
    The Hamlyn Publishing Group 1980
    ISBN 0-600-30439-6

The Cambridge Atlas of Astronomy
    Edited by Jean Audouze & Guy Israël
    Cambridge University Press 1994
    (First published in 1983 in French)
    ISBN 0-521-43438-6

Bound for the Stars
        Saul J Adelman and Benjamin Adelman
        Prentice-Hall Inc. 1981
        ISBN 0-13-080390-1

Encyclopaedia of Astronomy
        The Hamlyn Publishing Group 1979
        ISBN 0-600-30362-4

Manned Spacecraft In Colour (Revised Edition)
        Kenneth Gatland
        Blandford Press 1976
        (First published in 1967)
        ISBN 0-7137-0756-9

Ξ  **The Martian Enigmas A Closer Look**
        **Mark J. Carlotto**
        **North Atlantic Books 1991**
        **ISBN 1-55643-129-1**

**Race to Mars: The ITN Mars Flight Atlas**
        **Roxby Productions Limited**
        **Macmillian London Limited 1988**
        **ISBN 0-333-46177-0**

Final Frontier
Mars: America's New Frontier
        Article by Robert Zubrin
        June 1995 issue, Pg.42-46

## ASTROLOGY

The New Compleat Astrologer
        Derek and Julia Parker
        Mitchell Beazley 1984
        (First published in 1971)
        ISBN 0-85533-783-4

## ATLANTIS & RELATED TOPICS

Atlantis From Legend to Discovery
        Andrew Tomas
        Robert Hale & Co. (publishers) Editions Robert Laffont
        (Copyright) 1972
        (No ISBN No. provided)

The Mystery of Atlantis, The Eight continent?
        Charles Berlitz
        Souvenir Press Ltd.
        ISBN 0-285-62211-0

Mysteries From Forgotten Worlds, Secrets of Lost Civilizations
        Charles Berlitz
        Souvenir Press Ltd. 1991
        (First published in 1972)
        ISBN 0-285-62929-8

Timaeus
        Plato
        Translation by J.M. Dent & Sons Ltd 1965
        Everyman's library No. 493

**The Dialogues of Plato Vol. III Fourth Edition**
        **Oxford University Press 1953**
        **(First published in 1871)**

## BIOLOGY

**A New Science of Life, The Hypothesis of Formative Causation**
        **Rupert Sheldrake**
        **Blond & Briggs Ltd. 1985**
        **(First published 1981)**
        **ISBN 0-85634-198-3**

## HISTORY

A Man on the Moon, The voyages of the Apollo Astronauts
      Andrew Chaikin
      Penguin Books 1994
      ISBN 0-14-024146-9

Atlas of Ancient and Classical Geography
      J.M. Dent & Sons and E.P. Dutton & Co. 1912
      (First published in 1907)
      (No ISBN No. provided)

The Conquistadors
      Hammond Innes
      Collins 1986
      (First published in 1969)
      ISBN 0-00-217531-2

**Herodotus, The Histories**
      **Translated by Aubrey de Sélincourt**
      **Penguin Classics 1982**
      **(First published in 1954)**
      **ISBN 0-14-044-034-8**

The History of Herodotus
      Robert Maynard Hutchins, Editor in Chief
      William Benton, Publisher
      Encyclopaedia Britannica Inc. 1952
      (No ISBN No. provided)

Other Origins, The search for the giant ape in human pre-history
      Russell Ciochon, John Olsen and Jamie James
      Victor Gollanncz Ltd 1990
      ISBN 0-575-05151-5

**Chaos, Making a New Science**
      **James Gleick**
      **Abacus 1995**
      **(First published in 1988)**
      **ISBN 0-349-10525-1**

**HUMAN AURA & RELATED TOPICS**

**Hands of Light**
>   **Barbara Ann Brennan**
>   **Bantam New Age Books 1988**
>   **(First published in 1987)**
>   **ISBN 0-553-34539-7 (pbk.)**

How to Heal with Color
>   Ted Andrews
>   Llewellyn Publications 1993
>   ISBN 0-87542-005-2

Specialist Medicine (Magazine)
Melatonin: a neuropsychiatric profile
>   Article by S Daya,
>>          BSc(UDW),
>>          MSc(Rhodes),
>>          PhD(MEDUNSA)
>   September 1994 issue, Pg. 58 to 64.

**MYTHOLOGY AND RELIGION**

Encyclopaedia of Religion and Ethics
>   Edited by James Hastings
>   Volume 7 and Volume 2
>   T&T Clark, Edinburgh 1964
>   (First published 1914)
>   (No ISBN No. provided)

**Pentateuch & Haftorahs, Second Edition (The Old Testament)**
>   **Edited by former Chief Rabbi Dr J H Hertz**
>   **(Hebrew Text, English Translation & Commentary)**
>   **The Soncino Press, London 1937**
>   **ISBN 0-900689-21-8**

Who's Who in Mythology
>   Alexander S Murray
>   Bracken Books 1994
>   ISBN 1-85891-068-4

## NIKOLA TESLA, FREE ENERGY SYSTEMS, MAGNETO-GRAVITICS AND RELATED TOPICS

Persons interested in Nikola Tesla can receive a book and video catalog on numerous works (some better than others) either on, by or related to work done by Tesla, as well as on several other interesting topics, from:

Tesla Book Company
P.O. Box 121873
Chula Vista
California
91912

In Search of Nikola Tesla
        F David Peat
        Ashgrove Press 1993
        (First published 1983)
        ISBN 1-85398-020-X

**Prodigal Genius, The life of Nikola Tesla**
        **John J. O'Neill**
        **Neville Spearman 1968**
        **(No ISBN No. provided)**

Ξ  **The Philadelphia Experiment, Project Invisibility**
        **Charles Berlitz and William Moore**
        **Souvenir Press 1979**
        **ISBN 0-285-62400-8**

## PHILOSOPHY AND GENERAL SCIENCE

The Rebirth of Nature, The greening of Science and God
        Rupert Sheldrake
        Century, an imprint of Random Century Group Ltd. 1990
        ISBN 0-7126-3775-3

**Forbidden Science,**
**Exposing the Secrets of Suppressed Research**
> **Richard Milton**
> **Fourth Estate, London 1995**
> **(First published in 1994)**
> **ISBN 1-85702-302-1**

Space Warfare and Strategic Defense
> David Pahl
> Bison Books 1988
> (First published in 1987)
> ISBN 0-86124-378-1

New Scientist
Faster than the Speed of Light
> Article by Julian Brown
> April 1, 1995 issue, Pg. 26-30

## PYRAMIDS & RELATED TOPICS

Ξ **The Great Pyramid Decoded**
> **Peter Lemesurier**
> **Element Classic Edition 1993**
> **(First published in 1977)**
> **ISBN 1-85230-088-4**

The Great Pyramid *Your Personal Guide*
> Peter Lemesurier
> Element Books 1987
> ISBN 1-85230-016-7

Pyramid Power
> Max Thoth & Greg Nielsen
> Aquarian/Thorsons 1988
> (First published in 1974)
> ISBN 0-85030-840-2

The Pyramids of Egypt
        I.E.S. Edwards
        Penguin Books 1979
        (First published in 1947)
        (No ISBN No. provided)

The Orion Mystery
        Robert Bauval & Adrian Gilbert
        Mandarin 1995
        (First published in 1994)
        ISBN 0-7493-1744-2

Omni
A Modern Riddle of the Sphinx
        Article by Professor Robert M. Schoch
        August 1992 issue, Pg. 48, 68 & 69.

Geoarchaelogy Vol. 7, No. 6
Seismic Investigations in the Vicinity of the Great Sphinx of Giza,
Egypt.
        Article by Thomas L. Dobeki & Robert M. Schoch
        December 1992 issue, Pg. 527-544

KMT
The Sphinx Controversy, Another Look at the Geological Evidence.
        Article by James
        Harrell Summer-Fall issue 1994, Pg. 70-74

KMT
        Letter by A.M. Dodson, regarding the coffin fragment
        found in "Menkaure's" Pyramid (3rd Giza Pyramid) by
        Vyse's workforce.
        Summer-Fall issue 1994, Pg. 2-3 Letter No.1

KMT
        Letter by Robert M. Schoch, replying to the article by
        James Harrell in the same issue of KMT.
        Summer-Fall issue 1994, Pg. 5-7

Geoarchaeolgy Vol. 10, No. 2
Geologic Weathering and its Implications on the Age of the Sphinx.
        Article by K.Lal Gauri, John J. Sinai,
        and Jayanta K. Bandyopadhyay.
        1995 issue, Pg. 119-133

**Fortean Times, No. 79,**
**The Great Sphinx Controversy.**
        **Article by Prof. Robert M. Schoch**
        **February-March, 1995 issue, Pg. 34-39**

## —V—

## —W—

## —Z—

# BOOK   THREE

# 13 An Overview of *The Face on Mars* as a whole

As I already mentioned in the introduction to this second edition, in the main the original work stands up well to scrutiny, particularly in its main themes. As I had also mentioned in the original book, right at the start of chapter one (page 6), given the vastness of the topics covered, I had in any case painted the picture in broad strokes, and this has served the overall book well.

Despite this, in a period of two decades, it would be very unusual if some new information did not come to light on the topic of the real origins of human history, and this is indeed the case.

The new information can be split into two types: the details and the broad concepts. In the details the changes are tiny or of little consequence for the most part. A perhaps perfect example of this might be footnote 7 in the original Chapter 2 (page 24). The original reads:

> DiPietro and Molenaar are both engineers working at Goddarrd's Space Flight Center, who conducted their work in a responsible, technical, and honestly scientific manner, but as a result of their work being of a 'controversial' nature they have been forced to publish their discoveries independently of NASA and the orthodox 'experts', and without endorsement from the organization that employed them for their efforts.

At the time I wrote those words, sometime in early 1995 if not earlier, that was the best information I could gather concerning these two individuals. Today, a five minute search on Google would probably result in that footnote to be adapted to say, something like this:

> Vincent DiPietro was a computer scientist with NASA's Goddard Space Flight Centre. Greg Molenaar was a computer scientist with Lockheed Martin on contract with NASA at the Computer Sciences Corporation. Working Independently, they discovered the "misfiled" image 70A13 in the giant archive of Viking images from NASA and produced in 1980 a

short monograph entitled "Unusual Martian Surface Features", where they set out to explain the then cutting edge imaging techniques they had used to identify the Face.

Materially, very little changes for the reader regarding the events that took place. The slight correction in the actual chain of employment concerning Greg Molenaar, whilst potentially significant in terms of his ability to be a "spokesman" for NASA, is ultimately not the point. And while it is true that he was on contract to NASA and as such not directly employed by them and as such, not really in a position to either be a spokesman nor to be "silenced" by NASA officially, and thus, my original zeal may have been physically partially misplaced, it certainly was not misplaced morally, and even then, still applies to Vincent DiPietro, since he was directly employed by NASA. In other words, while it can certainly be stated with some clarity that I was not completely correct in what I wrote in footnote 7 as it originally stands, I think it is also fair to say that the error is relatively insignificant with respect to how it plays out in the larger context of the Face on Mars story. Nor was there any intent to deceive in my original footnote. It was merely the result of information travelling a little slower and with more steps between sender and receiver back in 1995 than now in 2014, and this too, I think, is quite obvious.
Such tiny divergences from truly precise facts are not, in essence, catastrophic to the overall theory, either alone or even taken as an aggregate whole. They are of course potentially quite important to the individuals concerned, the specific details of a scientific theory that might be furthered or hindered by such small misrepresentations and so on, but it is not really going to alter the overall points I made in my original work.

Nevertheless, where time, effort and personal knowledge permitted (all of a limited nature, sadly) I have attempted to set the record straight in the next chapter.

Where the new information instead affected broad concepts is much more easy to point out and remedy. In succinct form, the biggest change to the original work is in the details of antigravity technology and its origins. Despite these being important facts, and despite the fact that these origins necessarily bring into question (in the most severe way) the intentions, ethics and aims of the current users of antigravity technology, the point remains that as a whole, these facts do not alter the basic theory concerning Mars and the original Martians.

It does bring into question our current situation on planet Earth, and ongoing and recent discoveries of how things really work on our planet bring into sharp focus such issues.

In particular, the discoveries exposed by brave men such as Edward Snowden and Gary McKinnon go on to further inform us of the reality we are really in, as opposed to the somewhat matrix-like *faux-reality* we are subjected to with aggressive vehemence, but they do nothing other than further confirm the overall points and theorems I laid out in the original work.

In other words, all that twenty years of time has done is bring into sharper focus the original image I painted. If before I was using a Schiaparelli-type telescope and sketching my Martian Cities with a little imagination to fill in the gaps (though, in my defence, I tried to be very clear I was doing so when I was), now, two decades later, we have better lenses and more powerful telescopes. The Martian Cities are still there, and paradoxically both easier (technologically) as well as harder to see (psychologically – the brainwashing is going full tilt after all) for anyone who cares to do so. To continue with  the metaphor, my having being wrong about a detail of whether a specific feature on Mars was a highway or a canal, remains mostly irrelevant to the main point: We had cities on Mars! And Antigravity machines exist and are being used right now, as well as back then, to do fantastic things! Including, sadly, destroy whole planets, which is why you are being kept busy looking at American Idol and "reality" TV, and playing Angry Birds on your phone instead of hyperspacing to Aldebaraan in your own private scout ship.

And that is why I re-issued this book in its original format. It still retains all of the painstaking research and for the most part it all remains very accurate, and the tiny places where it's not 100% are easily spotted and not really an issue that might "throw you off" the whole giant path this topic lies within. It also retains something else. A level of concentration of mind that is more difficult and rarer these days. The Face on Mars was never written as an "easy read". It was intentionally written with the purpose of helping the reader think about life in general, our origins more particularly, and technology and ethics in some details. For most people, doing such things is almost painful. No one has the time left. No one can really do anything about it all anyway right? So why bother. But this is not true. You knowing this information *is* important. And it *does* make a difference. More than ever, I am convinced that morphic resonance (and even that term feels inadequate) matters. The more people are aware of something, the more things must change.

A brilliant essay by Václav Havel: *The Power of the Powerless* best expresses in detail my point with respect to awareness. This essay (which you can find free on the internet), which had a huge impact on Eastern Europe as a whole and one might say was if not instrumental, certainly prophetic with respect to what was eventually the relatively peaceful transition of power from the communist party to a parliamentary republic in Czechoslovakia, some ten years after it was written, basically explains that any ideology that dehumanizes and represses humans, in the end dehumanizes and represses everyone, including the so-called rulers, and that the shift of power never comes from expected and organised quarters, but rather from the unexpected and spontaneous bursts of consciousness that invariably erupt from the masses of the powerless.

Havel's essay really also explained and foretold the collapse of the whole Soviet regime though, and this at a time when the Soviet regime was thought to be the most impregnable fortress. So do not despair dear reader. Your educating yourself, your knowing consciously what is really going on, or at least knowing as much of it as you can and seeing the flash and the bang you are exposed to daily as the distractions they are, is in itself useful.

Life evolves not just physically, but psychically, and this psychic or psychological evolution is the precursor to the physical, not the other way round, as the theologian, philosopher, scientist and palaeontologist Pierre Teilhard de Chardin masterfully explained in his posthumous book *The Phenomenon of Man*.

So do not despair, do not feel despondent and powerless. If nothing else, awareness of how the world really works, gives you a certain level of immunity from the nonsense you are being fed daily in multiple ways, and it allows you more freedom to do and pay attention to those things that really do matter, such as your family and your interactions with others around you; but more importantly, it also offers a very real level of protection from real-world consequences of the financial misdeeds perpetrated by the banksters "elite". Actions that make no sense to a person who is unaware of the absolutely total veil of lies we have been wrapped in and the realities behind our completely fake system of finance, which is merely a shadow-world for the real activities going on behind the scenes. The reality behind our fraudulent banksters include activities that necessarily must and do include antigravity technology.

Becoming clear of the lies, especially the pernicious lies perpetrated on us all since the end of the Second World War, is salutary to your own individual mental sanity, but also to that of your neighbours and friends, even if you do not discuss these ideas with them in any detail. The knowledge itself invariably will bubble to the surface and cause an irrepressible number of spontaneous evolutions of consciousness that cannot help but influence the material reality we all share. So be content in your life and begin to see things in this different life, so that when those spontaneous evolutions happen, you are as prepared as possible to make the transition a non-painful one for yourself and those you love.

The next chapter will more specifically describe both the details as well as the broad concepts that may need some adjustment with respect to the original work. Finally, some final thoughts and potential actions you might take (to replace the original **What Now?** chapter in the original perhaps) are included at the very end.

# 14 Chapter by Chapter updates and changes

## Chapter 1

No real errors present nor are any changes or updates required to this chapter. The only exception might be that perhaps the concept of antigravity propulsion could have been introduced here too had I had the information then that I have now.

## Chapter 2

In some ways this might be the chapter with which new readers might find the most to object with. The Face evidence in the form of imagery has been now quite comprehensively "dismissed" thanks to the "new" and "better resolution" images released by NASA/JPL of the Face and supposedly surrounding areas. It will be beneficial here to discuss these new images in some detail if we are to try to really understand what is going on with regard to images of Mars (and the Moon and other objects in space too).

Shortly after The *Face on Mars* was published in December 1995, I began to look at images from the Clementine mission, which had run in January and February of 1994. The website that had been set up at the time was rather clunky and forced you to at first use a special image viewing software to see the images. This was not ideal because you had to save screenshots rather than be able to view the image directly in its native raw format, as I had been able to do with the images of Mars. Nevertheless I took notes regarding certain specific images that I found interesting. A few weeks later I returned to the site and tried to find the images again. When I did, the images had been radically changed. Some did not even resemble the originals at all, while others seemed to have been passed over with some kind of blurring tool.
By this time the idea that NASA/JPL and whichever other government department was involved would filter images was no

longer just idle speculation. I toyed with the idea of documenting these changes in some detail, but I was getting tired of the whole scenario and I let the idea lie for a while. When I returned the next time to the Clementine images the site had changed as a whole, a different software now had to be used, and the images I had noted down earlier were no longer even available.

I distinctly recall the sensation I had back then, I was 27 years old, had just spent about three years researching Mars and about a year writing the book and then getting it to print and I was rather exhausted by the whole process. My personal life was also about to take one of its periodic shifts, which in my case usually tend to be the rough equivalent of a planet's sudden pole shifts, and I left well enough alone. I honestly felt there was little I could do as an individual against the resources and tactics of the US government.

Nevertheless, a few years later, the Mars Global Surveyor (MGS) was tasked to send back images of The Face and like everyone else interested in the subject, I waited eagerly. The only difference being that I thought it was never going to be really released, I thought they would try and simply make the face "disappear" but I was interested in finding out *how* they would do this. The problem was that extensive work had been done on these images by people like Mark Carlotto and Vincent DiPietro, and by then, many amateurs as well as several imaging professionals. And some of us had published books on this subject, including Carlotto, who was the epitome of a true scientist, having documented his work thoroughly in his book (and later updated them too). There was also a relatively large following of laypersons who had read these studies and books and felt interested enough in the subject to have become significant in terms of numbers.

As it turns out, the methodology used to discredit The Face was genial, and even more-so because at first I did not recognise it even as it happened.

The MGS imaging team eventually released what became known as the "catbox" image of the Face in 1998. The image in question can be seen on my site at:

**www.gfilotto.com/mars-images**

The reason why this image received the nickname "catbox" is because it was clear to anyone that had been following the subject at all that the image had been tampered with. It had been stretched, taken at a low sun-angle and then flipped so it was the mirror image of what was actually imaged. It had also had little contrast, being of a generally monotonic character.

All of these issues were explained away as due to either the requirements of taking the image being "different" due to the location of the Face on Mars, which was utter nonsense, given the orbital nature of the imaging satellite, or, even less excusable as required by the process of data interpretation that occurred in the "complicated" and "scientifiky" process of acquiring the "complicated" and "scientifiky" data and then "converting it" into images that could be finally shown and seen by us "non-complicated", "non-scientifiky" plebeians.

Except that in the people who had taken an interest in the images of Mars there were more than a couple of people that were as "scientifiky" as you like.

Stanley McDaniel was at the time a Professor Emeritus at Sonoma State University who taught subjects such as the philosophy of science, logic, critical thinking and ethics. His book, The McDaniel Report, remains an unparalleled documentation of NASA lies. The clear and inescapable conclusion that these were intentional, conscious deceptions is the only one any sane person could reach after seeing the extremely well-documented McDaniel Report. Similarly, the work of Mark Carlotto with respect to imaging is unquestionably of the first order from a purely scientific perspective. Mark, by the way, was no stranger to classified projects and information. We met in London in 1996 and again a year later, and we struck up a friendship that continues to this day even if sometimes years go by without any contact between us. He told me at the time that he had worked on a project that was trying to achieve the storing of data on crystals. At around that time, there had been an item in the news that a new "light-chip" had been developed and that this new optical system was several orders of magnitude faster than a Cray supercomputer, and a chip the size of a small modern mobile phone had many times the processing power of one of these still room-sized behemoths of data-crunching. The problem was that the data for processing by such a "light-chip" would have to be stored optically, that is in crystals. Mark told me that not only had he succeeded in developing a way to store data on crystals, but due to the nature of crystals, these devices could hold huge amounts of data. Because data stored in one plane of the crystal would not interfere with other data that was stored in a slightly different angle of the crystal structure, you could "fill" the crystal with many "facets" of data in slightly different angles one from the other. I recall that he mentioned to me at the time that the data sizes were on the order of petabytes for tiny crystals of a centimetre cube or so. He told me this back in 1996, when having a 512Mb hard-drive was pretty awesome, and he had done this work some years earlier than that. The work he did was (unsurprisingly to me) taken over by some military types and he never heard of it again.

As an aside, this was my jumping off point for an aspect of my science fiction series *The Overlords of Mars* which is set in 1999. The computers used by the-powers-that-be in it, have exactly this kind of technology in them. But returning to science facts and the Face on Mars, the point was that any trickery of "light and shadow" by NASA would soon be put to the test, and indeed it was. Merely days after the image was released by NASA, it was nicknamed the "catbox" image, not by the experts who analysed it, but by their many followers and people like me who whilst not officially scientists, at least understood enough about the topic to know when we were being deceived. Initially then, NASA's attempt to discredit the Face seemed juvenile and botched to me. But I was young and naïve back then I suppose, and I didn't realise that what seemed obvious to me, who was interested in the topic and had some knowledge of the details, was certainly not obvious to the average person in the street, which by the way was subsequently bombarded by the mass media effect of "the Face on Mars turns out to just be a bunch of rocks". So pervasive was this mass media effect that when you started to point out how the image had been intentionally tampered with in polite conversation, about half the people present would assume you were a tin-foil hat wearing lunatic that believed in conspiracy theories. It was really at this time, the mid nineties, that the whole "conspiracy theory nut" persona became "a thing". The other half of the polite dinner party who did not immediately discount you as a crazy person and actually listened to the points being made split into two basic group, an interested but overwhelmed part, which felt powerless to do anything about what you had just told them, became visibly distressed if the subject was delved into at depth, and eventually ended off with comments like: "Oh I don't want to think about it!" Literally in some cases just saying that it made them uncomfortable and they were just going to focus on their day-to-day life and not think about such things. These people used to distress me the most. I could not for the life of me understand them. I tended to watch them with fascination, on some level probably thinking to myself something along the lines of, *well, look at that... they look like humans, they seem to act like humans, but what are they? What are they really? Monkeys? Human-monkeys? Only worrying about the next peanut?*

I am happy to say that as I have aged, my view has improved at least a little. These people are not bad or stupid, they are just busy trying to survive as best they can, and occupying their minds with things they cannot (as far as they see) affect, would just make their daily struggle to wade through life on this planet and be as happy, caring and good at being wives or husbands or parents that much more difficult. I really can't blame them, because no one told them that they actually do make a difference, and that just keeping these ideas alive in their heads is not just important, but will eventually have a positive effect on their lives. I will return to this point later.

Finally, the last quarter or so of the people would take some initial active interest, but over time, feeling powerless to change anything, would gradually fall into the camp of those that basically couldn't do anything about it and stopped caring.

In other words, within a short space of time, the interest in the idea of artificial objects on Mars had been effectively crippled at a global level. It did nothing to affect those of us that had more information and detail, but by effectively neutralising the vast majority of interest, and by keeping up a relatively constant bombardment of ridicule and mass media "debunking" of the very idea of ancient alien artefacts on Mars, the people actively working on this topic were relatively quickly isolated.

But let me share with you some of the reactions these professional scientists had to the initial "catbox" image:

**Mark Carlotto**

When I contacted him asking his opinion about the new image I had already formed my own. I was no imaging expert, but I had picked up enough to learn to do basic manipulations of images and it was clear from the way the small crater at the bottom of the Face was stretched out into an oval that the image angle was either very strange, or the image itself had actively been stretched. I also knew how to play with contrast and I quickly became aware of what seemed to be a general "fuzzyness" on the image which at first I honestly thought was artificially placed, but later, it was pointed out to me by someone more scientific than myself, that this was not anyone airbrushing the image, but rather it was a real effect of the image itself and actually was some kind of moisture or cloud in the atmosphere of Mars. The image had been taken in early morning apparently, and I would guess on purpose, so that the evaporation that might happen at those latitudes at that time could further reduce the quality of the image. Or it could have been coincidental of course. But I know which way I'd bet.

Anyway, I asked Mark what he thought whilst saying in a very restrained manner that as far as I could see, it still seemed artificial to me. Up until that point Mark had not yet expressed what he thought about the Face in terms of artificiality. He had given a presentation in 1996 in London, at which I was present, where I recall that his conclusion was that the Face exhibited enough artificiality (as measured by a variety of means he weighted in acceptable statistical fashion used by scientists every day) to be artificial to a 96-98% probability even allowing for standard deviations. After that talk I spoke to him privately at dinner and asked him what was his personal view, explaining I would keep it to myself. He told me that he had no opinion. He just did the work and measured and observed and reported what he did, but he had not really got an opinion either way. This was fascinating to me and I recognised that Mark was possibly the most objective scientist I had met to date. When we met again a year later at the same conference, he had by then read the first version of the book you now hold and he was enthusiastic. At a dinner with just him, myself and Professor McDaniel, Mark mentioned that he had really enjoyed my book and that they (meaning himself, Stanley and other professional scientists) should be more bold and take a hint from my own approach. Considering I was 27 at the time and did not even have a degree, (just a higher National Diploma in Civil Engineering) I was more than a little flattered. But even then, he specified that whilst personally he now hoped and thought the Face was probably artificial, he expressed that professionally, and ethically, he could only say that all the data he had pointed to artificiality to something above 95% likelihood. His excitement was more personal, that is, he, I guess, started to think more about the consequences of artificiality.

After my email about the MGS "catbox" image, the reply I got from Mark was something I had not expected. He wrote "It is definitely artificial." He also explained a bit more why, but this was not just his personal view. He based this statement on what he could see in the image after it had been adapted. I did not mention this point publicly for almost two decades for a number of reasons. Mark is a professional scientist and I knew for a fact that other professional scientists had been unable to express their professional opinions about the Face or risk their careers. One such person was a very prominent European scientist, whose instruments have been aboard rockets of the ESA. He had spoken to me in private and told me that I was right in my assertions and ideas and that he would love to be able to show support for my ideas in public but that if he tried to do that he would very quickly be out of a job. This was not a random crazy person from the audience. I researched him after we spoke and he was indeed who he had told me he was.

Mark had never specifically asked me to keep anything he told me private, but there are unwritten codes amongst professionals and amongst honest men, and I knew (because he had told me) that he had not appreciated the fact that he had done some work that had later been used by others to support their own theories without so much as a warning to him. Giving the impression that Mark was fully behind their pet theories, which was not actually the case. To take his comment to me and use it as proof he now thought the Face was artificial without his specific permission seemed unethical to me, and apart from that, my own life was rather complicated at that point and I had many other things on my mind. But my point stands. One of the giants of image processing now saw the "catbox" image the same way I did, contrary to the popular belief being driven at the point of a pitchfork by the mass media, this image only supported artificiality instead of detract from it once it was correctly exposed and the intentional warpings of NASA removed from it.

**Vincent DiPietro**

DiPietro is also a very experienced imaging analyst and along with being the original discoverer of the Face, if you read his work, you once again get the impression of a true scientist. Here is what he had to say, in his own words, about the MGS image:

## NEW MGS FACE IN CYDONIA ANALYSIS

The image of the Face in Cydonia taken by the MGS spacecraft has been analyzed by this writer to have the following shortcomings:

> 1. The initial image presented to the public and the news media which printed it, show the image presented as the "flipped" version. In other words if the image had annotation on it, the annotation would be read as the mirror image.

> 2. The side that we have always seen in sunshine on the Viking images 35A72 and 70A13 is now in darkness, and therefore, very little detail can be seen unless a contrast stretch in this darkness is done to reveal some detail, but like the Viking images, the detail in the dark areas could be overly stretched to produce artifacts which are really not there. The tall spiked peaks on the dark side of the Face are just such artifacts.

> 3. The image was taken as an angular shot to the side at a displacement of 45 degrees to the vertical. The images of the Face by the Viking spacecraft were taken directly overhead at 90 degrees to the vertical. The Face appeared at almost dead center in 35A72 (that is why IMHO the Face looks best in 35A72), and the Face appeared off to the upper right corner of 70A13 (that is why IMHO the Face is slightly different in perspective ratio as 35A72). Most all of the images of Viking, and ALL of the images of MGS were taken directly overhead, not off to the side at 45 degrees to the vertical as this ONLY image of the Face taken by MGS. The first clue to this angular perspective was seen in the oblong shape of the small crater to the

lower left side of the Face. In the Viking images, this crater is as round as a "soup bowl", and the shallow curvature of the terrain in the center of the crater was seen as a rolling smooth curved surface. In the MGS photo, this oblong crater appears to have a stack of peaks in its center. IMHO I do not believe that these peaks are actually there, and the crater is round not oblong. Likewise, many of the peaks in the central area of the Face, may also not be peaks, but might be artifacts of the MGS imaging.

4. The side that was in shadow in the Viking images, and that which we have never seen is now lighted in the MGS images, and is only a narrow portion representing less than 25% of the Face object on the right. By "rotating" as the Malin group has done, these pixels on the lighted side are effectively stretched out. Hence the image that was later presented by the Malin group. But now all of the peaks are present but none of the valleys, and all of these peaks are stacked up like a deck of cards and stretched out. Mark has also done further "rotation" to give the Face a more correct geometric look. The fact is that in much of the space on the lighted side there have been substituted data of the "peaks" and not the "valleys", so it is somewhat difficult to see the rolling contour of the surface features between the peaks.

5. A similar study of this type of angular type of "stacking the deck" can be seen in this scenario: If you have ever seen a baseball game on television where a picture of the pitcher, the batter, and the catcher taken from the outfield are shown in the scene, you will notice that all three of the players are "stacked up" like a deck of cards, and nothing in the "valleys" between them can be seen, only some "peaks". Obviously, there is a lot of grass and some roll in the ground which is obscured by the mound. In addition, most of any point on each of the players is returned to the camera as nearly a parallel line of sight; there is no perspective

view. This causes further distortion in aspect ratio, because some features may be sized differently from others, but from this "parallel projection", the size ratio cannot be determined.

6. There have been a number of details found in the "head dress" area of the Face just above the eyebrows. These include a "lazy X" which is centered above the nose, a vertical object bisecting the lazy X, and other "ornaments" in this area.

7. Dr. John Brandenburg and I have called Dr. Dan Goldin's office at NASA headquarters, and asked for two more shots to be taken directly overhead; one in morning sunshine, and one in evening sunshine.

8. I firmly believe that the image made by MGS at the 45 degree angle renders very little in additional data of the central portion of the Face, and nothing of the side we have never seen. The image is useful only in verifying the bilateral symmetry of the Face in the verification of the profound hairline around the Face, and the symmetry of the objects found on the forehead. If further images are not produced by this MGS or other spacecraft, it is my declaration that the Face cannot be "proved" or "disproved" as being "artificial" or "natural".

DYNAMIC RANGE OF MGS - The single image made by the MGS of "the Face on Mars" lacks the dynamic range of gray shades rendering an apparent larger pixel size than the 4.32 m/pixel as advertised. At 14% of the 256 gray scale range, it is hardly acceptable to resolve any detail, with only 40-45 gray values made available, especially with one side in darkness, reducing further the gray shade spread of values. Pixels of slight variations would be represented by the nearest rounded off value, thereby losing the definitions of gray shades which would define an edge or other detail. The result would be a very large ratcheted pixel.

It would be desirable for two more shots to be taken directly overhead by MGS; (90° instead of 45°), one in morning sunshine, and one in evening sunshine. The gray scale dynamic range setting should be greater than 50% of the 256 gray step capability of the A/D converter in the camera which is adjustable by increasing the gain of the amplifier of the camera. The picture that was made by MGS with the reduced gain of the op-amp on the camera does not allow definition of any details in the dark regions which is most of the area nearest the camera of the dark side. It is possible that the teeth would again be visible if another picture from directly overhead were made using a higher gain to the camera amplifier. The single image of the object taken at 45°, seems to have prevented a fair assessment of the object. I believe that taking two more images as I have described might be able to put this two decades controversy to rest. Whether the object shows work of intelligent design, or just an ordinary mesa, (I would gladly say either), the issue could be resolved if these images will be made and the results confirm this information one way or the other.

.     STRANGE PHENOMENON - The element - XENON129 was first disclosed by Dr. John Brandenburg. The second generation of a radioactive component produced when nuclear fission occurs is found in abundance in Mars' atmosphere.    Fission such as that from a reactor or bomb produces IODINE129 which has a half life of 17 million years releasing beta particles and XENON129 gas which is stable and lasts forever. These very same elements are present when there is nuclear fission. The accepted scientific explanation for why Mars even has any XENON129 is only a theory: Billions of years ago before the sun and solar system formed, there may have been a  supernova which produced the IODINE129 which later transformed into XENON129.   Detection of these elements are found all over the solar system, and on the solar system bodies that were probed. The mystery to this theory is that there is a much larger quantity of these elements found on Mars. Relative to Xenon-130, there is 3 times more Xenon 129 on Mars than on Earth! Where did all of the extra XENON129 come from?? and why on Mars?

Sincerely,

Vincent DiPietro

May 17, 1998

The "Mark" in his point number 4, is Mark Carlotto, who did the initial fixing of the mess that was the "catbox" image in order to try to make some sense of it. I think if you just read DiPietro's words, you can sense that he is being rather restrained in his description of what the Malin group released. Mike Malin was the individual placed in sole charge of pretty much all/any images coming from Mars. He has also made more than a few attempts at "debunking" any theory concerning artificiality of objects on Mars. To put it bluntly, he is not what I would consider an honest man, and he has a specific purpose in his position, and it is not the advancement of science in my opinion. You can read more about Mike Malin in Professor McDaniel's report.

If you had not read this whole book first, his last comment concerning Xenon129 might seem rather strange, but given the idea I presented here about how Mars was intelligently destroyed using gravity assisted asteroids as weapons, it actually only adds a shocking support to my theory. It would seem there is evidence of a widespread nuclear conflagration having taken place on Mars, which by the way, fits exactly with my view of almost 20 years ago of how that cataclysmic war must have happened (see pages 166 to 170). In other words, twenty years later, the only thing that has happened is that we have found more support for the idea that Mars used to harbour life and that it was destroyed by asteroid bombardment using antigravity capable spaceships and that the defence of the planet in all likelihood also involved the use of nuclear weapons to try and break-up the descending meteorites. My "crazy" theory actually is supported by scientifically verifiable facts, and is only "crazy" purely as a result of massive and constant *opinions* expressed to the wider population and by virtue of their repetition *ad nauseam*.

But let me continue onto more recent images of Mars. And also go on to explain why the disinformation campaign concerning the Martian objects has been so successful.

Between the MGS images and the later higher resolution images of Mars that "finally" proved the Face was just a hill of beans, a few years passed, and in these years a couple of web-sites went up, dedicated to the detailed analysis of various anomalies on Mars. An excellent one is www.marsanomalyresearch.com, which provides the official images and commentary and analysis on them. It is really a quite scholarly work and it has become clear to me at least, that the images we receive from the Malin group at JPL and their offspring, are quite clearly doctored. It is my considered opinion that filters have now been implemented on the large data sets of images and algorithms designed to brush over anything that is non-fractal in nature are actively in place.

In other words, the same kind of software used and developed by Mark Carlotto to analyse the non-fractal nature of objects (remember the military used this technology at first to find camouflaged items in satellite images) is now being run on images to identify the "problem" areas, prior to another program automatically airbrushing, merging and smearing the existing pixels until they are no longer non-fractal. It would not be hard to write such software in a way that it blended gradually with the nearest surrounding fractal areas (i.e. natural formations) and blended the non-fractal areas into rough averages of the surrounding fractal areas). The whole process can thus be automated and I would guess a further human check would result in any image which resulted in a level of artificiality above a certain threshold. The same general system is used to "flag" terrorist conversations or communications being captured by the NSA. And by the way, the recent revelations of Snowden concerning the total mass surveillance of every electronic communication on Earth, is not really news to us "Martian Conspiracy Theorists" (aka real scientists). Software programs such as Echelon and Carnivore have been known of since the 90s. And I have pretty much lived with the knowledge that, as Robert DeNiro's character in the film HEAT put it "Assume they got our phones, assume they got our houses, assume they got *us*! Right here, right now as we sit, everything. Assume it all."

In other words, to recap, what happened when they released the "catbox" image it was not a bungled attempt at a further cover-up. It was just stage one. It served the purpose of reducing mass interest on Mars.
Reducing the number of people who would pay attention to this and keep these kind of ideas in mind to a tiny fraction of what it was in the early 90s.

A recent reviewer on Amazon, a Mr. S. Bernardo (from Portugal) in his review of the book *Secret Mars: The Alien Connection*, by M.J. Craig, put it best when he said in part of his review (I have corrected minor spelling errors for clarity, and the bold emphasis in the second half of the review, on the next page, is mine, not in the original):

I first came across this subject about 25 years ago long before the internet allowed for mostly debunking attacks on it. I've been following this since mid-eighties, I've read a lot of crap about it pro and con but I always thought that what was missing on this fascinating mystery was a good introductory book on the theme.

The market now has a really good share of good books on it , but mostly are too eager to try to present the "evidences" in the most scientific way possible in a few pages and so most books tend to be a bit "artificial" (haha) when they should be about what the public is looking for. And what the public is looking for is not abstract mathematical proof that could only be verified by experts but we want to see peculiar recognizable "day-to-day" anomalies instead. Fortunately nowadays cool anomalies is something that is not lacking on Mars. True, there's still a lot of ridiculous crap among all the stuff that circulates on the internet but if you dig real well and spend some time within this topic you might be surprised to find out that the initial premise is not as ridiculous or farfetched as the mainstream (mainly American mainstream) scientific community wants to keep depicting it.
**The problem with this Martian theme nowadays too is that it has so much info and cross-referenced data available that for someone who comes right into this subject without any prior background only has two choices...either you go back to 30**

years + of independent analysis and research on it; read all the books, papers and articles (and even some wacky websites too), start to cross reference yourself separating what's interesting from what is purely ridiculous and sometimes pure disinformation and end up where I am now with an opinion based on actually doing a serious following of all the controversy over the years... OR you turn on the news or watch the Nasa/Mainstream sponsored typical documentary where someone, usually "an expert" tells you that it's all a big fantasy and conspiracy theory for lunatics' entertainment.

If you go for the first option, you have a lot of catching up to do, if you go to the second option then there's nothing more here to see and this book is not for you.


Most of us just do not have the time to that that kind of research in our spare time, and so, the idea that our real history is a lot more interesting than you have been told is throttled at birth. The subsequent release of the "final" images of the Face in 2001 (see my site at: www.gfilotto.com/mars-images) were then the final hammered nail in an already pretty tight coffin, and leaving the intervening years with nothing but more of the same (2006 and 2007) has finally driven the point home. There is nothing to see here, move on.


Besides, the powers that be are pretty good at making people believe whatever they decide. In 2001 they managed to blow up three buildings in the middle of New York, and part of the pentagon and make people believe the official story of how these buildings collapsed from a fire. One of them not having been hit by anything at all yet falling at freefall speed. And really, when you give up completely on well established laws of physics, reality and observation, well, you can be made to believe anything, and compared to the Face on Mars, the idea that 9/11 was perpetrated the way the official government story tells us it took place is a far more blatant lie. Yet, a lot of people believe it. Despite literally thousands of engineers around the world signing a petition against the fiction we have been bombarded with by the mass media and official, government approved stories.

So this is why the small "paranoid group of conspiracy theorists" is actually not at all "converted" to the idea that there is nothing interesting to see on Mars. On the contrary, we are more sure than ever that Mars has some *very* interesting things on it. And I don't mean pretty rock formations. I mean artificial objects of an ancient race of long-dead beings (but not as dead as you might think, because I think there is a good chance some of their genes carry on here on Earth in one way or another).

I hope I have given you a little food for thought, but keep in mind the following points. After the first version of this book was published various points of my thesis have only been strengthened. I present below a few items I had written about on 7th December 1998 concerning this:

- The pyramids I mention on p72 as being rumoured to exist in the Shensi province in China are actually a fact that I have been able to verify recently. Articles on them were carried by Amateur Astronomy and Earth Sciences magazine (recently relaunched as Quest for knowledge) as well as several others. The largest of the group is indeed a mammoth of a pyramid, beginning to approach the sizes of the Martian pyramids, being some 300 plus meters high and much larger at the base and in terms of volume than any of the Egyptian pyramids.

- Robert Bauval seems to have now taken the position that the Great Pyramid is indeed much older than 2600 B.C. and he has been deeply involved with trying to get as much information about the Great Pyramid out to the world at large as possible. In view of this, and the efforts he has undergone in this endeavour, (as well as the duplicity demonstrated by certain officials he had to deal with) my comments regarding his timidity on this point in chapter 3 of my book should be followed by the acknowledgement that I had misjudged his intentions. I now think Mr. Bauval is probably "rocking the boat" more honestly and efficiently than most.

- The apparent lack of a Martian magnetosphere in the past was given as one of the conditions in my book. However, I did indicate that Mars should have had one in the past if it was to have supported humanoid life. Well, as it turns out, it seems Mars DID in fact have a much stronger magnetosphere in the past than it does at present! This was the conclusion borne out by a study done by Joseph L. Kirschvink of Caltech and his fellows (Science, vol.

275, no. 5306, page 1633). So it seems that yet again, a piece of evidence that seems to be supportive of my general theory, has cropped up after the basic ideas have all been put down in black and white. With respect to my ideas as to why Mars may have lost its magnetosphere, I hold that while my ideas on that particular subject may still have validity (pg 144 to 148), a better theory might indeed have come along. My friend Robert Myles pointed out to me very sensibly (as mechanical engineers are wont to do) that if you hammer a magnet, it will lose its magnetism, and as those of you that will read my book know, Mars was most certainly "hammered" by massive asteroid impacts.

- A particularly ego-boosting occurrence for me was that a Belgian astronomer of considerable repute (his credentials are quite impressive in themselves, as he was part of the team that designed the Mir-Space Shuttle coupling as well as other projects) has taken the position that my ideas concerning how Mars was destroyed is correct. I refrain from mentioning him by name as a professional courtesy.

There is also the fact that Pyramids have continued to pop up in unexpected numbers and location, including Visoko in Bosnia, in Egypt and at least one other place I am busy trying to investigate independently first. The Chinese pyramids are now rather well-known even if access to them has remained practically non-existent.

In other words, the only reason you might be sceptical about chapter 2 of the original book today, is not really related to actual science, but rather propaganda. I am not here to try to convert you, I am merely presenting facts. Facts you can investigate for yourself, and given what is now in this addendum to chapter two, I urge you to satisfy yourself with a little more research. The Mars anomalies website mentioned above in this section is a good place to start, and this book itself is an excellent resource to begin the original trail of the information we had available to us twenty years ago, which turns out to have been more free of deception than most of what we have now. But as you will see a little later in this book, there are reasons for all of this.

Lastly, a few days ago, in February of 2014, I wrote Mark Carlotto an email asking him about the Face and his views now that we have all had opportunity to look at the latest images of the Martian "hill" that used to be known as the Face as well as make up our own minds as to the quality and origin of those images. I also explained that whatever his view, if he wanted me to keep it to myself I would do so. His comment to me was simply this:

> Nice to hear from you. I've kind of checked out of the Mars scene but do still follow what is going on.
>
> Almost forty years after it was first imaged, after much analysis, critique, and criticism, I believe the evidence supporting the idea that the Face on Mars is an artificially constructed object is greater than that it is enigmatic geology – an interesting looking natural rock formation.
>
> And you can quote me on that.

# Chapter 3

To a certain extent, this chapter is perhaps the one I would want to be clearest about. The dimensions, factual observations and so on provided by Peter Lemesurier concerning the Great Pyramid remain the best researched and most factual ones and are in my opinion impeccable. As to the *purpose* of the Great Pyramid however, I am today convinced it had a very real and practical purpose and it most certainly was not a mere tomb for Khufu, nor do I believe it necessarily has much to do by way of predicting humanity's future, though, there is an aspect of pretty esoteric and advanced physics that may well allow for that very idea to actually have some basis in fact too.

Some of the best work in respect to the purpose of the Great Pyramid has been done by Chris Dunn and I highly recommend his book *The Giza Power Plant*. Chris is a mechanical engineer and a pragmatic man and he has made some amazing discoveries concerning some of the factual aspects of how the pyramid was constructed and the tools used to do so.

A somewhat similar conclusion about the Great Pyramid being a practical piece of equipment rather than a ceremonial one, is the one arrived at by Dr. Joseph P. Farrell, in his trilogy of books on the subject called the *Giza Death Star*. Dr. Farrell is a brilliant researcher and although he does indulge from time to time in highly speculative ideas, he does make it very clear when he is doing so and it is easy to separate the facts from his guesses and opinions. The good Dr. Farrell is also very knowledgeable about physics, and in this respect we are somewhat similar in that I too am rather conversant with some very advanced ideas in physics. My problem in that respect is that my knowledge lacks the formal mathematics preferred by the Western physicists, and hinges quite a lot more on a form of logic that is more prevalent in Russia and the ex-Soviet scientists. It was in fact, not unknown for Soviet physicists asking for help from American ones when trying to provide the mathematical proofs for their ideas, ideas which were sound but based in a form of logic that deals more with ratios and patterns and is not as easily reducible to mathematics as the Western physicist's training is geared towards.

The good Dr. Farrell proposes what at first seems an unlikely idea, and that is that the Great Pyramid may have been a scalar weapon. In order to understand what a scalar weapon is you do need to know a little about physics, or at least be familiar with the work of another researcher, retired Army Lt. Colonel Tom Bearden. If you want your mind expanded about what kind of technology is really possible, as well as get to the root of why the physics you are taught in school has some fundamental flaws in it dating back to the simplification of Maxwell's equations, go visit his website at www.cheniere.org
I don't necessarily agree (or understand) all of what Bearden says, but the parts that I do understand and that I have looked at myself, do make sense to me, and his description of how electricity actually works is something I have printed out and keep in a file to this day, because it finally explained something I had been wondering about since highschool that never made sense to me about voltage and how it was described in ways that made it very ambiguous and confusing to me.

After you begin to understand what scalar weapon technology is, and you investigate it further and then find, with an increasing sense of dread and the hair on the back of your neck getting progressively straighter that it is not only real, but has been developed by DARPA and in fact has its roots all the way back to Nikola Tesla and his still classified work, you begin to see that Dr. Farrell's idea is far from crazy, and that the Giza plateau may well host the remains of what could have been an interplanetary weapon. Something that could have been used to destroy asteroids perhaps, as a form of protection from the kind of war that destroyed Mars (this is my idea, not his) or perhaps used more nefariously to cause titanic, perhaps planet busting explosions on other planets or areas of space that house enemy ships. In fact, the physics involved is so strange that the idea of this being a weapon that could affect other planets might not even be limited to planets in this solar system. Potentially, scalar weapons of the type the Great Pyramid might embody might well be used for the destruction of planets around other stars. And in a final twist, potentially, perhaps, depending on how they might actually work, you could use such a weapon to turn a sun into a supernova and destroy the whole system.

I know right? Sounds utterly fictional. Well...here's a little item of scientific fact for you:
Champagne Supernovae are not supposed to exist. Based on the apparently well-established and known laws of physics, supernovae are not supposed to occur in stars that have more than about 1.4 Sol masses, however, we now have observed some that go supernova after going above this limit by quite a bit, say 2 solar masses. Now this may seem like a simple error in physics, and it could well be, however if it is, then the alterations to physics proposed by people like Tom Bearden are not radical at all. They are relatively minor, because for the physics to be wrong enough that Champagne Supernova become a natural phenomenon, the overhaul of physics would have to be quite a bit more drastic.
We're talking not a shift from the donkey cart to the model T Ford, but from the donkey cart to the Ferrari Testarossa. There is however another reason that Champagne supernovae might be happening, and to give credit where credit is due, I first read of it in a science fiction book by Adam Roberts called Jack Glass. In his book, it turns out that the explosions are caused by star-busting weapons. And scalar technology is just the thing that would be able to do that.

Given these points, is it any wonder that the efforts to keep a lid on the Martian artefacts is so colossal? I don't think so. As you will see later in this book, while I have radically changed my mind to a certain extent about the real origins of antigravity technology, and why the implications in this are dark indeed, I do still believe that the weaponisation of these technologies is so dangerous that having some serious limitations of its use is the right thing to do. One of the easiest ways to avoid such weapons from proliferating in an unhealthy fashion is quite simply if no one believes they are even possible to construct in the first place.

The down side is that the people who *do* control this technology may well be very far removed from what we might hope are honourable, just and good men and women (See the addendums to later chapters and particularly to those of Part II of the book). I do hope though, that at least some of them are. If they are human, some of them must be. Look at Ed Snowden. He looks like a mousy accountant, and the least likely heroic superspy or James Bond that you might put on a cinema screen, but the reality is that Ed Snowden single-handedly did one of the bravest acts a man can do, and he has had done so probably more effectively than anyone in a few centuries. He has completely and irrevocably put his whole life on the line in the interest of telling the truth. He might not look like James Bond, but he certainly has a moral fibre that is actually far tougher than that of the fictional 007. And Ed Snowden worked for the NSA, the very definition of sneaky, evil, unchecked, power-hungry demons in human shape in my opinion. Made all the more demonic by the completely unfettered access and unchecked power that this enormous bureaucracy has. And yet, out of that pit of vipers, comes Ed Snowden. The human spirit is continually surprising, and it appears in the oddest places, so I have some hope that at least some people that *do* have access and control of this incredibly powerful technology, *do* have a sense of love, of compassion, of care for the entire human species and not just a few elite pieces of it. And I sincerely hope that such people get together and get rid of the Nazi ones. And then begin to allow this planet to become educated in a way that is conducive to humanity joining the other intelligent species that must exist out there, and that in some way, have found it possible to mostly avoid the type of conflagrations that create Mars as we find it today, or... even worse, that create Chamapagne Supernovae! And if not, if there are alien races out there as corrupt and insane as some of us humans here on Earth, and if they are actively creating Champagne Supernovae, then I hope those good guys get organised enough to get large numbers of us off-planet and busy colonising other planets sometime soon. As well as defend us from such beings. And if that is the case, or even just partially the case, you'll excuse me if I, personally at least, would like to be more involved in that self-defence of my home planet, thank you very much. Whether it be from merely human Nazi-like predators, or potentially alien Nazi-like predators. And for that to take place, some truth, some honesty, and perhaps some expansion of the organisations will be required (and in my view already is). And perhaps, some moves in this direction are actually beginning to take place.

## Chapter 4

The original data, as well as my contention that Mars shows clear signs that it was destroyed by *intelligently controlled* asteroid bombardment from space, using anti-gravity capable vehicles, remains unchanged. If anything, the theory has only grown stronger in the intervening two decades. My theory that Mars was destroyed by intelligent beings that had advanced technology and used it to direct asteroids to the planet surface is the first (and to my knowledge only) time, that such a theory has been put forward not as a result of some dubious pseudo-scientific ideas, but merely by looking at the facts that are still in evidence all over the geology of Mars as well as noting the very unusual astronomical and geological features of its two tiny moons, or as I refer to them, "unused bullets."

Many other works, websites and books have been dedicated to show certain "anomalies" on Mars, and while I do keep an eye open for such stories, I no longer follow these in any high level of detail. The basic theory remains strong and although some of the efforts at proving Mars is not quite as we are told (and in fact is quite far from it) are admirable, my sense of it is that it is all a little pointless. The political motivations for antigravity technology (and therefore the realities on Mars) remaining the highest secret on Earth are so overwhelmingly powerful that I doubt we will see any sort of clarity on it in my life-time. That is not to say I don't think the people making these efforts should stop doing so. Quite the contrary, but my personal battlefield is different; and I believe I have, in a way, "already served", with this book and its update. As I grow older and unfortunately my personal life also gets more complex, I feel my energies are best spent in understanding how to change the situation at a global scale. Not that I think I can do that on my own, but perhaps, I could point the way, much as I hope this book did originally, and, if the e-mails I occasionally still receive concerning this book, which was written almost 20 years ago and had a print run of only a thousand two hundred books or so are any indication, I think any efforts in this direction are always worth it.

There are very serious practical realities we need to face as individuals before we can hope for any truly global change to take place, and many of these begin with educating ourselves about the political realities (as opposed to the absolutely false ones that are *constantly* bombarded at you in a million ways a day) of our planet. It certainly *sounds* like a dystopian reality. After all, if I am right, consider what this means:

- You live on a planet where almost everything you learnt is a lie
  - The laws of physics taught at school are at best incomplete. At worst, the truth about them is actively obscured by a massive effort of a few who know, and are making sure the overall masses (including the teachers) do not question the status quo because "it's so obviously correct" (never mind if it fits reality or not). *Keep watching the reality TV show. Look at the shiny lights. That's right, now where were we? Oh yes...light...nice stuff huh? Nothing can go faster than it. Believe that!*

  - The history of humanity as you have learnt it is an utter lie and far less interesting that the truth of it, which necessarily must involve spaceships, beings or people from other star systems and genetic manipulation of our planet's creatures at the minimum. And quite possibly concepts that we don't even have a name for. Never mind terra-forming...we're talking of solar-system forming. Adding things like the Moon to the Earth-Moon system, destroying the odd planet here or there in a bit of a war, or if you prefer "ethnic cleansing" on a scale that makes the Nazis look like play-school amateurs, and weapon technology that literally can destroy whole planets and possibly even whole star systems in the blink of an eye.

o   Every other significant human event in history, be it the world wars or the Babylonian myths, the Bible and so on, are all twisted after-effects of a larger story; and therefore, also, strictly speaking, lies; no matter what hidden gems of truth may also hide inside them as a larger whole.

o   There is (and always has been) a class of banksters-elite that have access to knowledge and technology that is beyond that of most of us, and their understanding is (and has been throughout the ages) such that they feel the best thing to do is use this knowledge to their exclusive advantage and keep the rest of the population much as a farmer keeps sheep. Sadly, odious as this idea is, the actual release of all the real information and technology throughout the population would almost invariably result in the extinction of the human race. Mars lies as the uncomfortable dead corpse in the cupboard of our history, to remind us just of what we do with unlimited power. As an aside, there is some evidence that the ancients actually had a primitive form of radio that allowed the priest class to create events within the temples of the Gods that gave the people no alternative but to believe in the manifest reality of these supernatural forces. See my article on this at:

http://www.gfilotto.com/ancient-radios

•   The technology being kept from us would allow this planet to become a Garden of Eden with no pollution, almost free energy for everyone and the capacity to travel to other stars. Unfortunately any idiot with a car-sized antigravity machine can reduce the whole planet to a bunch of dead rocks floating in space. And possibly our sun to go supernova, even if it's not supposed to be able to. So it's kind of a bummer. *YOU would never do that right? But that neighbour of yours...the one with a different take on things, not as educated as you, not as smart, not as compassionate...you know...maybe if we just got rid of HIM then we would be fine...* Starting to see how the banksters control your thoughts yet?

- The lie is huge and massive, but everyone has bought into it because of how much energy has gone into creating it, and sadly, because our neurology has not been on our side for about at least 200 years or so. Evolution moves slowly unfortunately (but perhaps providentially up to now). You think you would not have been a Nazi if you lived in Germany in the 30s? Think again. Most people do not actually THINK. They react to emotions. Our biology is only beginning to change now as a result of modern conditions, but for millennia, (possibly a couple of million years actually) emotions have had some survival strategy that was valid. Mostly, in the post-industrial revolution world, they do not. As a small child I wondered how a whole country could become so mentally unstable as to actually believe the insanity of Hitler. I was seven at the time and I couldn't make sense of it. Today at 44, I still can't. But I know something I didn't know then. I am a borderline Asperger's with a quite high IQ. That's normal people-speak for: my brain actually does logic before it does emotions as a general rule. If you think that has given me some advantage in life, you would be right, but they are not as many, nor of the kind you probably think. I know I will never go hungry for very long. Whatever country, environment or situation you put me in, I will eventually get myself into a comfortable enough position to take care of my basic (and some luxury) needs. And it doesn't really matter if you start me off naked in the jungle. Unless something kills me, I eventually will achieve the minimum level of success to survive in short order. So that's the plus side. But on the negative side? My relationships have always been a mess. My relations to other humans is very limited because I can't process a lot of the behaviour they do as being sane, and they can't process my laser-like ability to see logical flaws in their thinking, their work, the way they go about  trying to do most things. It becomes intolerable very quickly for both people when what to me would be obvious errors that need correcting to them are just "Tuesday." Trying to explain it to them more often results in an argument than a resolution, probably because my natural mode of processing this type of situation is to simply point out the logical flaw, which gets received as somewhat "lacking in compassion" from their perspective, to put it mildly. ("Impossible bastard" seems popular. Also you'd not realise how "Yes, you're always right!" could be an insult, but I assure you it can be. Even when true.) Having discovered the Asperger's condition and understanding not just that I process things differently but WHY and HOW I do, has helped a lot, but I am still new at it and if it has helped me understand things better, it has also made it clearer how distant I am from most people and how difficult it is to be able to give and receive that quality of love that we all need. In a way, if you have read this far, and understood what you read, you too are an unusual type of human. It is not that we lack compassion. Quite the contrary. We have plenty of it. We just need more people who can THINK to share it with, because as long as this planet functions on monkey-like emotions, ugly and sad as it is, the safest thing is to actually keep most of us in the sheep-pen. But it's a circular notion. Eventually the sheep outnumber the "farmers" and something will break. And break in a bad way for all of us. The only way forward is to consciously become aware of these things, and elevate ourselves as human beings. To become able to process our emotions in a rational context based on facts. Certainly not to give up compassion or love or a sense of connection with others, that way only another Hell lies, but all the  same, to relegate emotions to their proper place. Which as far as I am concerned is when you have positive ones. Good emotions are good. Let's keep them. Bad emotions? Not so much. But not having bad emotions lies in the ability not just of ourselves, but those around us, to be able to THINK better too. This book, is merely an effort to show some of the realities we need to process in order to be able to be trusted with a much higher level of accountability for all that we do. And no; it doesn't start with any fictional thing called "society". It starts, begins, lives and ends with one person. And one person only. You.

So if I am right, the only way forward is for you personally and individually to choose to consciously evolve. Become aware of the reality behind the façade. The massive lie is actually surprisingly easy to see once it's been seen the first time. It' a bit like those drawings that show two or more things. At first maybe you can only see one thing, but after it's been pointed out, you can't un-see the other aspect of the image. A classic example is the one making fun of Sigmund Freud, a drawing of the bearded "psychologist" and Sigmund look-alike and the naked woman it also shows in the outline of his face. It's one drawing that looks like a man's face or a naked woman reclining. Once you see both aspects, you can't un-see them. In this case it's quite obvious, but the principle is essentially the same. Once you see that the whole financial system is a massive, artificial and utterly unbelievable lie, that the petroleum based economy of this planet too is a fiction and that antigravity technology is very much a reality, the cracks in the illusion are too broad and fundamental for you NOT to see how things really work. Suddenly events that made no sense or were confusing become immediately clear. Little side-wars that could break out into WW3 like the current Ukrainian crisis become far easier to understand and you start to follow the money-trail, the Nazi ideology and the political realities, instead of the unreality shows produced by the through-feeding mechanism you have installed at your own cost in your lounge; possibly with HD quality and whatever the latest improvement on Dolby sound is.

As for what to do about it all, well, do not despond yet my dear reader (also, read the rest of this book as there is more on this in the addendums to later chapters).
Firstly, there is some pretty massive scientific evidence that just by you changing yourself and becoming aware of such things, you are actively helping to change others too (morphic resonance has been proven time and again in the last 20 years and is no longer really in

dispute, much as has telepathy by the way, though it's been a quiet revolution in science and mostly unreported by the mainstream (no surprises there then!))

Secondly, there are practical things in your life you will automatically change once you are aware of reality to a higher degree. And these changes themselves cascade and eventually avalanche, even without any directed conscious effort by yourself specifically. Of course, we also can and should take specific actions wherever and whenever possible, but it is best to be clear on your aims and the dangers of it before doing so. As we have seen, the indiscriminate dissemination of antigravity technology on Earth is NOT the smart way to go. But on the other hand, neither is the perennial cretinization of the 99.9% of humanity the answer. We may be sheep for now, but let us begin by learning to do math properly, and then let's see what happens when we begin to use that new idea for the benefit of ourselves and the other sheep around us.

## Chapter 5

No real errors present nor are any changes or updates required to this chapter. Other than the fact that on any one of the topics I touch upon, literally a couple of books could be spent digging a bit deeper. But you know how it is… so many interplanetary wars and exploded planets and so little time.

## Chapter 6

Once again, no real amendments are required to this chapter, but I would like to add a couple of points. Where I describe "magic" as potentially "possible" only insofar as it relates to higher technology or higher states of consciousness, I would like to make you aware of the incredible work of Peter Gariaev (his name has various spellings as he is Russian and due to the Cyrillic alphabet, names are generally spelt phonetically. My own name for example, when written in Russian would look like DJUZEPPE if translated letter for letter, so this scientist has also appeared on the internet as Pieter/Piotr and Garjaev/Garaiev and possibly a few other permutations). I communicated with him a little by e-mail with the help of a Russian friend to translate between us in 2011, and I confirmed that his work was in fact real by (quite coincidentally as it happens) also being in touch with the only American scientist that I am aware of that replicated his work. I wrote more about this in my second non-fiction book: *Systema – The Russian Martial System*, because certain aspects of Systema, do relate to the unconscious use of our body, and training the body and mind to act instinctively, which produces results so effective that for some time, the ability of martial artists trained in Systema seemed to be the stuff of myth and legend, but in fact, are anything but (although some people are not above aggrandising certain effects for personal gain, which was part of the reason I wrote the book. I wanted the truth and science of Systema to be better known and understood).

Peter Gariaev's work is far too complicated for me to go into any detail, however, there are a number of things he has proven. One of them is that DNA is mutable as a result of information being focussed at it. He did it with a complicated apparatus of scattered laser light, but the amazing thing is that he proved not only that he could repair DNA that had been damaged by radiation and resulted in non-viable seeds, but that he did so, merely by "flashing" the "image" of a healthy seed at the damaged seed.

The point is though, that the "image" of the DNA he read from the healthy seeds remained suspended in a liquid solution even after the lasers were switched off. There are words too complicated for me to have mastered, and concepts related to them that are beyond me, (try looking up solitonic waves) but what I understood of it is that DNA can be affected not just by lasers, but by thought. Gariaev essentially explained why things like hypnosis, precognition, telepathy, race memory, morphic resonance (see chapter 7) and so on work. And if you are thinking I just went off the deep-end in that last sentence, you'd be wrong.

Precognition as a human skill is a fact. It has been proven in a meta-study that spanned some 50 years of precognition studies and though small, the effect is very much statistically significant. But if you don't trust the meta-study don't worry, the American scientist I told you about that worked with Gariaev's work? His name is Dr. Rollin

McCraty and I was in touch with him too in 2011. Again, this is discussed in more detail in my book on Systema, but the reason I was in touch with him is because he has proven in a lab that humans can predict an event, especially if dangerous or emotionally significant, some time before it happens. If you think this is pseudo-science, you need to know that aside from the fact that the man is a real scientists, doing real science, he is also contracted by the US military for this very research now.

In other words, they want to see if he can teach soldiers how to develop this ability, because you can see how it would be kind of useful to know where to be about a second or two before something that would kill you is going to occupy the same sector of the space-time continuum as you.

One last little bit…as far as I can tell, if I understood the good Dr. Gariaev properly, it seems that our DNA is actually essentially capable (and does) produce tiny mini-worm-holes and that these get information both from the past and the future, to create our present, and as such, our thoughts on the matter (but especially our "unconscious" thoughts) make a huge difference. In effect, the hypnosis "trick" (but which works, I know, I am a trained hypnotist and I have treated dozens of people for various things) of "changing your past" to improve your future, is actually something that produces results because, in a way, you really are changing the past, and creating a new and different future too when you do what is sometimes called "timeline therapy".

If we can improve and fix and change our own DNA merely by focussing our thinking and/or our unconscious mind, and indeed our own lives, and when you consider that some 90% + of our DNA is considered "junk" by most scientists, and serves no theoretical good purpose, I am willing to bet there is a whole bunch of other stuff we might be able to do once we begin to "activate" more of that so-called 90% "junk" DNA. Which by the way…seems to already be happening to some children in some admittedly very early reports from the front-line of edge-of-science publications. It seems more of our "junk" DNA is "coming online" so to speak, and no one is really quite sure what it means or what it will result in.

One last point, that is, a personal one I have come to conclude, and that is that the rapid increase in Aspergers and autism is not in fact a bad thing, but may well be the next step in human evolution. Unfortunately, evolution, as any biologist knows, is a blind bitch, and for every eventual beautifully winged swan, she may produce 50 stump-wearing half-frogs with feathers, so the process is messy and anything but pretty, but without a doubt, the next step for humanity, if we are not to blow ourselves up altogether, is to develop a more logical and at the same time more compassionate outlook.

The abandoning of irrational emotions —more prevalent in the female of the human species for very real biological reasons, which developed over about 2 million years of hominid history, so, not something you are going to just toss aside in a mere 100 years or so, no matter how politically incorrect it might be to say so— is a much and urgently needed next step. But at the same time, logic without the higher awareness that we call "love" or "compassion", but is really just a higher level of conscious awareness in general, can easily become an abomination. In essence, logic without love becomes an ideology and any ideology, no matter how supposedly noble at first, invariably results in mass killings. If you need to understand this better, I suggest you read the essay written by Václav Havel, the Czech dissident (and later president!) on The Power of the Powerless.

All ideology becomes an abomination in due course. And love without reason behind it is just as tragic and leads to acts of "passion" that can and do become just as repulsive as the ones of unfettered "logic".

In this respect it is the male of the species that can learn from the female, though, in fairness, at present, the males have already made great strides in this department considering how dire the situation was.
Sadly, the females are now lagging behind.

Yes, it is "impolitic" to say so out loud in polite conversation, but it remains a fact. A generalisation, to be sure, but a fact nonetheless. And before you go labelling me a misogynist, know I have a little daughter. And I fully hope she goes on to take over this planet and the Galaxy beyond it.

Women are behind in the human evolution game for a simple reason. Men's survival pressure was the objective universe for most of human history. We soon learnt that the Universe does not especially care too much individually and we need to adapt to it or we will often die. Women on the other hand, had mostly other women as their survival pressure for most of human history, so deception, a general level of superficial co-operation that reduces the potential for violence in general (but that also does not allow the level of do-or-die co-operation men can have) and most important of all, the biologically instilled frame of mind that she can NEVER admit wrongdoing. Think about it... a female in the caves admitting to stealing some food or another female's food source and protector (a big strong male) would be shunned or killed by the other females in short order. And whilst men killed each other with abandon already in general, between men out hunting, the admission of guilt and acceptance of ritual punishment added respect and even honour. A man who admitted his guilt and took his punishment could be relied upon. Certainly more than one who lied about it or made implausible excuses even when caught red-handed. And men and women did not really argue with each other in the sense that we do now. The big violent male monkey would crush skulls that irritated him too much, so the female monkeys might bicker with each other, but not too loud and not too close to the big male monkey, in case he got upset and went all violent ape on everyone. Come the modern world and now males do not kill each other as much over the silliest thing. Shooting your waiter because the coffee is cold is frowned upon these days, no matter how strongly you might feel about it.

The female of the species however, still has her monkey-brain relatively intact and whilst this may have been a survival aid in the past, it's not doing anyone any favours now. Her emotions in the caves "mattered" because they were a social display that was designed to help convince the other females (and the males even more-so) and that had a function within that setting. Her taking everything subjectively was an important part of her survival. If she did not put herself first, her children might not get enough food, she might not get the best mate, she would eat less and be protected less. Her being indirect and leading-on, and testing the emotions of a potentially suitable mate for stressors made sense then. The biggest, meanest monkey in the group would not fly off the handle at the first tantrum she threw, because he had the pick of any female he wanted and everyone knew it. And a female tantrum was nothing serious. Every male monkey knew that, and if they got to loud, well, a couple of slaps quieted them down. Now the male monkeys have generally learnt that slapping the females down is frowned on almost as much (or more-so) than the shooting of slow waiters. Unfortunately, there really hasn't been much of an incentive for the female monkey brain to develop out of the "stress-testing" of the male, or of the seeing the Universe as personally responsible to her and for her, nor for her to realise that in the main, emotions per se are useless most of the time, and especially so when you can't tell the difference between an emotion generated by your now faulty (because we don't live in caves anymore) monkey wiring, or an emotion generated by a real intuition as a result of some unconscious level of data acquisition.

Think I am making this up?
In the tests of Dr. McCraty, the women would generally predict an event with emotional significance (violence, sexual content of a graphic nature, etc.) earlier than the men. Interestingly enough, when the men were trained to become calm, yet relaxed and aware, a mental and emotional state that can actually be measured when you are hooked up to the right equipment, a state that is understood to combat veterans or trained martial artists as a kind of Zen mode, where you are both calm yet ready, the men seldom reacted to images, regardless of content. The women on the other hand still generally reacted. This, to me, makes perfect sense. A man's best weapon is whatever tool he has fashioned as such, especially in the cave-monkey times. But if he jumped at every shadow, he would soon be exhausted and more likely to make a mistake when a real danger appears and thus less likely to survive. The female's best weapon however, was quite simply a man. So better be safe than sorry and wake his lazy monkey butt up! Regardless of whether it was a shadow or not! And that takes an extra second or so of alarm…

What does all this talking of male and female monkeys have to do with Asperger's you ask? Well… in its milder form, Asperger people are more capable of seeing reality, particularly the more logical aspects of an issue clearly. Their discomfort with other people stems from the fact that what is so clear and obvious to them, often does not compute in the same way in the more monkey-fied brains of other humans. This makes no sense to the Asperger's sufferer and even if they sometimes have high IQs, they cannot understand the irrational and absurd statements, emotions, choices etc. being made by the other person. This cause anxiety and eventually frustration that can lead to depression, rage, anger etc. In general an Asperger's will try to remove themselves from such a situation, but if they can't they will become so uncomfortable that eventually they will react in some way.

Imagine how ugly this gets when you have a slightly Asperger's man and a very emotional, feminine, woman. It's a recipe for disaster and is reflected in the ever-rising divorce rates. People (both men and women) actually WANT to be married. All the statistics on this say so. Yet less and less of them stay together. It's my opinion that this is not so much as a result of "more choices" being available, but the result of the male and female roles changing so rapidly and without a roadmap ahead, that they went from complementary (however brutal the reality of this complementary situation was) to competitive.

Men still have a long way to go, but they have made great strides in conquering the monkey-brain, and they are making steps towards the space-man brain, and now, ladies, you need to catch up. And make no mistake, fighting the monkey brain is the hardest thing any human will do, but it is necessary. I would tell you more, but that really is another book, and yes, I will write it. I need to.
But for now, just be aware that human evolution is happening right now! And you are in it, and YOU matter!


## Chapter 7

There could be endless additions here, but I covered the main ones in my addendum to chapter 6 above.

## Chapter 8

Again, I could have volumes to add here, but in the main, the original remains essentially on the right track, as do the appendixes to part I of the book.

## Chapter 9

This is the one chapter where I certainly was missing some of the puzzle pieces. The information in it is accurate (and the video analysed by Hoagland is now on Youtube under the label STS-48, which is the designation NASA used to give to shuttle missions) but I had missed a whole important aspect of antigravity technology.

I was, wrongly as it turns out, under the idea that antigravity technology was mostly "an American thing."

After all, T.T. Brown, Tesla, Einstein, all those geniuses, and even people like Bob Lazar, the huge Area 51 which I considered just your perfect place to invent new military technology and so on, it all pointed to America being where it probably all had begun. Although, many pieces of this theory did not actually seem to fit with realpolitik in the world even back then and it was certainly unclear if there was any kind of global consensus or if there was infighting between the various "factions" behind closed doors. And as much as I could make sense of it, the general "vibe" of the powers that be that controlled this technology seemed to range the scope from darkly murderous towards any researcher that got too close, to almost benevolently whistleblowing enough information to give people like me an important clue (or a bit of guided misdirection, as the case may be).

Even back then in 1995 I thought I was missing something, but I also thought it was inevitable I would not be able to have a clearer idea of exactly the who and how and why behind the use of this most secret of technologies. I ascribed such secrecy in general to essentially the military mindset, which concerns itself with practical safety before most other considerations. And in this at least I was generally correct, if also terribly naïve.

Luckily, others have filled in what had been a blind spot for me. The real origin of antigravity technology are now much more clearly understood and well documented, and these origins are not

American, but rather, unquestionably German. That is, antigravity technology has its roots in Nazi Germany.

The political realities of that fact, if like myself in 1995 you did not know this, make for a very rude awakening. And more than a little reading. Very uncomfortable reading.

In this respect, the work of Dr. Joseph Farrell, and in particular his books *Nazi International, The SS Brotherhood of the Bell* and Reich of the *Black Sun.* Have been instrumental in providing much of the documented research. Similarly, the work of Henry Stevens entitled *Dark Star,* which concerns itself with weapon advances and secret Nazi bases is again a very good place to dive in at the deep-end.

Such books deal with the real history and practicalities of the Second World War, and it soon becomes very obvious that a massive ideological cover-up has taken place at the end of that war. This is where it gets shaky. The book that first highlighted to me in a way quite like no other the real nature of the Nazis was a book called *The Spear of Destiny*, by Trevor Ravenscroft. This book was heavily criticised for being variously filled with inaccuracies, lies and "fantasy". And admittedly when Ravenscroft begins to delve deeply into some of Rudolf Steiner's ideas towards the end, he did lose me. But whatever failings the book may have had, it brought to light for me something that really needs to be better grasped in general.

The nature of the Nazis was essentially cult-like. This was not a political party per-se. It was essentially a cult. One steeped in mysticism, the occult, secret societies and, like Peter Levenda (author of the also excellent [if disturbing] *Unholy Alliance – a History of Nazi Involvement with the Occult*) says in various seminars and some of his books, to try and lighten the mood of a very dark subject, "They were the best-dressed cult." Considering it was a death-cult, I suppose he's right, though I am certainly not partial to the fashion myself.

It is hard enough to deal with the realities of what the Nazis did, both during and "after" the Second World War. Because once you look into it, it becomes quite clear that many of the top players in the Nazi party are really unaccounted for. Including Bormann and Hitler, but more interestingly perhaps, people like General Hans Kammler (who was in charge of all "wonder weapons" development, and a mass-murdering Nazi of the worst sort, being very much responsible for the slave labour at various facilities) and others related to the secret technology of the Nazis.

There are essentially two threads here. The first, and easier to follow if strange, fantastic and disturbing enough, is that towards the close of the war, when it was clear Germany would lose, the top ranks of the Nazi regime made plans to continue from outside Germany and by other means. And this effort itself was broadly speaking split into two itself.

a) One "branch" would deal with the massive outflux of billions in gold, gemstones, currency and whatever other assets could be removed from Germany. Mostly this went to Argentina where essentially the Nazis had enough land that they had their own country-within-a-country. We are talking country-sized "estates" where you did not go in or out without having an SS ID. And the perimeters were manned by armed SS. All under the protection of the Peron government. In fact, Argentina joined the war just days before the end mostly as a pretext for loading up Nazi scientists and materiel and flying it back to the Nazi's "home away from home". This side of the equation set up thousands of shell companies and German firms and using the looted billions essentially put in place a vast network of financial sustainability. The Nazis had basically decided that as they had lost the war on the ground, they would simply continue, with their ideology unchanged, in the boardrooms of finance. The evidence for this is actually quite blatant.

b) The "weapons" side of the equation it seems really did in fact extract Hitler and Eva Braun, and possibly Bormann too as well as some other top Nazis to Argentina. This is not really news to many Argentines, but for us Westerners sounds like science fiction. Hitler died in the bunker surely, right? The problem is… which Hitler? There were by some accounts up to six bodies that were supposed to be Hitler. And none of them were identified by anyone other than...ummm...other Nazis that were in the bunker. Who also identified different bodies at different times as being Hitler. But no problem right? We had dental records back then. Yes, we did. And guess who was given the task of "proving" the identities of dead Nazis by the use of dental records, for the most part? A Nazi dentist.

The story of Hitler was perhaps most comprehensively told in a relatively recent book called *Grey Wolf – The Escape of Adolph Hitler*, by Simon Dunstan and Gerrard Williams, though it draws heavily also on a previous work by an Argentine titled Hitler Murió en Argentina (Hitler died in Argentina) written in the 70s by Manuel Monasterio who used the pseudonym Jeff Kristenssen.

The second part of the "weapons" side of the "defeated" Nazis, seems to be that they made various deals, both with Russia and America, in exchange for various technologies and/or promises of technology and/or involvement in some of the huge financial restructuring that was orchestrated by Dr. Hjalmar Schacht who was really a key person in the war, being as all the financial motivations related to it have him at the centre of it, along with some other Anglo-American banksters (see Antony Sutton's masterful *Wall Street and the Rise of Hitler*). The result was a US government so compromised that by the time they realised that they had not perhaps, in fact been given the keys to the treasure horde (despite re-housing and looking after some 3,000 Nazi scientists under project paperclip) it was too late to turn back. The rest we "know" right? Wernher Von Braun was not really that bad a Nazi, and he helped create the Apollo program. Cue applause. Well... not quite. By some very credible accounts, the Nazis were dangerously close to dropping the Atom bomb on America, and in fact did their first atomic weapons test on October 10-11 1944. At the end of the war, the bombs dropped on Japan, it seems, had components that had been developed by the Nazi scientists. As well as some of the enriched uranium required, which, inexplicably, the Manhattan Project suddenly had in abundance, when a few months before it had not produced enough in the previous 18 months for a single bomb. The theory goes that a U-boat that was documented to have been allowed to reach US shores delivered the missing Uranium, along with perhaps other things and people of interest. But people that if they were Nazis, certainly did not see any kind of trial or jail or length of rope like some of their colleagues in Europe.

c)   There is one more brief but important element of the "weapons" side of the not-quite defeated Nazis. They seem to have kept the most advanced technology of antigravity for themselves long enough at least to frighten the Americans into making some deal. Consider that Admiral Byrd, a man that by all accounts was a serious naval officer not easily shaken, was put in charge of an "exploratory force" of some 5,000 men, including warships, destroyer and airplanes, to go and "take a look" at Antarctica. The mission was supposed to last months and this was very shortly after the close of the war. Why spend such huge sums to scout a barren land of ice? Perhaps it has something to do with the persistent notion that the Nazis had huge U-Boat bases there. Once you begin to look into this, it really gets into the rabbit hole, because between the absurd (but very real) Nazi realities and the outlandish exaggerations of some, it becomes difficult to establish fact from fiction. But Operation Highjump itself is a fact. And Admiral Byrd, on returning from Antarctica after a mere few weeks, and having lost men and equipment on the mission, spoke to a journalist of *El Mercurio* whilst still aboard one of his ships. In this interview, before he was sanctioned by the Navy back on US soil, Admiral Byrd stated explicitly that America should prepare to defend itself from aircraft that could fly at unheard of speeds, and that they would be of Polar origin. Which seems pretty unambiguous to me. It is interesting to note that all of Operation Highjump's details are still classified, as are Admiral Byrd's journals. He never spoke of Operation Highjump again. It is also interesting to note that at first the Roswell crash of the now famous "alien" craft was in fact reported as being "alien" as in "of German origin," And lastly, the first UFO flap, in

Washington, where the term "flying saucer" was coined, as described on page 360 of the original version of this book, could easily be seen as a warning to the White House. "Look here, we can bomb you right at home if we want to, and there is nothing your Air Force can do about it. Better stick to the deal we made." And so, the US government went to bed with the Nazis. Nazis with their ideology intact, and no doubt feeling quite smug at having pitted the two superpowers against each other as a kind of distraction from themselves, the real third force behind the scenes. As a last aside, in the 1950s, some atomic weapons were exploded in Antarctica for reasons that remain classified and obscured even today. Really it makes no sense. Unless there was a pesky base of Nazis still out there possibly. And yes, we now have entered the realm of speculation, but not as wild as you might have though. And if you research this a bit yourself, from principal sources, you will see that the picture of the world as it really is, is a very different one from the one you have been brainwashed into accepting.

When I say you have been brainwashed, I truly mean it. And it's no fault of yours. The Nazis invented many advances in warfare, one of them, which is an actual word in German was "world-view warfare" which gets reduced to "propaganda" today, but the real meaning of this word is far more insidious and frightening. If you control what people think, you really don't need to kill them off. They essentially become resources. When you start to understand some of the experiments in this area that were done by the CIA and others, it quickly becomes obvious that the whole NSA "spying on everyone" has nothing to do with "terrorism" (other than perpetuating it I mean). The motives are financial and the idea is to have totalitarian control. Do you really think the huge resources of the NSA have to do with some rag-tag Islamic fundamentalists living in caves in Afghanistan? Do you realise how absurd that even sounds when you think about it in the cold light of day? Think about the ability to hear and record EVERY conversation, and film EVERY event that is near any kind of camera on any specific person you want. What would you use that for? What COULD you use it for? The answer is pretty much "anything I want to". And how easy is it to manipulate whole societies and nations merely by adjusting a little bit how certain things are presented?

It is also beginning to become clear that most of the uses of "direct action" of the NSA (aside from possibly some drone strikes, though it's doubtful the NSA did much directly there) were mainly to direct and control financial markets and perhaps help syphon off those trillions of dollars that just seem to disappear into "black projects". (See the addendum to chapter 11 below). A secondary use was to pass on information to other agencies to criminalise America Citizens under the auspices of the "the war on drugs", which might more properly be labelled as stage 2 of "command and control" and getting your population to generally accept ever increasing doses of police-state authority. That is how brainwashing is done. Slowly and thoroughly. Do you realise that many everyday actions the average American could take in the 70s are now considered criminal? Or that the discourse both online and outside of the virtual world has become dominated by an insidious version of "political correctness" that bears no brook with individual thought, but instead forces (usually by the expedient means of "social shaming") us all to think the same thoughts about the same things. Reality has to give way to the happy-land of unicorns where we all get along just fine. And of course, you might say, "Well, what's wrong with that? Why shouldn't we all get along?" We should, but that's only going to happen once we actually converse, speak and understand each other's differences at the root, and evolve those differences in which we are personally deficient and help the other evolve where they are. Peace is never going to be achieved by a police-state mentality that criminalises free expression of thought in some very real cases. And arbitrarily criminalises other behaviour. These are not policies based in reality, they are policies based in controlling people and mechanising them. Strictly speaking, a very Nazi idea.

So we have gone from UFOs to politics. Which we would, when the UFO technology is steeped in Nazi ideals. So what to do? Well, personally, a few years after I wrote The Face on Mars, I came across an amazing piece of writing. On reading it, I realised that the author had verbalised in the most exact way, my very own view of how to live life with regard to all things legal and all things political. That piece of writing was an essay called Natural Law, by Lysander

Spooner. You can easily find it free on the web. I read it when I was essentially making a transition from atheism to agnosticism, (which is at least a more honest position when you observe the world around you openly for long enough) and at the time I had no idea what a Deist was, but even then, Lysander's use of the word God never disturbed me. His logic was, and remains, unassailable. And the use of the word God as a fundamental axiom was no more disturbing if he had used the word Nature. Indeed, it can be taken in that context. "the state of things as we find them, whatever their origin" is a rather cumbersome phrase really. I mention this because non-believers are liable to dismiss anything that has the word "God" in it. I understand such people well as I used to be one of them. And no, I have not had a conversion to any specific religion, though I have had an evolution, and my agnosticism has indeed changed (only in the last year or so I would say) to more properly align with Deism more than Agnosticism. Deism is basically the converse to an atheist with a large dose of Agnosticism, and might be summed up as:

*After careful consideration of all the facts, and all the experiences I have had, and all the reasons I can see and all the  best actual science I can do, it seems clear to me that there is something more to all this than mere machinery. What that something is, I cannot say. What its qualities are, I cannot really precisely define, yet, like a fog of unknowing, it seems to permeate all around me. And for want of a better word, I might use God to refer to it.*

It is a constant revision of our facts and a constant re-assessment of our theories that is the real nature of science. If after careful consideration we become aware of more facts, or more relevant facts, or at least of some "pesky" signals and indicators that somehow do not fit the theory snugly, we need to adjust. We need to say "this is not quite right yet." We might  not ever get it quite right, but the idea is to do the best we can, and the best we can do, invariably has to take into account that we really need to put our emotions aside. Emotions are not a reliable way to do anything, except maybe make love (and even then…you wanna be careful!) but they certainly are no way to think. Emotions need to be relegated to the nurturing of loved ones, to the compassion for our fellow man, but not as a short-hand for reasoning. Reasoning is something different from having an emotion and then trying to justify it in the face of a reality that quite obviously contradicts our emotional expectations.

Reasoning is seeing reality as close to how it really is as possible. And woe betide the fool who says "we all have our own truth". No. No we really don't, you journalistic excess, you waste of oxygen you, there is only one truth at any specific moment in time about any specific perspective. Even if we take the classic Einstenian idea of two ships moving close to the speed of light, one heading towards a star and one heading away from it, yes it's true that each captain aboard each ship will see the star as being of a different colour, but so what? There is nothing "relative" about it, once you understand how light works. The star is neither "blue" nor "red" or could be thought of as both at the same time depending on your position in relation to it, but none of this is "open to interpretation", and beware anyone who speaks that way. There is only one truth, and it is the same one for all of us. Always. Whether we see it or not, reality is not our playground. At least, not until we accept and realise there is such a thing as objective reality. Then…mysteriously, once we accept this, we may actually begin to change the reality we have for a better one. But first we need to understand how this one works.

In the context of this book then, understanding the reality of how this planet actually works, what the real motivations behind the powers that be are and so on, is all tremendously valuable in shaping the future. So take care what ideas you let in. And scrutinise them well. And never stop doing so. The future needs a lot of *reasonable* people. That is…*people that can reason*, which is really not necessarily connected to being a mealy-mouthed, well-behaved drone that never disagree with anyone for fear of social ostracism.

So keep your guard up, and remember: There are Nazis out there!

# Chapter 10

Although the information in this chapter is not wrong, there could be much more added to it. In this respect, the work of Dr. Farrell, already mentioned in the addendum to chapter 9 above, is invaluable. Particularly his book *The SS Brotherhood of the Bell – The Nazis' Incredible Secret Technology*, which details the early history of how the Nazis rediscovered and re-developed this ancient technology. Similarly, the book *Dark Star – The Hidden History of German Secret Bases, Flying Disks and U-Boats*, by Henry Stevens is an amazing resource. Once you have had these as starting points, you will be on the right track to discover more and also be able to better differentiate facts from fiction with respect to not only the origins of antigravity technology, but sadly, the ideology that informed its beginnings, as well as the sad fact that such ideology is far from being removed not just from the world, but from the very top corridors of power of those who control and have access to antigravity technology. The Nazis did not really lose the Second World War. Not by a long shot.

# Chapter 11

Once again, this chapter could comprise several books worth of information, and that's only if we limit ourselves to demonstrably egregious falsehoods perpetrated by governments with regard to covering up anything related to antigravity technology.

It is interesting to note that the hacker Gary McKinnon, who has been in legal battles to fight his extradition to the USA for over a decade, is guilty of supposedly breaking into the Pentagon computer system and accessing files that gave detailed lists of personnel aboard ships in space that could carry upwards of 300 people at a time.

For this to be the case, it needs to be true that a huge shadow-economy that is potentially at least a significant size of the global economy, if not equal or larger to it, needs to be ongoing. You can't just build a bunch of huge spaceships without some kind of organisation and even if you do it off-planet (be it on Mars or the Moon or wherever) the first stages of such a long-term plan would have required huge black budgets on Earth. Do we have evidence for this? Actually yes. Plenty of it. So much so that even classic scholars of economy and physics (because physics has quite a bit to do with economy, a lot more than you might think unless you are very familiar with the work of Joseph Farrell) are starting to realise it. And even write official papers that begin to expose such systems as actually existing. See for example the work titled "The interrupted power law and the size of shadow banking" produced by Professor Davide Fiaschi (Dept. of Economic Sciences at the University of Pisa), Professor Imre Kondor (Department for the Physics of Complex Systems – Eotvos University, Budapest, though this paper was worked on via the Parmenides Foundation in Germany) and Dr. Matteo Marsili (Abdus Salam International Centre for Theorethical Physics). Although Shadow Banking does not specifically refers to hidden black budgets put aside for the construction of off-planet ships and other such things, it does refer to a system of financial commerce that is essentially separate and hidden from the one you and I are privy to or have anything to do with. And what *IS* the size of this "shadow banking" that goes on behind closed doors? Well, according to the Financial Stability Board (a body created to coordinate the unholy mess of international finance, so likely to have as many snakes in the grass as an Indiana Jones film on crack) estimated it at $26 Trillion in 2002, rising to $62 Trillion in 2007 and $67 Trillion in 2011 (this despite the global financial crash of 2008). But Fiaschi and the other two academics decided to use raw data and crunch the numbers and see what they could find. As it turns out, it seems that most of the shadow banking is concentrated in the top 13% of financial institutions, and unsurprisingly, that the estimates released by the FSB were...shall we say "conservative" to put it mildly, since Fiaschi and friends found the size of the shadow banking sector to be about $90 Trillion in 2011 and rising to over $100 Trillion in 2012. Because Fiaschi's figures are raw data, as opposed to studied cherry-picking from within the industry itself, it is more likely that his figures are correct. Now keep in mind that *ALL* of this $100 Trillion is basically finance that sort of "disappears" into a "shadow banking" black hole at the very top of a huge pile of finance. Where it moves, how it moves, who it goes to, is all unknown. All we can say really is that it disappears. Oh yeah... and that it is larger than the whole global economy which we *CAN* see. I know. It really sounds absurd doesn't it. It *is* absurd. But it's also very much reality on planet Earth.

The point is quite simply that once you look into it, it is clear that not only is all of this possible, but in fact all the structures for it to exist are factually in place. Furthermore, once you have created enough of a breakaway civilisation off-planet that has enough antigravity machines, you effectively also can break away financially. There is literally nothing stopping you from mining asteroids and other planets (at much higher rates of yield than you can do on Earth given the advanced technology you can use) to the extent that the economy off-planet becomes the dominant one and the one on Earth literally becomes just an appendix to it. The only limit to this would be the various competing factions who may also be already off-planet.

What the composition of such factions might be is difficult to determine, but given global events on Earth, it seems at least quite likely that the so-called Western banking Elites are on a different side to the Russian banking Elites, and as it turns out, it might be the Russians that are approaching more the idea of being "the good guys". Once you are aware of these realities, some of the political machinations displayed on Earth that are mystifying to most people become pretty easy to read and the subtext of certain actions so obvious that it is a clear message to those in the know and in actual power. In this respect, though no one is foolish enough to think Vladimir Putin is a soft cuddly bear, it needs to be said that he genuinely seems to have acted in ways that have —without any doubt about it— resulted in the saving of many thousands of lives by avoiding war in Syria and more recently by his cautious but firm defence of Russia's interests in the Ukraine. More importantly, if you listen to the speeches Putin has given in public to the world leaders at various functions he has been directly and clearly stating that the homogeneous one-world government policy of the USA is essentially one that results in wars and the death and suffering of millions and that such a unipolar philosophy cannot result in anything else as it has to crush any dissension from the ranks by force. A position that he personally, and the Russian government as whole, find inhuman and distasteful. You know you are doing something wrong when the ex-head of the KGB tells you your methods are too brutal.

# Chapter 12

This is without a doubt the most dated of the chapters and I smile at how simple and naïve where the propaganda efforts back in the almost unbridled freedom of 1995-6. Misinformation and propaganda has now achieved heights that even Goebbles and Hitler could not have imagined. To a certain extent, the "world-view warfare" has been won, and you and I reader, have lost.

The good thing about it is that as the losers we no longer have to "fight" for the truth. We can just see it. Quietly figure it out and then begin to naturally see the lies and illusions; illusions that nevertheless cause untold very real misery, death and depravation, for what they are. We may be prisoners on this planet, but we are not going to think we are not. And once a whole bunch of us see it and begin to disengage from the false reality, a shift begins. They say Rome was not built in a day, but what everyone forgets is that neither did Imperial Rome collapse in a day. It crumbled. Slowly, but unstoppably. And that is what the banksterism of today is headed for. Something else will replace it, and it will not be "good" either, but hopefully it will be "better" enough to give the broader humanity a chance and a better chance of getting off-planet too. We belong amongst the stars. Just not as the uncivilised and violent savages we are today. So let's hurry up and evolve. I might not get out there, flying in a spaceship and seeing the rest of this Galaxy, but I'd like my little daughter to one day be able to do so, if she wants to.

# Epilogue

Most of the information here remains relevant, except the imaging of Mars. It is my current belief that publicly released images of Mars are heavily doctored and automatically pass through highly advanced algorithms designed to "photoshop out" any non-fractal components. Furthermore, highly fractal or "problem" images would then receive individualised attention and either be suppressed or further altered to produce less artificiality as their final output.

There is now strong and very real evidence that most of the images taken in the Apollo landings were doctored and adjusted in pretty severe ways (ways that are now becoming exposed, ironically enough, mostly by the advances we have made in image manipulation techniques since the crude attempts of the 70s. Crude for us now, but very advanced at the time they were done) as well as that some of these images were probably staged and possibly some of the videos too. Although I personally am sure we did go to the Moon, it is clear to me that we did not do so in the manner presented. My theory is that antigravity technology was involved and used in the LEM ascent module. The staging of some videos and photographs on a studio on Earth, much as the film *Capricorn One* depicted for a Mars landing, is not inconsistent with such a view.

In other words, both things are true. Yes we did go to the Moon, and yes, NASA lied about it. A lot. They lied about the how, the why and the what they found there, and in my opinion continue to lie about our not having gone back there (though to be fair I think that is a different branch and not really NASA). Logic suggests that we would not just have undoubtedly gone back, but probably that we have permanent bases there by now. As for who I mean by "we"... well...that is a little complicated, but if you read the addendums to chapter 9 you will have a better grasp of it. If you also read some of Joseph Farrell's work it will become even clearer. But saying "the Americans" would just be simplistic and probably only partially correct. You might also wish to read my science fiction series *Overlords of Mars*, which though began as pure fiction, has had the uncomfortable effect of being closer to reality than even I first suspected. I stress, it is fiction, yet... elements of it are pretty close to the mark. Uncomfortably close to it I think.

## What Now

This section is no longer really applicable. In fact, even back in 1996, after my life took one of its periodic sudden shifts, I never cashed any of the cheques I received. Today, you can visit my website at www.GFilotto.com, where I do try, as my very limited time permits, to keep you up to date on what I am up to, but barring my becoming independently wealthy, receiving huge amounts of funds for further research and possibly a small but comfortable private island in which to undertake such research, my efforts are generally limited to what one man can do in his almost non-existent spare time.

# Final Thoughts

There have been a number of other indicators and pointers over the years that have consistently added weight to the idea that this Solar System was previously inhabited. Anomalies on the Moon and Mars are plentiful, and indicators of large fleets of spaceships being present off-planet are also more than mere rumours.

But there are other, socio-political elements that make it quite clear that the so-called "vast conspiracy theory" that could only be true in the fevered imagination of a tin-foil hat wearing insane geek, is actually a very real, very disturbing reality.

We now know beyond a shadow of a doubt, thanks to Ed Snowden for the most part, that yes, actually the global surveillance network is such that you are literally being monitored every time you make a call, send an e-mail or are in front of any camera anywhere. Including the ones on your own laptop. We now know as an undeniable fact that Google was sponsored by an R&D type firm that is a subsidiary of the CIA. We know that social networks are just another way of making sure you are trackable, recognisable and forever digitally followable. Face recognition software present in your phone is tagging images you take and if anyone at the NSA ever wants to know where you are they can literally find you live anywhere on the planet probably within seconds. The old film with Will Smith in it, *Enemy of the State* was not fiction even back then in 1998.

The financial reality of how this planet works have become a little bit less opaque even to the common man, and anyone willing to do a bit of research can very quickly come away with a nauseated view of bankers in general and the even more absurd reality that props up even most bankers.

Even if you only read the four books written by Michael Lewis on the financial world, (*Liar's Poker, Boomerang, The Big Short* and *Flashboys*) which are funny and brilliant aside from informative, and which do not even remotely delve into UFOs or any kind of conspiracy, you will still begin to realise just how artificial the decisions that affect the lives of millions around the globe are, and how a very few people hold the power of life and death over the masses. None of this is new, but when you begin to realise there are also trillions of dollars missing that no one can account for, and black projects that seem to have endless funds funnelled to them in untraceable ways, then the idea of a "vast conspiracy" is no longer just a theory. It is a fact. You might never be able to identify the specific individuals involved in it, but the fact the conspiracy exists is still knowable, and knowable to a high enough degree of probability that we can call it a fact. Everyone knows about Area 51 now and we also know it is a huge complex, with thousands of people and large underground facilities. Try to find any kind of money trail as to how it was built. And yet all that equipment doesn't just appear magically out of nowhere.

But what is the point of knowing all this stuff if you can't do anything about it anyway?

There is an important point in all this.

It seems to be true that whatever we unconsciously allow to reside in our mind, shapes reality. A book that explained this quite nicely was *Reality Transurfing*, by Vadim Zeland. But generally it has also been the refrain of self-help gurus that con you out of hard-earned cash for...well...a long time. Snake-oil never goes out of fashion, it just changes form. But the point is that hard science seems to now be confirming it. And no, it never works the way the self-help gurus tell you. You will NOT become a millionaire by chanting positive affirmations. If you become one it will be because you are either very lucky or because you work your butt off, but the point is, still, in some ways, the way you think, the deep thought patterns of you unconscious do shape reality. In a subtle but perceptible way.

The global efforts to control your underlying thoughts is the most effective way to keep you completely under control. In essence, the film The Matrix was so successful precisely because it metaphorically exposed a deep truth. Reality is not as you think it is. Knowing these things, figuring them out, seeing them for yourself immunises you to a great extent from the continuous brainwashing that you are being subjected to. And once that stops working so well on you, then your thoughts change. And you change. You feel free-er inside. You strangely, begin to ignore the news that is fed to you and begin to recognise patterns in the fog of lies. You begin to be able to predict what will happen next, because once the veil of stupefaction is lifted, the game is actually not that complicated to see or understand. You see the "bad" guy (presently Putin for example) is actually not as "bad" as he is presented since he has to date prevented a couple of wars and is trying to resist being pulled into another one in Ukraine (if he will succeed or not remains to be seen). You will begin to care more about your immediate family and friends and learn to ignore global events. You begin to have a growing certainty that you will no longer be a sheep trained to think as your government, or the people directing them want you to think. And in this there is power. An incalculable power. The best expression of how this works, as I have already mentioned a couple of times, is explained in the essay *The Power of the Powerless*.

Ideology dehumanises. All ideology ultimately does, because all ideology is ultimately reductive and human individuals are not machines, nor are they parts of "society". They are forever and irreducibly individuals. Even if they act in concert, to force them to do so at the expense of their humanity ultimately will always fail. And once you begin to realise that much of what you thought was reality is actually just an ideology that has been imposed on you without your knowledge, then that ideology, and the false reality it supported, begin to collapse. With such collapse comes a raising of human consciousness and generally also an increase in the humanity. An increase in those things that make being human worth living. Compassion for your fellow man, love, kindness, understanding, co-operation amongst strangers.

So knowing these things, knowing human history is vastly different from what you have been told, knowing there are other intelligences in the Galaxy we live in, knowing we already have the technology to go to the stars, it's important. It matters. It will change the way you think and the way your children will think and it will give humanity a better chance of escaping the vicissitudes of a life lived only from the worm's eye view of the material world.

There is another world. Many of them amongst the stars, but I refer to an internal world, one that is based ultimately on love. All religions try to point to it but get lost and twisted over time, but there is another way. I think perhaps in older times, when we were closer to nature we had more communion with it. Sadly we were also more ferocious then. Now we might be a bit less ferocious (physically at least, and even that is arguable if you take into account say the American police state today) and perhaps, if we reconnect to that other state of being, that plane on which telepathy and love and empathy are not just ideas but reality, we might be better able to sustain living in it more and more of the time we are awake.

Perhaps we will begin by living there in our sleep at first, which is not a bad thing, because the unconscious is what ultimately leads us where we will go. And if we all dream the same loving dream, well…what a world might that make.

But don't get me wrong. This does not mean you need to be a docile flower all the time. There are Nazis out there and they seem to be running things. At times, holding your resolve and fighting back might be required. If so, be true, be loyal, protect the innocent and the weak, be just and be powerful in your actions and sure in them. And, as well, be ready to *not* act if that is the best thing to do. But act from what is highest in you. Act not for yourself alone but for the good of all.

These things are easy to say. Less easy to do. Be someone your children and your friends and even your enemies can one day look at and say "There. That one is for Justice. For Truth. For Love."

I wish you, my long-suffering reader, a long, happy, joyful life. May your troubles be brief and your laughter long and heartfelt.

Giuseppe Filotto
11th May, 2014
London